Lecture Notes in Geoinformation and Cartography

Series Editors: William Cartwright, Georg Gartner, Liqiu Meng, Michael P. Peterson

Jiyeong Lee · Sisi Zlatanova (Eds.)

3D Geo-Information Sciences

Springer

Editors

Dr. Jiyeong Lee
University of Seoul
Department of Geoinformatics
13 Siripdae-gil, Dongdaemun-gu
Seoul 130-743
Korea
jlee@uos.ac.kr

Dr. Sisi Zlatanova
Delft University of Technology
Section of GIS Technology
Jaffalaan 9, 2628 BXDelft
The Netherlands
s.zlatanova@tudelft.nl

ISBN: 978-3-540-87394-5 e-ISBN: 978-3-540-87395-2

DOI:10.1007/978-3-540-87395-2

Lecture Notes in Geoinformation and Cartography ISSN: 1863-2246

Library of Congress Control Number: 2008938587

Typesetting: Camera-ready by the editors

Cover design: deblik, Berlin

Printed on acid-free paper

9 8 7 6 5 4 3 2 1

springer.com

Preface

In recent years 3D geo-information has become an important research area due to the increased complexity of tasks in many geo-scientific applications, such as sustainable urban planning and development, civil engineering, risk and disaster management and environmental monitoring. Moreover, a paradigm of cross-application merging and integrating of 3D data is observed. The problems and challenges facing today's 3D software, generally application-oriented, focus almost exclusively on 3D data transportability issues – the ability to use data originally developed in one modelling/visualisation system in other and vice versa. Tools for elaborated 3D analysis, simulation and prediction are either missing or, when available, dedicated to specific tasks.

In order to respond to this increased demand, a new type of system has to be developed. A fully developed 3D geo-information system should be able to manage 3D geometry and topology, to integrate 3D geometry and thematic information, to analyze both spatial and topological relationships, and to present the data in a suitable form. In addition to the simple geometry types like point line and polygon, a large variety of parametric representations, freeform curves and surfaces or sweep shapes have to be supported. Approaches for seamless conversion between 3D raster and 3D vector representations should be available, they should allow analysis of a representation most suitable for a specific application.

In the last decades research and industry have addressed a variety of technical issues related to 3D data acquisitions, processing, visualization, data management and spatial analysis aiming at a fully developed 3D geo-information system. A variety of new sensors (space, airborne, terrestrial) have contributed to accelerating the research in 3D reconstruction. As demonstrated by some authors in this book, integrated utilisation of sensor data will allow for fully automatic approaches for 3D data reconstitution at various levels of detail. However, due to the rapid developments in sensor techniques large amounts of 3D data have become available. This poses new challenges to storage and management. Emerging visualisation applications such as 3D globe based interfaces, navigation systems presenting a 3D perspective, etc, have greatly contributed to acceptance and understanding the benefits 3D systems. These visualisation environments have to be further developed towards spatial queries, simulations, etc.

The critical topic in 3D remains the modelling process. In the last decades a large number of geometry and topological models have been researched and tested. 3D models can be maintained in TEN (Tetrahedral Network), constructive solid geometry (CSG) models, TIN boundary representations, 3D layered/topology models, voxel based models, 3D polyhedrons-based models, but only a limited number of solutions are available on the market. 3D topological models are still only in the research labs. They are expected to boost 3D analysis and 3D simulation techniques, which will enrich the 3D spatial applications. In the last years much attention has been given to 3D indoor building models and the possibilities of integrating construction designs with existing 3D geo-information. Several attempts have been already reported in the drive to bridge 3D formal models as used in architecture, construction and engineering (ACE) and GIS, which opens new exciting areas for research and developments.

These dynamic developments require intensive dialog between researchers, developers, data providers and users. It is also very important to document appropriately the progress made. This book attempts to achieve this goal. It comprises twenty-six full papers prepared from the contributions of sixty-nine authors. The papers are organized in two parts: Part I contains three chapters that are the invited keynotes related to 'From Map to GIS and Virtual Geographic Environments', 'Introduction to Korean Land Spatialization Program', and 'Introduction to CityGML'. Part II contains twenty-three chapters organized in five themes: 3D Data Models, 3D Database Management, 3D Data Acquisition and Processing, 3D Data Analysis and 3D Geo-Visualization.

The papers in this book are selected from the Third Workshop on 3D geo-information, 13-14 November, Seoul, Korea (http://3DGeoInfo.uos.ac.kr), all of which have been thoroughly reviewed by three members of the international programme committee. The authors of the best and most original contributions were asked to submit revised versions based on these comments. Two similar previous events took place in Kuala Lumpur, Malaysia, on 7-8 August 2006 and Delft, on 12-14 December 2007. The selected papers from the first workshop were published in 'Innovations 3D Geo Information Systems' edited by Abdul-Rahman, Zlatanova and Coors, Springer-Verlag, 2006 and the papers from the second were published in Van Oosterm, Zlatanova, Penninga and Fendel (eds) 'Advances in 3D Geoinformation Systems', Spinger-Verlag, 2008.

The editors of this book would like to express their thanks to the local organizers June Hwan Koh, Chul-Min Chun, Yunsoo Choi, Impyeong Lee, Sung-Gil Cho, Jay Hyoun Kwon and Ki-Joune Li for the pleasant support. Furthermore, we are grateful to the all the members of the scientific committee who have helped in the review process and selection of papers. We would like to thank all the authors for their original contributions. The editors are also grateful for the support provided by University of Seoul and the two projects, which are 07KLSGC04 'Indoor Spatial Awareness' supported from Cutting-edge Urban Development - Korean Land Spatialization Research Project funded by Ministry of Land, Transport and Maritime Affairs, and C6A2104 'Advanced Education for Global u-spatial Information' funded by Ministry of Education, Korea.

Seoul,
September 2008

Jiyeong Lee and Sisi Zlatanova

Organisation

Program chairs

Jiyeong Lee	University of Seoul (South Korea)
Sisi Zlatanova	Delft University of Technology (the Netherlands)

Local organizing committee

June Hwan Koh	University of Seoul (South Korea)
Chul-Min Chun	University of Seoul (South Korea)
Yunsoo Choi	University of Seoul (South Korea)
Impyeong Lee	University of Seoul (South Korea)
Sung-Gil Cho	University of Seoul (South Korea)
Jay Hyoun Kwon	University of Seoul (South Korea)
Ki-Joune Li	Pusan National University (South Korea)
Alias Abdul-Rahman	University of Technology Malaysia (Malaysia)

Scientific committee

Andy Hamilton	University of Salford (United Kingdom)
Alias Abdul-Rahman	University of Technology Malaysia (Malaysia)
Masatoshi Arikawa	The University of Tokyo (Japan)
Guo BinXuan	WuHan University(China)
Roland Billen	University of Liege (Belgium)
Lars Bodum	Aalborg University (Denmark)
Seongkil Cho	University of Seoul (South Korea)
Woosug Cho	Inha University (South Korea)
Jinmu Choi	Mississippi State University (USA)
Jin Won Choi	Yonsei University (South Korea)
Yoon Soo Choi	University of Seoul (South Korea)
Volker Coors	University of Applied Sciences Stuttgart (Germany)
Claire Ellul	University College London (United Kingdom)
Robert Fencik	Slovak University of Technology (Slovak)
Andrew Frank	TU Wien (Austria)
Yi FuZhong	Bureau of Surveying and Mapping in HeiLongJiang Province(China)
Christopher Gold	University of Glamorgan (United Kingdom)
Norbert Haala	University of Stuttgart (Germany)
Joon Heo	Yonsei University (South Korea)
Daniel Holweg	Bentley (Germany)
Tu JianGuang	WuHan University (China)
Chulmin Jun	University of Seoul (South Korea)
Byung-Guk Kim	Inha University (South Korea)
Eun Hyung Kim	Kyungwon Uinversity (South Korea)
Junehwan Ko	University of Seoul (South Korea)
Thomas Kolbe	Technical University Berlin (Germany)
Marc van Kreveld	Utrecht University (the Netherlands)

Mei-Po Kwan	Ohio State University (USA)
Jay Hyoun Kwon	University of Seoul (South Korea)
Hugo Ledoux	Delft University of Technology (the Netherlands)
Impyeong Lee	University of Seoul (South Korea)
Jiyeong Lee	University of Seoul (South Korea)
Ki-Joune Li	Pusan National University (South Korea)
Zhilin Li	Hong Kong Polytechnic University (Hong Kong)
Hui Lin	Wuhan University (China)
Wu Lixin	North Eastern University(China)
Mario Matthys	High School Science and Art Gent (Belgium)
Martien Molenaar	ITC Enschede (the Netherlands)
Peter van Oosterom	Delft University of Technology (the Netherlands)
András Osskó	FIG/Budapest Land Office (Hungary)
Chris Parker	Ordnance Survey (United Kingdom)
Wanning Peng	ESRI (USA)
Norbert Pfeifer	TU Wien (Austria)
Clemens Portele	Interactive Instruments (Germany)
Jonathan Raper	City University London (United Kingdom)
Siva Rayada	Oracle Corporation(USA)
Massimo Rumor	University of Padova (Italy)
John Wen-zhong Shi	Hong Kong Polytechnic University (Hong Kong)
Ryosuke Shibasaki	University of Tokyo (Japan)
Sungwoong Shin	ETRI (South Korea)
HongGyoo Sohn	Yonsei University (South Korea)
Uwe Stilla	Technical University of Munich (Germany)
Jantien Stoter	ITC Enschede (the Netherlands)
Rod Thompson	Queensland Government (Australia)
Ho-Lin Tsay	National Applied Research Laboratories (Taiwan)
Frank Van den Heuvel	CycloMedia Technology B.V. (the Netherland)
George Vosselman	ITC Enschede (the Netherlands)
Peter Widmayer	ETH Zürich (Switzerland)
Peter Woodsford	1Spatial / Snowflake (United Kingdom)
Keiji Yano	Ritsumeikan University (Japan)
Lai Zhibin	Reijing Peking University ChinaFront High Technology (China)
Qing Zhu	Wuhan University (China)
Alexander Zipf	University of Bonn (Germany)
Sisi Zlatanova	Delft University of Technology (the Netherlands)

List of Contributors

Alias Abdul-Rahman
Department of Geoinformatics, Faculty of Geoinformation Science and Engineering, University Technology Malaysia, Malaysia, e-mail: alias@utm.my

Reza Aghataher
Head of GIS department in National Geographic Organization, Iran, e-mail: Reza_aghtaher@yahoo.com

Yusuf Arayici
Research Institute for the Built and Human Environment (BuHu), University of Salford, UK, e-mail: Y.Arayici@Salford.ac.uk

Thomas Becker
Institute for Geodesy and Geoinformation Science Technische Universität Berlin, Germany, e-mail: becker@igg.tu-berlin.de

Lars Bodum
Centre for 3D GeoInformation Aalborg University, Denmark, e-mail: lbo@3dgi.dk

Pawel Boguslawski
Department of Computing and Mathematics, University of Glamorgan, UK, e-mail: pbogusla@glam.ac.uk

Cláudio Carneiro
Geographical Information Systems Laboratory, EPFL, Switzerland, e-mail: claudio.carneiro@epfl.ch

Jinmu Choi
Department of Geosciences, Mississippi State University, USA, e-mail: jc778@msstate.edu

Volker Coors
University of Applied Science Stuttgar, Germany, e-mail: Volker.coors@hft-stuttgart.de

Sagi Dalyot
Mapping and Geoinformation Engineering, Technion - Israel Institute of Technology, Israel, e-mail: dalyot@technion.ac.il

Mohmoud Reza Delavar
Center of Excellence in Geomatics Eng. and Disaster Management, Dept. of Surveying and Geomatics Eng., College of Eng., University of Tehran, Tehran, Iran, e-mail: mdelavar@ut.ac.ir

Jia Dongzhen
Institute of Satellite Navigation & Spatial Information System, Hohai University, Nanjing, China, e-mail: jiadongzhen@hhu.edu.cn

Yerach Doytsher
Mapping and Geoinformation Engineering, Technion - Israel Institute of Technology, Israel, e-mail: doytsher@technion.ac.il

Claire Ellul
Dept. of Civil, Environmental and Geomatic Engineering, University College London, UK, e-mail: clairee@ge.ucl.ac.uk

Markus Gerke
International Institute for Geo-Information Science and Earth Observation (ITC), the Netherlands, e-mail: gerke@itc.nl

François Golay
Geographical Information Systems Laboratory, EPFL, Switzerland, e-mail: francois.golay@epfl.ch

Christopher Gold
Department of Computing and Mathematics, University of Glamorgan, Wales, UK, e-mail: cmgold@glam.ac.uk

Mordechai Muki Haklay
Dept. of Civil, Environmental and Geomatic Engineering, University College London, UK, e-mail: m.haklay@ucl.ac.uk

Andy Hamilton
Research Institute for the Built and Human Environment (BuHu), University of Salford, UK, e-mail: A.Hamilton@Salford.ac.uk

Ya Hu
Institute of Space and Earth Information Science, The Chinese University of Hong Kong, Surveying Engineering Department, School of Civil Engineering, Southwest Jiaotong University, China

Ari Ikonen
Information Technology, Pori, Tampere University of Technology, Finland, e-mail: ari.ikonen@posiva.fi

Umit Isikdag
Delft University of Technology, the Netherlands, e-mail: egetera@superonline.com

GUO Jiateng
Institute for GIS/RS/GPS and Digital Mine Research, Northeastern University, China, e-mail: guojiateng@mail.neu.edu.cn

Baris M. Kazar
Oracle USA, Inc., USA, e-mail: Baris.Kazar@Oracle.com

Tor Yam Khoon
School of Civil and Environmental Engineering, Nanyang Technological University, Singapore, e-mail: cyktor@pmail.ntu.edu.sg"

Mahmud Shahrear Kibria
ESRI Nederland B. V., the Netherlands, e-mail: mkibria@esri.nl

Byung-Guk Kim
Korean Land Spatialization Group, Korea, e-mail: byungkim@inha.ac.kr

Erik Kjems
Centre for 3D GeoInformation, Aalborg University, Denmark, e-mail: kjems@3dgi.dk

Jan Kolar
Centre for 3D GeoInformation, Aalborg University, Denmark, e-mail: kolda@3dgi.dk

Thomas H. Kolbe
Institute for Geodesy and Geoinformation Science, Technische Universität Berlin, Germany, e-mail: kolbe@igg.tu-berlin.de

Gerhard König
Technische Universität Berlin, Institute for Geodesy and Geoinformation Science, Germany, e-mail: gerhard.koenig@tu-berlin.de

Ravi Kothuri
Currently at Innerscope Research, USA, e-mail: RaviKothuri@gmail.com

Sandra Lanig
Department of Geography, University of Bonn, Germany, e-mail: lanig@geographie.uni-bonn.de

Roosevelt De Lara Jr.
Universidade Federal de Santa Maria CESNORS-FW/RS, Brazil, e-mail: roosevelt@smail.ufsm.br

Jiyeong Lee
University of Seoul, Department of Geoinformatics, 13 Siripdae-gil, Dongdaemun-gu Seoul 130-743, Korea, e-mail: jlee@uos.ac.kr

Hui Lin
Institute of Space and Earth Information Science, The Chinese University of Hong Kong, Hong Kong, e-mail: huilin@cuhk.edu.hk

Wenshi Lin
Department of Atmospheric Sciences, School of Environmental Science and Engineering, Sun Yat-sen University, China

Tarmo Lipping
Information Technology, Pori, Tampere University of Technology, Finland, e-mail: tarmo.lipping@tut.fi

WU Lixin
Institute for GIS/RS/GPS and Subsidence Research, China Uni of Mining & Technology, China, e-mail: awulixin@263.net

Edson A. Mitishita
Universidade Federal do Paraná Curso de Pós-Graduação em Ciências Geodésicas, Brazil, e-mail: mitishita@ufpr.br

Eugenio Morello
SENSEable City Laboratory, MIT, USA, e-mail: eugenio@mit.edu

Claus Nagel
Institute for Geodesy and Geoinformation Science Technische Universität Berlin, Germany, e-mail: nagel@igg.tu-berlin.de

Pascal Neis
Department of Geography, University of Bonn, Germany, e-mail: neis@geographie.unibonn.de

Giwon On
Fraunhofer IGD, Germany, e-mail: giwon.on@igd.fraunhofer.de

Jari Pohjola
Information Technology, Pori, Tampere University of Technology, Finland, e-mail: jari.pohjola@tut.fi

Zhou Qi
LIESMARS, Wuhan University, China, e-mail: seven-laand@hotmail.com

Carlo Ratti
SENSEable City Laboratory, MIT, Cambridge, MA, USA , e-mail: ratti@mit.edu

Siva Ravada
Oracle USA, Inc., USA, e-mail: Siva.Ravada@Oracle.com

Sara Saeedi
GIS Division, Dept. of Surveying and Geomatics Eng., College of Eng., University of Tehran, National Cartographic Center, Tehran, Iran, e-mail: Saeedi@ut.ac.ir

Najmeh Samany
GIS Division, Dept. of Surveying and Geomatics Eng., College of Eng., University of Tehran, National Cartographic Center, Tehran, Iran, e-mail: nsamani@ut.ac.ir

Arne Schilling
Department of Geography, University of Bonn, Germany,
e-mail: schilling@geographie.uni-bonn.de

Yonghui Song
Research Institute for the Built and Human Environment (BuHu), University of Salford, UK, e-mail: Y.H.Song@Salford.ac.uk

Alexandra Stadler
Technische Universität Berlin, Institute for Geodesy and Geoinformation Science, Germany, e-mail: stadler@igg.tu-berlin.de

Rodney James Thompson
Department of Natural Resources and Water, Queensland, Australia,
e-mail: Rod.Thompson@nrw.qld.gov.au
Delft University of Technology, OTB, GIS Technology, The Netherlands,
e-mail: rthompson@tudelft.nl

Yixiang Tian
International Institute for Geo-Information Science and Earth Observation (ITC), e-mail: ytian@itc.nl

Jari Turunen
Information Technology, Pori, Tampere University of Technology, Finland
e-mail: jari.j.turunen@tut.fi

Muhamad Uznir Ujang
Department of Geoinformatics, Faculty of Geoinformation Science and Engineering, University Technology Malaysia, e-mail: uznir@utm.my

George Vosselman
International Institute for Geo-Information Science and Earth Observation (ITC), e-mail: vosselman@itc.nl

Hongxia Wang
Research Institute for the Built and Human Environment (BuHu), University of Salford, UK, e-mail: H.Wang@Salford.ac.uk

Bingli Xu
Institute of Space and Earth Information Science, The Chinese University of Hong Kong, Shatin, Hong Kong

Jie Yu
State Key Laboratory of Information Engineering in Surveying, Mapping and Remote Sensing, Wuhan University, China, e-mail: yujie2xw@126.com

Izham Mohamad Yusoff
Department of Geoinformatics, Faculty of Geoinformation Science and Engineering, University Technology Malaysia, e-mail: izham1@utm.my

Yeting Zhang
State Key Laboratory of Information Engineering in Surveying, Mapping and Remote Sensing, Wuhan University, China, e-mail: zhangyeting@263.net

Yunsheng Zhang
State Key Laboratory of Information Engineering in Surveying, Mapping and Remote Sensing, Wuhan University, China, e-mail: sheng_650@163.com

Zhong Zheng
School of Civil and Environmental Engineering, Nanyang Technological University, Nanyang Avenue, Singapore

Jun Zhu
Institute of Space and Earth Information Science, The Chinese University of Hong Kong, Shatin, Hong Kong
Surveying Engineering Department, School of Civil Engineering, Southwest Jiaotong University, China

Qing Zhu
State Key Lab of Information Engineering in Surveying Mapping and Remote Sensing, Wuhan University, China, e-mail: zhuq66@263.net

Alexander Zipf
Department of Geography, University of Bonn, Germany,
e-mail: zipf@geographie.uni-bonn.de

Sisi Zlatanova
Delft University of Technology, the Netherlands, e-mail: S.Zlatanova@tudelft.nl

Contents

Part I: Keynotes 1

1. **A Virtual Geographic Environment for a Simulation of Air Pollution Dispersion in the Pearl River Delta(PRD) Region** 3
Hui Lin, Jun Zhu, Bingli Xu, Wenshi Lin and Ya Hu

2. **Representing and Exchanging 3D City Models with CityGML** 15
Thomas H. Kolbe

3. **Research and Development Program for the Third National Geographic Information System Project in Korea: Korean Land Spatialization Program** 33
Byung-Guk Kim, Hoon-Sung Choi and Yeun J. Jung

Part II: Papers 45

4. **Construction Operators for Modelling 3D Objects and Dual Navigation Structures** 47
Pawel Boguslawski and Christopher Gold

5. **A Multilayered Space-Event Model for Navigation in Indoor Spaces** 61
Thomas Becker, Claus Nagel and Thomas H. Kolbe

6. **Towards Defining a Framework for Automatic Generation of Buildings in CityGML Using Building Information Models** 79
Umit Isikdag and Sisi Zlatanova

7. **Managed Objects for Infrastructure Data** 97
Erik Kjems, Lars Boduml and Jan Kolar

8. **Integrating Terrain Surface and Street Network for 3D Routing** 109
Arne Schilling, Sandra Lanig, Pascal Neis and Alexander Zipf

9. Using a B-Rep Structure to Query 9-Intersection Topological Relationships in 3D GIS – Reviewing the Approach and Improving Performance 127
Claire Ellull and Mordechai Muki Haklay

10. Query Processing in 3-D Spatial Databases: Experiences with Oracle Spatial 11g 153
Siva Ravada, Baris M. Kazar and Ravi Kothuri

11. Making Interoperability Persistent: A 3D Geo Database Based on CityGML 175
Alexandra Stadler, Claus Nagel, Gerhard König and Thomas H. Kolbe

12. Use of Finite Arithmetic in 3D Spatial Databases 193
Rodney James Thompson

13. Automatic Digital Aerial Image Resection Controlled by LIDAR Data 213
Roosevelt De Lara Jr., Edson A. Mitishita, Thomas Vögtle and Hans-Peter Bähr

14. Automatic Surface Patch Generation from a Video Image Sequence 235
Yixiang Tian, Markus Gerke, George Vosselman and Qing Zhu

15. Indoor 3D Modeling and Visualizing with a 3D Terrestrial Laser Scanner 247
Jia Dongzhen, Tor Yam Khoon, Zhong Zheng and Zhou Qi

16. Automatic Image Mosaic-Building Algorithm for Generating Facade Textures 257
Yunsheng Zhang, Qing Zhu, Jie Yu and Yeting Zhang

17. 3D Continuous K-NN Query for a Landmark-based Wayfinding Location-based Service 271
Najmeh Samany, Mohmoud Reza Delavar, Sara Saeedi and Reza Aghataher

18. 3D Geo-Network for Agent-based Building Evacuation Simulation 283
Jinmu Choi and Jiyeong Lee

19. Hierarchical Modelling of Multi-Geospatial Databases as Basis for Geo-Oriented 3D Analysis Capabilities 301
Sagi Dalyot and Yerach Doytsher

20. Solar Radiation over the Urban Texture: LIDAR Data and Image Processing Techniques for Environmental Analysis at City Scale 319
Cláudio Carneiro, Eugenio Morello, Carlo Ratti and François Golay

21. Creation and Error Analysis of High Resolution DEM Based on Source Data Sets of Various Accuracy 341
Jari Pohjola, Jari Turunen, Tarmo Lipping and Ari Ikonen

22. A Topological Analysis Method for 3D Geo-Entities Structured as Hexahedron Tessellations 355
GUO Jiateng and WU Lixin

23. Constraint-based Generation and Visualization of 3D City Models 365
Volker Coors, Karina Hünlich and Giwon On

24. GeoVEs as Tools to Communicate in Urban Projects: Requirements for Functionality and Visualization 379
Mahmud Shahrear Kibria, Sisi Zlatanova, Laure Itard and Machiel van Dorst

25. Producing 3D Applications for Urban Planning by Integrating 3D Scanned Building Data with Geo-spatial Data 397
Yonghui Song, Hongxia Wang, Andy Hamilton and Yusuf Arayici

26. 3D Dynamic Simulation and Visualization for GIS-based Infiltration Excess Overland Flow Modelling 413
Izham Mohamad Yusoff, Muhamad Uznir Ujang and Alias Abdul Rahman

Keyword Index 431

Part I:
Keynotes

Chapter 1
A Virtual Geographic Environment for a Simulation of Air Pollution Dispersion in the Pearl River Delta (PRD) Region

Hui Lin, Jun Zhu, Bingli Xu, Wenshi Lin and Ya Hu

Abstract. The rapid economic development of the PRD (Pearl River delta) has led to severe air pollution. Air pollution control is critical for sustainable development. From a geographic viewpoint, this paper will focus on how to develop a virtual geographic environment (VGE) to study environmental issues. Unlike existing work, our methods pay attention to dynamic 3D volume representation in virtual PRD terrain environment, which make air pollution transportation and dispersion easily understood to public and officers. Moreover, some GIS spatial analysis and real-time interactive operation with VGE will also be implemented to support the management and control of atmospheric pollution. A prototype system is developed to construct a virtual geographic environment for simulating the PRD region air pollution and dynamic interaction. Experimental results prove that the scheme addressed in the paper is effective and feasible.

1.1 Introduction

Urbanization in China has occurred most rapidly in the coastal areas, due to the stronger economic base and more developed infrastructure, as well as the greater abundance of natural resources. As a result, city clusters have arisen in coastal areas and nearby regions. The three largest city clusters – the Beijing–Tianjin–Bohai Bay, Yangtze River delta, and Pearl River delta regions – have become the forerunners of modernization in China. Over the past two decades, such clusters have played a leading role in China's economic growth, owing to their collective economic capacity and

---

Institute of Space and Earth Information Science, The Chinese University of Hong Kong, Shatin, Hong Kong;

Surveying Engineering Department, School of Civil Engineering, Southwest Jiaotong University, Chengdu, 610031, P.R. China;

Department of Atmospheric Sciences, School of Environmental Science and Engineering, Sun Yat-sen University, Guangzhou, 510275, P.R. China

interdependency [10]. However, the economic boom has led to a general decline in environmental quality. Currently, more than three-quarters of the urban population are exposed to air quality that does not meet China's national ambient air quality standards. Great efforts have been made to bring air pollution under control [6].

GIS (Geographic information system), visualization and atmospheric dispersion model have been integrated to support air quality management and air pollution control [3, 1, 2, 12, 9]. They mainly utilize GIS systems to realize the management, the inquiry, the analysis and the visualization of air pollution data. However, their research objects are mostly confined to a single city and don't consider the between-city influence in the transportation and the dispersion of the pollutant among city clusters. Atmospheric dispersion has obvious space-time dynamic characteristic. It can be represented as (x, y, z, t, a) , where (x, y, z) is spatial dimension, t is temporal dimension and a is attribute (it can include a lot of different attributes). The process of atmospheric pollution dispersion results from the dynamic change of spatial dimension, temporal dimension and attributes. Three-dimensional air pollution in existing researches is generally compressed into two-dimensional information, which is represented with isoline, isosurface and slice by means of graphic software tools such as GrADS and Vis5D. Due to the limitations of the 2D map space, these methods have to omit much useful information about air pollution. The intrinsic obscure tendency of map metaphor based information encoding and decoding in lower dimensions of space misdirects the communication between air pollution information and users.

Based on GIS and Virtual Reality (VR) technologies, a Virtual Geographic Environment (VGE) is a virtual representation of the natural world that enables a person to explore and interact, in cyberspace, with the vast amounts of natural and cultural information gathered about the physical and cultural environment [11]. Unlike the traditional data-centered GIS, the VGE has double-core data and model. The VGE pays attention to offer more intuitive and perceptive expressing methods and means, such as 1D speech or text, 2D map, 2.5D DTM (digital terrain model), 3D GIS, 4D animation, and xD thematic attributes. Unusual knowledge therefore can be easily discovered [7]. The utility of VGE enables the abstract and complicated phenomena to be more actual and intuitive, and then facilitates spatial decision-making support, such as exploration, explanation, prediction and planning. The multidimensional and dynamic analysis methods in the VGE have become the fundamental approaches for exploring spatial problems from all dimensions [4]. The interactive visualization of the three-dimensional models is of great importance for an in-depth analysis of the air pollution.

This proposed research will use the VGE method for dynamic modeling atmospheric pollution dispersion phenomena in city clusters. From a geographic viewpoint, this proposed research will focus on how to develop a interactive virtual geographic environment (VGE) to study environmental issues. Unlike existing work, we pay attention to dynamic 3D volume representation in virtual terrain environment, which make air pollution transportation and dispersion easily understood to public and officers. Moreover, some GIS spatial analysis and real-time interactive operation with the VGE will also be implemented to support the management and the control of atmospheric pollution.

The remainder of this paper is organized as follows. Section 1.2 addresses atmospheric pollution dispersion model for the Pearl River Delta (PRD) region in China. Section 1.3 discusses in detail how to construct a virtual geographic environment for atmospheric pollution dispersion. A prototype system is developed and rudimentary experimentations implemented in section 1.4. Conclusions and future research plans are addressed in the final section.

1.2 Atmospheric pollution dispersion modeling

Atmospheric dispersion modeling is the mathematical simulation of how air pollutants disperse in the ambient atmosphere. It is performed with computer programs that solve the mathematical equations and algorithms of fluid mechanics computation that simulate the pollutant dispersion. The dispersion models are used to estimate or to predict the downwind concentration of air pollutants emitted from sources such as industrial plants and vehicular traffic. Such models are important to governmental agencies tasked with protecting and managing the ambient air quality. A typical simulation scale of pollution dispersion in city clusters is from a dozen kilometer to several hundred kilometers and can use several types of air pollution dispersion models including the Lagrangian model and Eulerian model, as well as some hybrids of them.

This paper used the Eulerian model based on a modified mesoscale atmospheric model to set up a real-time three-dimensional dispersion model on the basis of simulation of meteorological field, called SYSUM [8]. SYSUM is a comprehensive, three-dimensional, multiscale, nonhydrostatic, Eulerian atmospheric chemical transport model, which is designed to simulate single air pollution dispersion in the troposphere. The model is modified from the mesoscale meteorological model PSU/NCAR MM5 [5], plus an equation of air pollution transportation and dispersion. The MM5 provides the time-dependent three-dimensional wind, temperature, pressure and specific humidity fields. The air pollution transportation and dispersion is driven "on-line" by the meteorological output of the MM5. The model has been developed for simulations of air pollutant dispersion in complex terrain. The time-varying trace gas concentrations are predicted by solving numerically a species continuity equation that includes advection, diffusion and deposition processes. The form of equation is:

$$\frac{\partial c}{\partial t}+\frac{\partial uc}{\partial x}+\frac{\partial vc}{\partial y}+\frac{\partial wc}{\partial z}=\frac{\partial}{\partial x}\left(K\frac{\partial c}{\partial x}\right)+\frac{\partial}{\partial y}\left(K\frac{\partial c}{\partial y}\right)+\frac{\partial}{\partial z}\left(K\frac{\partial c}{\partial z}\right)+Q-L \tag{1}$$

Where c stands for the concentrations of an air pollutant (g/m³), u、v、w is the velocity component in three dimensions (m/s), K is the eddy diffusivity, and Q and L are the emission and deposition fluxes, respectively. Only the dispersion of SO2 without any chemical reaction has been considered in this first model application. This paper does also not consider the dry and wet deposition of SO_2.

Our case study area is the Pearl River Delta (PRD) Region, with longitude between 21.43658°N and 23.95069°N, and latitude between 112.3945°E and 114.9579°E. The PRD region, with a total area of 42824 km^2, is located on the South China coast and is home to several large Chinese cities such as Hong Kong and Guangzhou, as well as a number of mid-sized cities. This region has experienced astonishing economic and industrial development in the past two decades. The domain PRD region has 91×91 grid points with a grid length of 3km. SO_2 emission sources in PRD region include

3179 point sources obtained in advance. The emission inventory is available at a horizontal resolution of 3km×3km which corresponds to the horizontal grid interval of the model domain. Apart from the PRD, SO_2 emission sources are neglected. For model to adequately resolve the boundary-layer processes, from the top to surface level, there are 45 vertical sigma levels. Fifteen layers were below 500 m height. The first level is situated 2.5 m above the ground. An experiment was initialized at 0000UTC on October 28th 2004 and integrated for 72 hours. The time step was 60 seconds for the domain PRD region. No chemical reactions have been considered because we focused solely on the transportation and diffusion of the air pollutant, SO_2 released from an elevated point source in the short simulated time (72 hours). Dry and wet deposition processes were also neglected as there is no rainfall in the short simulated time.

1.3 Modeling virtual geographic environment

The VGE in this paper aims to reach three main goals: 1) dynamic 3D volume representation for atmospheric pollution dispersion; 2) multi-scale, virtual terrain scene to meet different users' requirements; 3) dynamic interactive with virtual scene and implementing spatial analysis such as querying spatial position and overlaying thematic maps. From a technical point of view, the VGE system should implement real-time rendering and dynamic scene interaction, which allow users to interactively operate, explore and analyze scene models.

To reduce costs and save time, we used open source software and commodity graphics hardware. As the real-time graphic library, we used OpenSceneGraph (OSG). The OSG library based on OpenGL has many features, such as cross-platform (Windows, Linux, Irix), real-time optimization, wide range of input formats support, extensible through-plug-in architecture and built in support of paged LOD (Level of Detail) terrain generation and navigation. The OSG already provides a tool to generate hierarchical paged terrains from geo-images and digital elevation models (DEM). Moreover, it is easy to implement our algorithm into OSG framework.

1.3.1 3D volume visualization of air pollution

Pollutants can be classified as dynamic and fuzzy boundary volume objects, which are irregular in shape as opposed to regular objects that can be presented by points, lines, areas as well as small 3D elements. They are quite different from other kinds of information, for their unique properties of multidimensional structural and dynamic change. Particle system-based modeling is a well-known approach to modeling fuzzy boundary volume objects. The particle system refers to a computer graphic technique to simulate fuzzy phenomena, which has been widely used in many areas of computer graphics and scientific visualization.

After preprocessing air pollution data, they will be seen as a collection of many minute particles that together represent air pollution objects. Each particle has many attributes including position, shape, size, color, speed, direction and lifetime. Over a period of time, particles are generated into a system, move and change within the system, and die. In general, each parameter specifies a range in which a particle's value must lie. Normally, a range is specified by providing its mean value and its maximum variance. Stochastic processes are used to represent the fuzzy feature of pollutants.

In the virtual geographic environment, the most ideal situation is that all particles fill in the whole three-dimensional space. However, because of the computer's limitations, the particle quantity will directly influence the real-time efficiency of 3D scene rendering, especially in a large-scale and complicated scene. Visibility culling algorithms and creating LOD models are good ways to reduce the amount of particles. Therefore, we present a general purpose volume rendering algorithm that is based on the continuous level of detail idea. It maintains an octree to construct a view-dependent representation of regular volume data. After decomposing each leave node of the octree into tetrahedra these can be rendered efficiently.

Firstly, we design one data structure for hierarchical volume representation. Given a three-dimensional scalar field, which is defined by an array with 2^n+1 (n>0) grid points in each dimension, a hierarchical volumetric mesh is constructed by building an octree in a bottom-up fashion. Grids with a size other than 2n+1 have to be padded or re-sampled. Secondly, in order to perform a view-dependent simplification the octree has to be updated for each frame. During a top-down traversal of the octree, our approach calculates an upper limit on the local screen space error of each node. If the local error exceeds a predefined threshold the corresponding node is split into eight children. Thirdly, view frustum culling is used to speed up rendering. During the rendering traversal of the octree each node is tested against intersection with the view frustum. If a node does not overlap with the view frustum, it is invisible and can be discarded. The OSG can offer related functionality and be easily implemented to view frustum culling. Finally, imaged-based rendering techniques, such as impostors and billboard can also be used to improve rendering efficiency, which replaces a complex object by an image that is projected on a transparent quadrilateral. The particle system is used to represent close scene models, and imposter technique is used to represent distance scene models. Different LOD nodes can switch over automatically by means of one projection error of the screen pixel.

1.3.2 Virtual geographic scene

Virtual terrain landscape plays an important role in the virtual geographic environment, which includes DEM data, 2D imagery data, such aerial photography or topographic maps and 2D planning data, such as cadastre data or street networks. Real-time terrain scene rendering has to cope with handling large-scale terrains and related textures. VirtualPlanetBuilder (VPB) based on OSG can generate paged 3D terrain models from digital elevation data and imagery. VPB produces a hierarchy of terrain patches as a directory of .ive native optimized and compressed .osg format files. The built-in paged LOD engine of the OSG, takes care of loading the required terrain patches, according to the user selected view position and orientation; the frame rate is kept as constant as possible, while the background thread keeps increasing the detail of the loaded tile-patches.

OsgGIS is a GIS integration library for OpenSceneGraph, which can support the importing of GIS data and implement spatial analysis. Even if VPB can produce good quality paged terrains, it does not handle any scene data, vector information in GIS format (shape files) or 3D models. osgGIS makes it easier to access, visualize, and query, and manipulate GIS data from within an OSG application. In short, we can generate spatially accurate 3D geometry from GIS data layers while maintaining connectivity to the GIS data sources. It can make it easy to: 1) connect to and query various GIS data sources, 2) bring vector and raster data into OSG data structures for visualization, 3) deal with map projections in OSG, 4) access GIS feature attributes, 5) display 3D GIS data in conjunction with a VPB terrain model, 6) manage geometry for very large numbers of features, 7) automate the creation of 3D data layers from GIS data.

Moreover, we can use the OSG to implement interoperability among different scene models. The osgDB library of OSG allows applications to load, use and write 3D databases. It provides support for a wide variety of common 2D image and 3D model file formats using plugin architecture. The osgDB maintains a registry of and oversees access to the loaded OSG plugins. There are 45 plugins in the core OSG distribution, which can help us to import and display many other format 3D models.

1.4 Prototype system and experiments

Using Visual Studio 2005 .NET, OpenGL and OpenSceneGraph, a prototype system was developed to construct a virtual PRD geographic environment. The experimental results have been measured on a Hewitt Packard (HP) Compaq 6910P Notebook PC equipped with a 1.8 Ghz Intel core2 processor, 2048 MB of memory and a ATI Mobility Radeon TM X2300 graphics accelerator with 128 MB of graphic memory. The computer can support the real-time walkthrough for large scale terrain and real-time simulation for dynamic atmospheric pollution dispersion.

The PRD terrain data in this study includes remote sensing data, digital elevation models and surface structure data, which have different kind of resolution types. Fig. 1.1 shows three different LOD layers terrain models (e.g., PRD region with DEM data of 100m resolution and ETM images of 30m resolution, Hong Kong region with DEM data of 10m resolution and RS images of 10m resolution, the Chinese University of Hong Kong campus with DEM data of 1m resolution and aerial photography photo of 1m resolution).

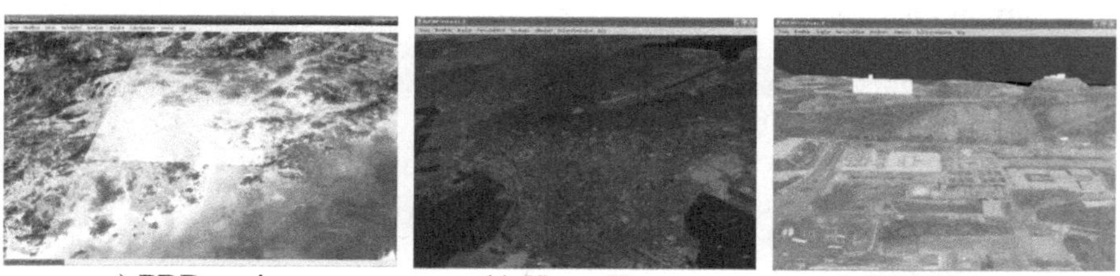

a) PRD region b) Hong Kong c) CUHK campus

Fig. 1.1. Multi-scale representations of PRD region terrain scene

Fig. 1.2(a) shows a terrain scene rendered with a wireframe generated with LOD screen projection error of 1. The rendering speed is 60 frames per second. Fig. 1.2(b) shows a terrain scene screenshot of a DEM overlaid by Landsat Thematic Mapper (TM) image texture (terrain model size is 10GB).

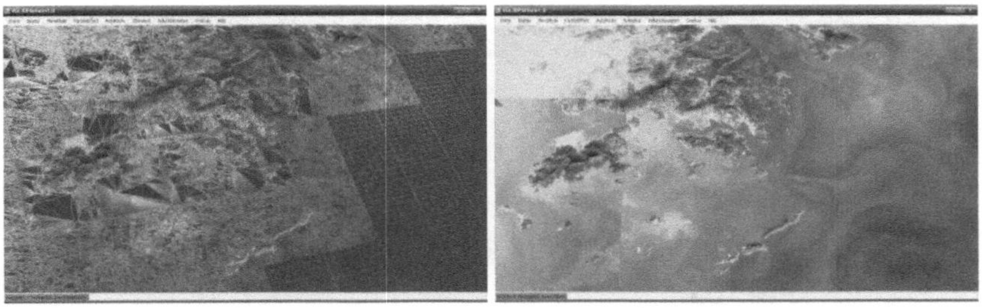

a) Wireframe terrain scene b) Terrain scene after using texture

Fig. 1.2. Terrain visualization experiment results

Our prototype system can offer some basic scene navigation functionalities (e.g., flying, walking). Some GIS's functionalities were developed to support geo-analyst. Fig. 1.3 shows virtual terrain overlaid thematic maps, which can display the geographic distribution of road traffic and buildings. We can use the mouse to investigate 3D coordinates of objects in virtual space and select one or several 3D objects to implement basic scene editing. These functions include move, zoom in/ zoom out, rotation, copy, delete and save. Fig. 1.4 shows a simulation of an imaginary pollution source on the CUHK campus site, based on the particle system technique.

A high performance computer was used to compute the atmospheric pollution dispersion modeling and exported a series of results data files according to the time step. Each file includes information on all spatial grid points, such as spatial position, simulation time, layer and pollution intensity. Based on the first data file, the particle system for volume rendering was initialized. Subsequent results of calculation data dynamically restrained changes in particles' attributes. Fig. 1.5 shows different effects including virtual terrain, air pollution effect and add the corresponding scene effect with fog.

Fig. 1.3. Thematic maps overlay

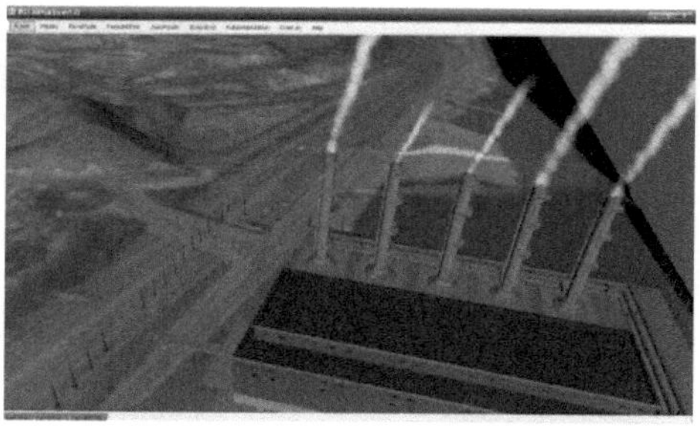

Fig. 1.4. Simulation of pollution source

a) Virtual terrain effect b) air pollution effect c) fog effect

Fig. 1.5. Different visualization effects

Fig. 1.6 shows simulation effect; first, at 20 minutes and subsequently at 60 minute intervals, which clearly shows spatial and temporal change of air pollution dispersion in the virtual geographic environment. The process of atmospheric pollution dispersion results from the dynamic change of spatial dimension, temporal dimension and attributes. Our prototype system can use dynamic 3D volume representation in a virtual geographic environment to explore air pollution dispersion from all dimensions. This approach leads to a better visualization of air pollution transportation and dispersion and greater understanding of air pollution transportation and dispersion in the PRD region for government officials and the general public.

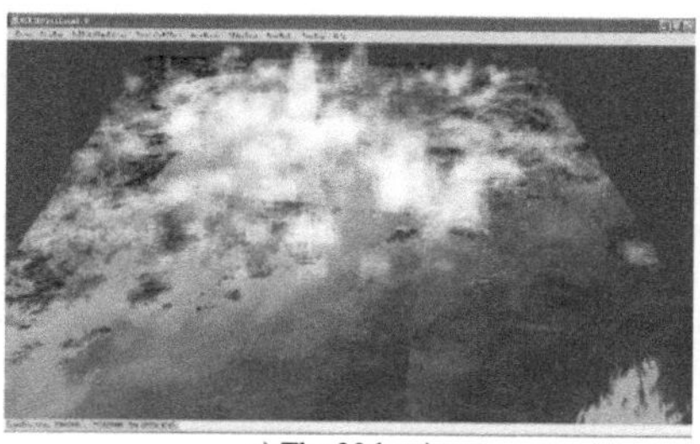

a) The 20th minute

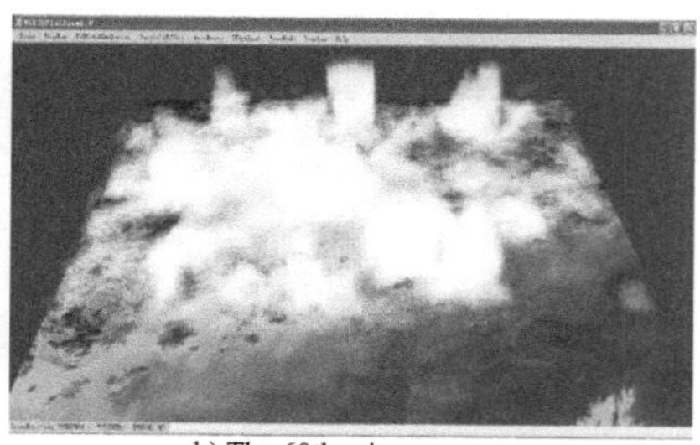

b) The 60th minute

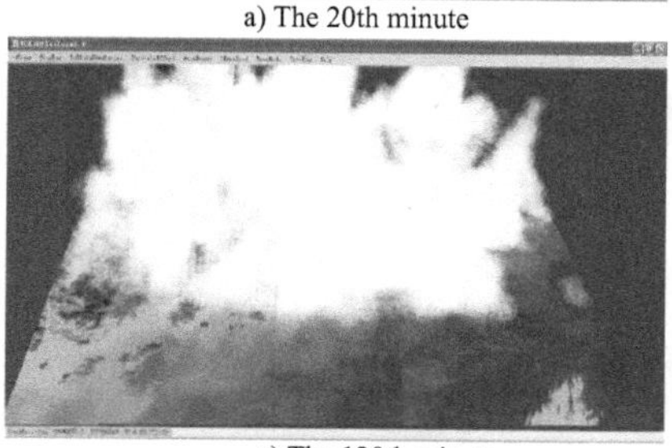

c) The 120th minute

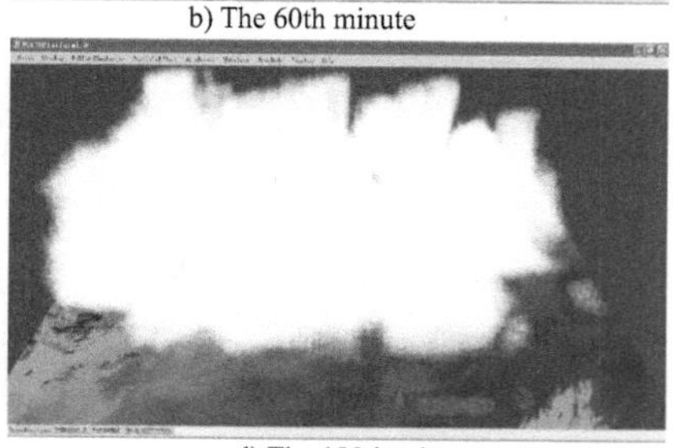

d) The 180th minute

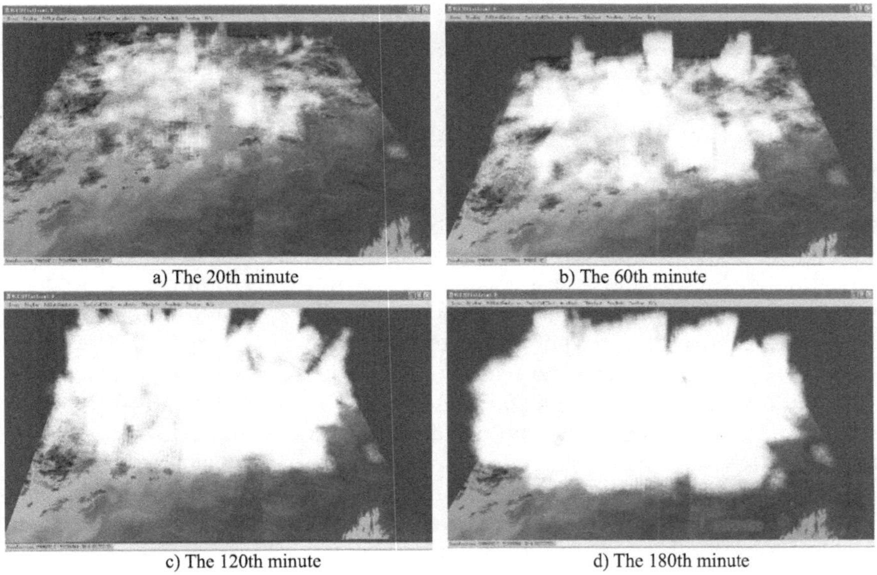

Fig. 1.6. Air pollution dispersion in the virtual geographic environment

From a view of visualization effect of simulation, our approach has better image quality. The virtual reality 3D volume rendering effects in the VGE can improve the overall appearance and realism of a simulation. Meanwhile, the rendering speed can be not less than 15 frames per second and well support interactive operations well, which allows users to interactively operate, explore and analyze models in the VGE scene. GIS spatial analysis and real-time interactive operations can be offered to support the management and the control of atmospheric pollution. The above experimental results prove that the scheme addressed in the paper is effective and feasible.

1.5 Conclusions

The utility of VGE enables the abstract and complicated air pollution dispersion to be more actual and intuitive, and facilitates spatial decision-making support, such as exploration, explanation, prediction and planning. Therefore the VGE method is good way to support the management and the control of atmospheric pollution. Air pollution control is critical for sustainable development because the rapid economic development of the PRD region has led to an increase in air pollution. China is expected to quadruple its GDP by 2020 (using 2000 as the base year for comparison) and, consequently, will face even more serious environmental challenges. As research background of the PRD air pollution, this paper pays attention to integrating the geographic database with a numerical meteorological model and air pollution dispersion model. Using Visual Studio 2005 .NET, OpenGL and OSG development package, a prototype system has been developed to construct the virtual PRD region environment, which can support real-time rendering and interaction of large-scale virtual scene models, such as 3D terrain and air pollution phenomena. It can offer an intuitive, efficient, and interactive visualization environment through which participants can explore complicated spatial air pollution information and implement GIS spatial analysis and inquiry.

Unlike existing research work such as GrADS and Vis5D software, our approaches pay attention to dynamic 3D volume representation in virtual PRD terrain environment. It can explore air pollution dispersion from all dimensions including spatial dimension, temporal dimension and attributes. The multidimensional and dynamic modeling methods of the VGE can lead to increased understanding of the air pollution issues in the PRD region. The VGE system can efficiently offer intuitive and accurate information to support decision-making in practice applications for the management and control of atmospheric pollution. For example, the virtual reality 3D volume visualization makes air pollution transportation and dispersion easily comprehensible to public and officers. It is a very significant factor in protect the atmospheric environment, reducing the atmospheric environmental pollution. Based on supercomputer clusters and WAN networks, future work will integrate geo-collaboration into our VGE system to build a collaborative virtual geographic environment, where geographically separated users can share a digital landscape environment and conduct collaborative work for the study of PRD air pollution.

1.6 Acknowledgments

This research is supported by the National High Technology Plan (863) of the People's Republic of China, Project No. 2006AA12Z207, CUHK RGC Project No. 447807 and the Ph.D. Programs Foundation for New Teachers of Ministry of Education of China (No.20070613005).

References

1. Chen HM, Wang XQ, Chen CQ, et al. (2005) GIS-Based Urban Atmospheric Dispersion Model – A Case Study in Fuzhou. Geo-information Science (In Chinese) 7(4): 101-106
2. Cinderby S, Forrester J (2005) Facilitating the local governance of air pollution using GIS for participation. Applied Geography 143-158.
3. Elbir T (2004) A GIS based decision support system for estimation, visualization and analysis of air pollution for large Turkish cities. Atmospheric Environment 38:4509-4517
4. Gong JH and Lin H (2000) Virtual Geographic Environments and Virtual Geography, Proceedings 9th International Symposium on Spatial Data Handling 28-39.
5. Grell GA, Dudhia J, Stauffer DR, (1994) A description of the fifth-generation Penn State/NCAR mesocale model (MM5) , NCAR Technical Note, NACR/TN-398+STR, National Center for Atmospheric Research, Boulder, CO, 117
6. Hao JM, He KB, Duan L, et al. (2007) Air Pollution and its control in China. Frontiers of Environmental Science & Engineering in China 1(2):129-142

7. Lin H and Zhu Q (2005) The Linguistic Characteristics of Virtual Geographic Environments, Journal of Remote Sensing(In Chinese) 9(2):158-165
8. Lin WS, Du DS, Zeng LP, et al. (2008) Numerical Simulation of Sea-land Breeze and Plume Dispersion in the Complex Coastal Terrain of South China. Submitted to Environmental Modeling & Assessment.
9. Pebesma EJ, Jong KD, Briggs D. (2007) Interactive visualization of uncertain spatial and spatio -temporal data under different scenarios: an air quality example. International Journal of Geographical Information Science 21(5):515-527
10. Shao M, Tang XY, Zhang YH, et al. (2006) City clusters in China: air and surface water pollution. Frontiers in Ecology and the Environment 4(7): 353-361
11. Zhu J, Gong JH, Liu WG, et al. (2007) A Collaborative Virtual Geographic Environment Based on P2P and Grid Technologies. Information Science. Information science 177(21):4621-4633.
12. Wang XH (2005) Integrating GIS, simulation models, and visualization in traffic impact analysis. Computers, Environment and Urban Systems 29:471-496

Chapter 2
Representing and Exchanging 3D City Models with CityGML

Thomas H. Kolbe

Abstract. CityGML is an open data model and XML-based format for the representation and exchange of virtual 3D city models. It is based on the Geography Markup Language version 3.1.1 (GML3). Both CityGML and GML3 are international standards issued by the Open Geospatial Consortium (OGC). CityGML not only represents the shape and graphical appearance of city models but specifically addresses the object semantics and the representation of the thematic properties, taxonomies and aggregations. The paper gives an overview about CityGML, its modelling aspects and design decisions, recent applications, and its relation to other 3D standards like IFC, X3D, and KML.

2.1 Semantic 3D City Models

Virtual 3D city models have been used in the past mainly for the visualisation or graphical exploration of cityscapes. Nowadays, an increasing number of applications like environmental and training simulations, urban planning and facility management, disaster management and homeland security, and personal navigation require additional information about the city objects given in a standardised representation. Semantic 3D city models comprise besides the spatial and graphical aspects particularly the ontological structure including thematic classes, attributes, and their interrelationships. Objects are decomposed into parts due to logical criteria (and not due to graphical considerations!) which follow structures that are given or can be observed in the real world. For example, a building will be decomposed into different (main) building parts, if they have different roof types and their own entrances like a house and the garage.

The semantic modelling of cities requires the appropriate qualification of 3D data. This can be done by an automated process in some cases or by manual interpretation.

Institute for Geodesy and Geoinformation Science
Technische Universität Berlin
kolbe@igg.tu-berlin.de

Anyway, it increases the efforts needed to create and maintain the 3D city model. From an economic viewpoint, the semantic modelling of cities only makes sense, if the data (in particular the semantic information) can be used by different customers within multiple applications. This, however, would require to find a common information model over the different users and applications.

This is what CityGML is all about. The aim of the development of CityGML was to reach a common definition and understanding of the basic entities, attributes, and relations within a 3D city model. By providing a core model with entities which are relevant to many disciplines the city model can become a central information hub to which different applications can attach their domain specific information (see Fig. 2.1). Information exchange between different disciplines can then be aligned with the objects of the city model. Similar ideas were discussed in the towntology project [1] and the nD modelling project [2].

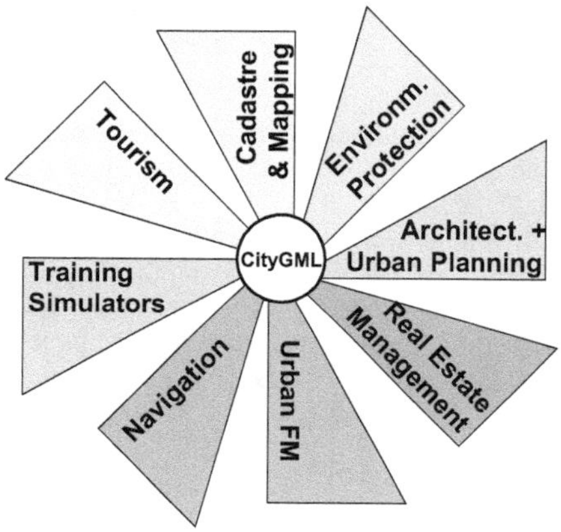

Fig. 2.1. The semantic 3D city model as the information hub which connects different disciplines by aligning information exchange with the city model entities.

2.2 CityGML and its Modelling Aspects

CityGML is an international standard for the representation and exchange of semantic 3D city and landscape models recently adopted by the Open Geospatial Consortium (OGC) [3, 4]. The data model behind CityGML is based on the ISO 19100 standards family and is implemented as an application schema for OGC's Geography Markup Language (GML 3.1.1) [5].

CityGML has been developed by the Special Interest Group 3D (SIG 3D) of the initiative Geodata Infrastructure North-Rhine Westphalia, Germany. The group consists of more than 70 members from industry, academia, and public administration. The different application backgrounds of the members include cadastre, urban planning, environmental and training simulations, civil engineering, architecture, computer graphics, geoinformation sciences, tourism, and telecommunication. Thus, a broad range of requirements, but also of expertise and experiences was brought into the consensus process for the definition of the data model. This process took a considerable amount of time. For example, it took more than 2 years with a regular full day meeting every six weeks of about 20 people to define the building model. Afterwards, the model was presented and discussed on international conferences and within the OGC. Much of the feedback was considered and led to further improvements. Although this does not guarantee to achieve a perfect data model, there is a good chance of not having overseen important issues. In general, all modelling suggestions (feature classes and attributes) have been rated with respect to their multifunctional use and were only considered if they were needed in various application domains.

CityGML represents four different aspects of virtual 3D city models, i.e. semantics, geometry, topology, and appearance. Above, all objects can be represented in up to five different, well-defined levels-of-detail (LOD0 to LOD4 with increasing accuracy and structural complexity). The following subsections will discuss each of these aspects as well as spatio-semantic coherence and extensibility of CityGML in more detail.

Since version 1.0.0 CityGML is functionally partitioned into modules (cf. Fig. 2.2). The vertical modules provide the definitions of the different thematic models like building, relief (i.e. digital terrain model), city furniture, land use, water body, and transportation etc. The horizontal modules (CityGML core, appearance, and generics) define structures that are relevant or can be applied to all thematic modules. This structure allows for partial CityGML implementations on the one hand and easy extension by further thematic models on the other hand.

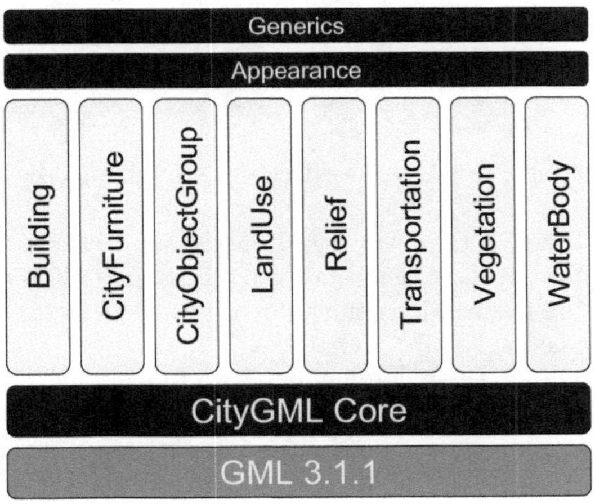

Fig. 2.2. Modularization of CityGML 1.0.0. Vertical modules contain the semantic modelling for different thematic areas.

2.2.1 Multi-scale representation

CityGML differentiates between five consecutive levels of detail (LOD), where objects become more detailed with increasing LOD regarding both geometry and thematic differentiation. Different LODs often arise from independent data collection processes and facilitate efficient visualization and data analysis. In a CityGML dataset, the same object may be represented in different LODs simultaneously, enabling the analysis and visualization of the same object with regard to different degrees of resolution.

The coarsest level LOD0 is essentially a two and a half dimensional digital terrain model (DTM). LOD1 is the well-known blocks model, without any roof structures. A building in LOD2 has distinctive roof structures and larger building installations like balconies and stairs. LOD3 denotes architectural models with detailed wall and roof structures, doors, windows and bays. LOD4 completes a LOD3 model by adding interior structures like rooms, stairs, and furniture. Beyond buildings, the LOD concept applies to the other object classes as well. Fig 2.3 illustrates the five different LODs in CityGML.

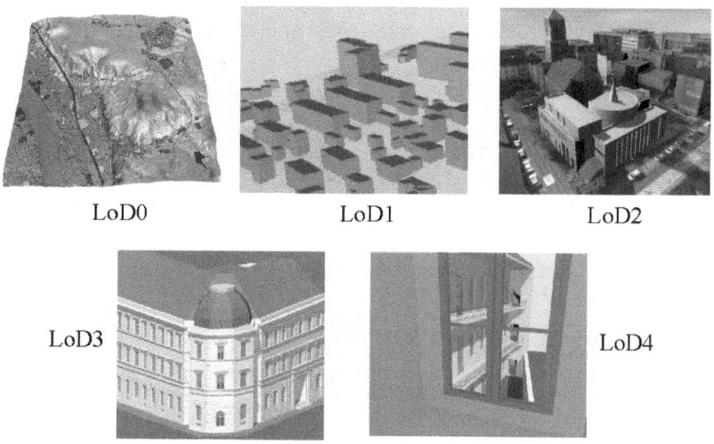

Fig. 2.3. Illustration of the five Levels-of-Detail (LOD) defined by CityGML.

The definition of the 5 LODs was based on previous work by different research groups on the usage of levels of detail in the representation of 3D city models [6, 7, 8]. LOD1 to LOD3 are inspired by similar classifications from [6] and [8]. LOD0 and LOD4 are originally defined by the SIG 3D.

The classification of models according to these 5 LODs may sometimes seem to be too coarse or restrictive (see [9]). However, the fixed number of LODs together with their specified information contents and accuracies establish quality classes. Just by naming the LOD of a CityGML dataset (or of objects within a CityGML dataset), one can quickly indicate the quality class for the 3D city model dataset. The LOD category makes datasets comparable and both the provider and customer / user get an idea of the data granularity, complexity, and accuracy.

2.2.2 Semantics

The semantic model of CityGML employs the ISO 19100 standards family framework for the modelling of geographic features. According to ISO 19109 geographic features are abstractions of real world objects [10]. They are modelled by classes which are formally specified using UML notation [11]. Geographic features may have an arbitrary number of spatial and non-spatial attributes. Object oriented modelling principles can be applied in order to create specialisation and aggregation hierarchies.

CityGML provides class definitions, normative regulations, and explanations of the semantics for the most important geographic features within virtual 3D city models including buildings, DTMs, water bodies, vegetation, and city furniture.

Fig. 2.4 depicts the top level class hierarchy of CityGML. The base class of all thematic classes is the abstract class *CityObject.* It inherits the attributes *name*, *description*, and *gml:id* from the GML superclass *Feature* and provides the additional attributes *creationDate* and *terminationDate* in order to model different object states over time periods (please note, that attributes are not shown in Fig. 2.4 for the sake of readibility). Above, every CityObject may be linked to objects in other datasets or external databases by an arbitrary number of *ExternalReferences*. These references may point to other representations of an object in cadastres, enterprise resource management systems or other application specific datasets. This is especially useful to maintain back links to original objects from which the 3D models might have been derived. *CityObjects* can be aggregated to build a *CityModel* which is a subclass of the GML superclass *FeatureCollection.*

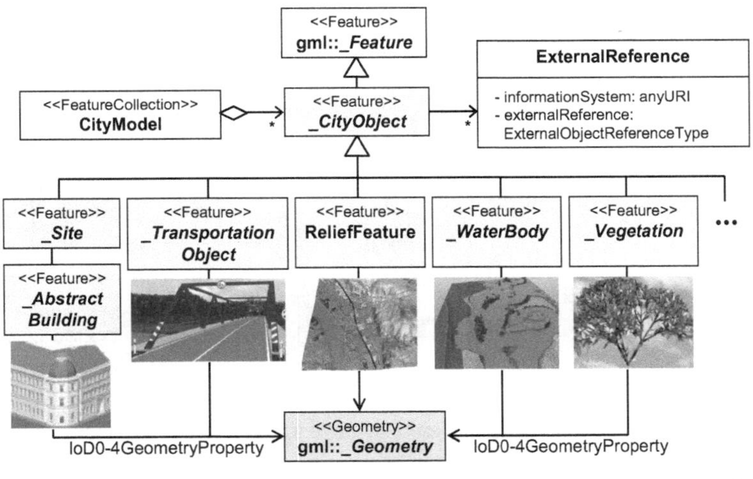

Fig. 2.4. UML diagram [11] of the top level class hierarchy of CityGML. All thematic objects are considered geographic features (following [10]) and their classes are derived from the abstract superclass CityObject. Attributes are omitted here for readability.

The subclasses of *CityObject* belong to different thematic areas and are defined within their corresponding modules (cf. Fig. 2.2). In the following, the building model will be discussed in more detail to illustrate the general modeling principles in CityGML. A complete description and normative reference is given in the CityGML specification document [3].

The pivotal class of the building model is *AbstractBuilding* from which the two non-abstract classes *Building* and *BuildingPart* are derived. These three classes follow the general composite design pattern [12]: a *Building* may contain *BuildingParts*, and as the latter class is also derived from *AbstractBuilding* a (recursive) aggregation hierarchy of arbitrary depth may be realized (see Fig. 2.5).

A *Building* or *BuildingPart* is described by optional attributes inherited from *AbstractBuilding* (not shown in Fig. 2.5): function, usage, and class, year of construction and demolition, roof type, measured height, number and individual heights of storeys above and below ground. *Building* and *BuildingParts* can be assigned multiple postal addresses.

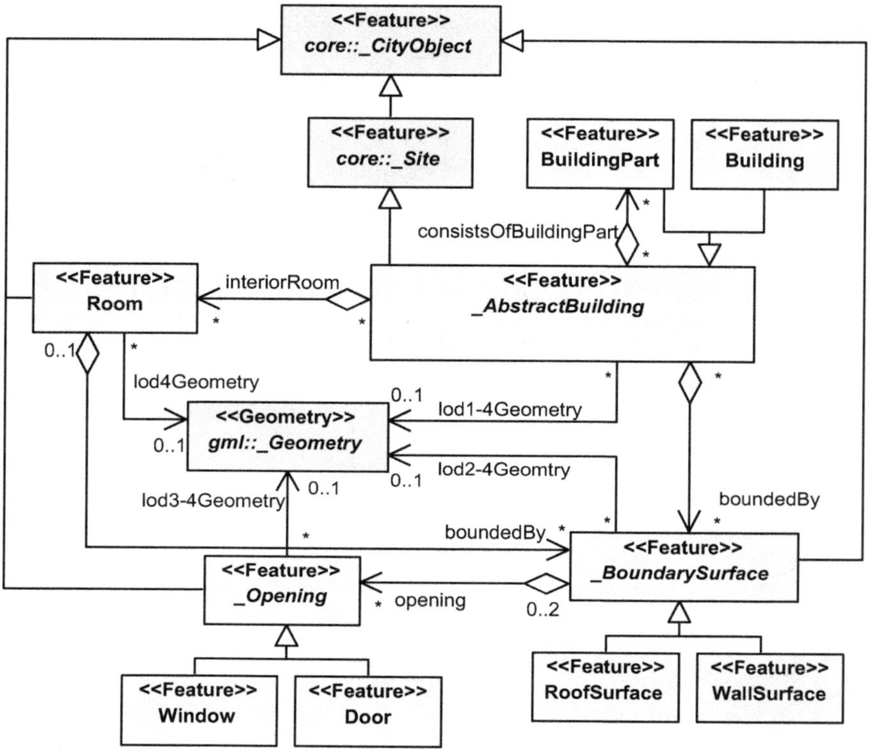

Fig. 2.5. UML diagram showing a simplified excerpt from the CityGML building model. Attributes are omitted for readability. Please note, that feature classes have multiple relations to geometry classes (e.g. *lod1Geometry*, *lod2Geometry* etc.).

Starting from LOD2 the boundary surfaces of *Buildings* and *BuildingParts* may be represented as semantic objects. *BoundarySurface* is the abstract superclass of these thematic objects having *RoofSurface*, *WallSurface*, *GroundSurface*, *ClosureSurface* etc. as subclasses (only the former two are shown in Fig. 2.5). As they are also derived from class *CityObject* they inherit its attributes and relations. In LOD3 and LOD4 openings in boundary surfaces may be represented by explicit thematic features of the classes *Window* or *Door*. Only in LOD4, buildings are allowed to have *Rooms*. Please note, that all feature classes have additional thematic attributes and further relations to different geometry classes not shown here.

The definition of this data model restricts the way buildings can be represented in a city model. In conjunction with the normative definitions of the feature classes this has an impact on the way how concrete data would have to be registered and structured. While this may be seen as restrictive concerning the freedom of modelling on the one side, it establishes a bindingness or liability of the producer with respect to potential users or customers on the other side. Users and their applications can expect buildings to be structured in a well-defined way. This makes it profitable to develop new applications that exploit the semantic information and structure of the *CityObjects*.

2.2.3 Geometry

CityGML uses a subset of the GML3 geometry model which is an implementation of the ISO 19107 standard [13]. According to ISO 19107 and GML3, geometries of geographic features are represented as objects having an identity and further (geometric) substructures. GML3 provides classes for 0D to 3D geometric primitives, 1D-3D composite geometries, and 0D-3D geometry aggregates. Composite geometries like *CompositeSurface* must be topologically connected and isomorphic to a primitive of the same dimension (e.g. *Surface*). Aggregate geometries like *MultiSurface* or *MultiSolid* do not impose topological constraints and thus their components can also permeate each other or be disjoint.

Volumetric geometries are modeled according to the well-known Boundary Representation (BRep, see [14]) where each solid is bounded by a closed surface (typically a *CompositeSurface*). In contrast to scene graphs or Constructive Solid Geometry (CSG) [14], all coordinates must belong to a world coordinate reference system (CRS) and no local transformations are allowed. 3D CRS refer to geographic or projected world reference systems. Above, GML3 also supports composite 2D+1D CRS with different reference systems for planimetry (x, y) and vertical coordinates (z). In general, the CRS define the frame for real world coordinates and they often differ for different countries, states, or even regions. However, the explicit association of each geometry with a CRS facilitates the integration of datasets by the application of the respective geodetic datum and coordinate transformations.

The advantage of having only absolute coordinates is that each geometry object belongs to exactly one fixed place in space. This allows to easily create and maintain spatial indexes in geodatabases or geoinformation systems. In fact, commercial and OpenSource RDBMS like Oracle Spatial or PostGIS both natively support (most of) the GML geometry model and different CRS. A big disadvantage is, however, that shape definitions cannot be reused like with CSG and scene graphs. For example, if a city model would contain 100 street-lamps or trees, 100 different, but equally shaped geometries would have to be created. In order to overcome this drawback, CityGML provides an extension to the GML3 geometry model called *ImplicitGeometry*. *ImplicitGeometries* refer to a shape geometry in a local coordinate system (to be reusable) and additionally provide a transformation matrix and an anchor point in a world CRS. The real world coordinates are computed by the multiplication of the local shape coordinates with the transformation matrix and then adding the coordinates of the anchor point. *ImplicitGeometries* are called like this, because the anchor point locates a place on earth where the associated shape 'unfolds' and thus implicitly describes the occupied space in real world. Features with implicit geometries can be retrieved from a GIS or spatial database by spatial selections on the anchor points.

In order to ensure a broad system support, CityGML is also restricted to non-curved geometries, as these often cannot be handled by GIS or spatial database management systems. Therefore, curved surfaces of real world objects must be approximated by facetted surfaces of planar patches.

According to ISO 19109 geographic features can be assigned more than one spatial property. This is generally being used in CityGML where most feature classes like *AbstractBuilding* or *Room* are assigned individual geometries for the different LODs by multiple associations (e.g. *lod1Solid*, *lod2Solid*, *lod3Solid*; c.f. Fig. 2.5) to the same geometry class (e.g. *Solid*).

For a given city model it is often not known, whether the data is topologically sound. For example, the boundary surfaces of buildings may not be closed or contain additional surfaces, which do not belong to the boundary of the volume (e.g. a roof overhang). For this reason, the geometry of nearly all thematic features can be represented by either *MultiSurfaces* or *Solids*. While *MultiSurfaces* are an unconstrained collection of surfaces in 3D space, *Solids* must be bounded by a closed (composite) surface (c.f. [3]).

2.2.4 Topology

For many applications topological correctness of the object geometries is crucial. For example, the surfaces bounding a building must be closed in order to be able to compute its volume. Also the solids of adjacent objects like *BuildingParts* must touch but their interiors are not allowed to permeate each other, because space can only be occupied by one physical object. In the application field of indoor navigation, rooms should be topologically connected as this facilitates the computation of a connectivity graph which then can be transformed to a 3D geometry network for route planning [15, 16].

In the past, different frameworks for the representation of 3D topology have been presented [17, 18, 19]. Nevertheless, most of them propose a geometric-topological structure where coordinates are stored only within the nodes or the points associated with the nodes. Higher dimensional primitives are then constructed by connecting primitives of lower dimensions. This way, nodes, edges, and faces can be shared by different primitives in order to construct complex shapes. By sharing primitives an explicit connection between higher dimensional shapes can be established and gaps or permeations can be avoided. This is illustrated in Fig. 2.6.

The topology model of ISO 19107 and GML3 also follows the line of full decomposition of *n*-dimensional topological primitives into (*n*-1)-dimensional primitives, which again are decomposed down to the level of nodes (0D). Please note that each primitive becomes an object (with ID). Furthermore, the topological representation in GML would require appropriate topological properties in the CityGML application schema. That means, that besides the geometry properties like *lod1Solid* to *lod4Solid* of class *Building* also corresponding topology properties like *lod1TopoSolid* to *lod4TopoSolid* would have to be added to the data model.

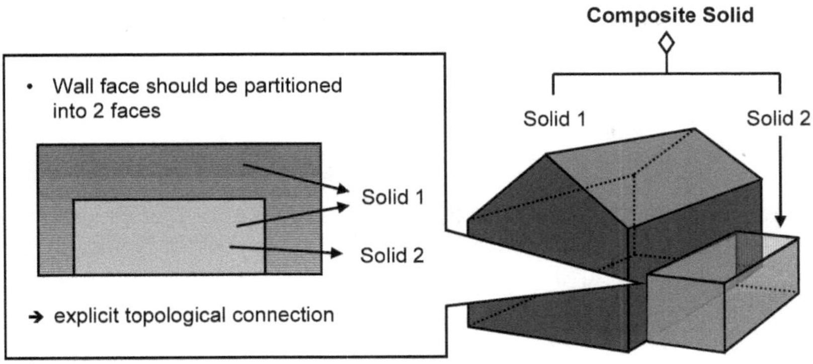

Fig. 2.6. A building with a connected garage. The topological adjacency between the solids representing the building extent and the garage extent can be made explicit by sharing the common wall surface.

As CityGML should be able to represent purely geometric models on the one hand and geometric-topological models on the other hand usage of the GML3 topology model would significantly increase the complexities of both the data model and concrete instances, i.e. datasets. If we take a closer look at Fig. 2.6 then we can also see, that it is not necessary to decompose the 3D building shape down to the level of nodes to establish the topological connection between the house and the garage. Indeed, it is sufficient to simply reuse the common wall surface of the house and the garage. This can be implemented in GML by providing the definition of the surface geometry inline within the specification of the solid geometry (bounded by a composite surface) of either the house or the building. In the representation of the other solid this surface is then included by reference (and not by value) which creates the connection between both solids. The following Listing shows an excerpt of a CityGML file which illustrates the referencing of a polygon 2.1 using the XLink mechanism [5]:

```
<bldg:BuildingPart>
  ...
  <bldg:lod2Solid>
    ...
    <gml:surfaceMember>
      <gml:Polygon gml:id="wallSurface4711">
        <gml:exterior>
          <gml:LinearRing>
            <gml:pos srsDimension="3">32.0 31.0 2.5</gml:pos>
            ...
          </gml:LinearRing>
        </gml:exterior>
      </gml:Polygon>
    </gml:surfaceMember>
  </bldg:lod2Solid>
  ...
</bldg:BuildingPart>
...
<bldg:BuildingPart>
  ...
```

```
    <bldg:lod2Solid>
      ...
      <gml:surfaceMember xlink:href="#wallSurface4711"/>
      ...
    </bldg:lod2Solid>
  ...
  </bldg:BuildingPart>
```

Listing 2.1. Establishing a topological connection between two solids by referring to the same polygon within the outer shells (composite surfaces) of the two solids.

2.2.5 Appearance

Information about a surface's appearance, i.e. observable properties of the surface, are considered an integral part of virtual 3D city models in addition to semantics and spatial properties. *Appearances* capture characteristics of surfaces as they appear to specific sensors like RGB or infrared (IR) cameras. Appearances are not limited to visual data but represent arbitrary categories called *themes* such as infrared radiation, noise or sunlight emission, or even earthquake-induced structural stress. Consequently, appearances can serve as input for both visualisation and analysis tasks.

Each surface is allowed to have assigned multiple appearances (e.g. one RGB texture for a building facade recorded in summer and another one from winter). All appearances must be assigned to themes like e.g. the "spring RGB theme" or the "winter IR theme". These themes are useful if one wants to visualize a 3D scene differently for winter or summer time. Switching of themes in a viewer client would mean a replacement of the surface materials and textures. By using themes, also multi-texturing can be supported by CityGML. One example could be the precomputation of shadow maps for each surface which would all be assigned to a theme that might be called "shadows". The shadow textures could then be mapped together with the RGB texture images to the corresponding surfaces.

The appearance data is given by texture images or by material definitions (adopted from X3D [20] and COLLADA). CityGML provides different ways how the raster data is mapped to the surfaces: *GeoreferencedTextures* can be used to describe the appearance of non-vertical surfaces. This way, e.g. an aerial orthophoto can be used together with its georeferencing information to easily describe the appearances of an arbitrary number of e.g. roof surfaces. It is up to a viewer to compute the proper extents of the textures from the 2D footprint of the corresponding 3D surface.

ParameterizedTextures use either explicit texture coordinates for each surface they should be mapped onto or a projection matrix. This can be computed from the exterior camera orientation and focal length associated with the texture image. 3D surfaces can simply refer to the picture in which their surface is best observed. The rest can be computed from these information by a viewer (or a 3D portrayal web service like the Web 3D Service [21]).

2.2.6 Spatio-semantic coherence

The ISO 19109 standard for the modelling of geographic features implies a dual structure of thematic object classes (the feature classes) on the one side and spatial properties (using the geometry and topology classes of ISO 19107) on the other side. In CityGML, objects like buildings may be decomposed both regarding their thematic structure into *BuildingParts*, *Rooms*, *WallSurfaces*, etc. and regarding their geometric structure into *CompositeSolids* consisting of *Solids* and these again be bounded by *CompositeSurfaces*.

In general, it is desirable to have a coherent thematic and spatial structuring of the objects. This means, that each complex thematic object is associated with a complex geometry object and each of the thematic components are also assigned to geometric components (being substructures of the original complex geometry). For example, a building may be assigned a solid geometry in LOD2. If the building is further decomposed into thematic surfaces like *WallSurface*, *RoofSurface* etc. their associated geometries should refer to those surface geometry objects which are part of the outer shell of the building's solid geometry. A fully coherent dataset has the advantage that each geometry object 'knows' what thematic role it plays and that each thematic feature 'knows' its location and spatial extent.

CityGML is flexible about the spatio-semantic coherence. It is possible to represent fully coherent datasets but also those types of data, in which either the semantic objects or the geometry objects are deeper structured. This can occur, for example, in a situation where the building geometry is reconstructed or acquired as a complex aggregation hierarchy which, however, is assigned to a simple Building object without any further thematic decomposition. A detailed discussion of spatio-semantic coherence and the typical cases for 3D city models is given in [22].

2.2.7 Extensibility

CityGML has been designed as a universal topographic information model that defines feature classes and attributes which are useful for a broad range of applications. However, in practical applications it is often necessary to store and exchange extra attributes or even 3D objects which do not belong to any of the predefined classes.

For these cases, CityGML generally provides two different ways of extension. The first is the usage of generic city objects and generic attributes, both defined within the module 'generics' (Fig. 2.2). Any *CityObject* may have an arbitrary number of additional generic attributes. For each generic attribute of an object the name, type, and value has to be given within the CityGML dataset. Supported data types are string, integer, real, date, and URI. *GenericCityObjects* may be assigned arbitrary geometries or *ImplicitGeometries* for each LOD. As they are derived from *CityObject* they may also be assigned generic attributes.

The second concept for extending CityGML are the so-called Application Domain Extensions (ADE). An ADE specifies systematic extensions of the CityGML data model, see Fig. 2.7. These comprise the introduction of new properties to existing CityGML classes like e.g. the number of habitants of a building or the definition of new feature classes. The difference between ADEs and generic objects and attributes is, that an ADE has to be defined within an extra XML schema definition file with its own namespace. This file has to explicitly import the XML Schema definition of the extended CityGML modules. The advantage of this approach is that the extension is formally specified. Extended CityGML instance documents can be validated against the CityGML and the respective ADE schema. ADEs can be defined (and even standardised) by information communities which are interested in specific application fields. More than one ADE can be actively used in the same dataset.

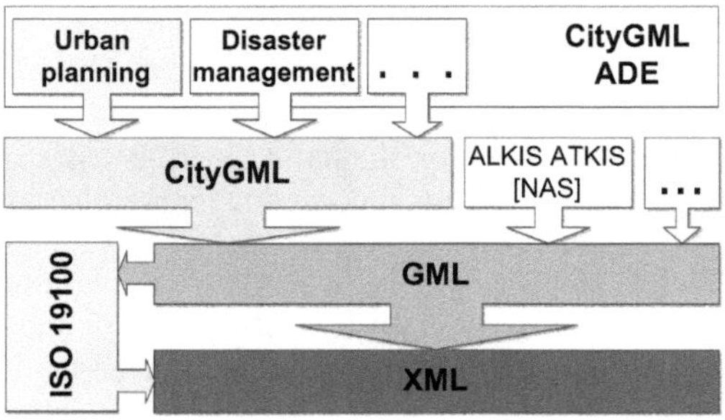

Fig. 2.7. CityGML Application Domain Extensions are XML schema definitions based on the CityGML schema. They extend CityGML by new feature classes and additional attributes for existing classes. CityGML is an application schema of GML (like others, e.g. the German national cadastre standard ALKIS).

2.3 Relation to other 3D Standards

Virtual 3D city models are not only investigated in the context of geoinformation systems. The field of architecture, engineering, construction, and facility management (AEC/FM) as well as the field of computer graphics provide their own standards for the representation and exchange of 3D models. The following subsections discuss the relations of CityGML to these standards.

2.3.1 Building Information Modelling / IFC

Building Information Modelling (BIM) means the semantic modelling of objects and processes in the the field of AEC/FM and CAAD. Like in CityGML, thematic objects are represented with their 3D spatial properties and interrelationships. Data are typically exchanged using the Industry Foundation Classes (IFC, see [23]), an ISO standard describing a product model and data exchange format for the built-up environment developed by the International Alliance for Interoperability (IAI).

IFC provides a very detailed semantic model for 3D building representations using constructive elements like beams, walls etc. Like in GML, IFC geometries are spatial properties of semantic objects. IFC has a very flexible geometry model (CSG, BRep, and Sweep representations), but does not provide support for CRS. Thus, georeferencing is not possible (yet).

Since the scope of IFC is restricted to buildings and sites, no topographic feature classes like terrain, vegetation, water bodies etc. are included. These would have to be modelled and exchanged in IFC as generic objects (i.e. IFC_Proxy).

IFC is a semantic model like CityGML, but with a different scope and at a different scale. IFC models can be converted to CityGML in different LODs preserving most of their semantic information [24]. CSG and sweep geometries have to be converted to BRep though. This way, IFC objects can be brought into the context of a city model within a GIS or spatial database and could then become subject to spatial and thematic queries (see [25]).

The derivation of IFC objects from CityGML data is a topic of future research, because from CityGML's surface models 3D volumetric components would have to be reconstructed. However, CityGML may be a good intermediate step in the (semi-)automatic acquisition of IFC models, because the CityGML object classes like *WallSurface*, *RoofSurface* etc. are much closer to photogrammetric observations or geodetic measurements (including laser scanning) than the component models of IFC (cf. [26]).

2.3.2 3D Computer Graphics and Geovisualisation / X3D & KML

CityGML is complementary to 3D computer graphics standards like X3D [20], VRML, or COLLADA and geovisualization standards like KML [27].

These standards address the way how to efficiently and nicely visualize 3D models and how to interact with them. Both in X3D and KML georeferenced coordinates can be used. Although it is in principle possible to also exchange semantic information using X3D (see [28]) or KML, both specifications do not standardise the way how to represent complex geographic features and their interrelationships.

CityGML should be considered a rich source format from which X3D or KML can easily be derived. CityGML is not optimized with respect to efficient visualization. However, the semantic information given by the explicit association of CityGML objects to thematic classes like buildings, doors, plants, and the provision of thematic attributes can be exploited to filter objects and to create 3D graphical shapes, appearance properties and materials accordingly.

2.4 Applications

CityGML is being employed already in different application areas. For example, the mapping of environmental noise pollution for the whole state of North-Rhine Westphalia in Germany is done based on CityGML data for about 8.6 million buildings in LOD1, road and railway networks in LOD0, and the digital terrain model in LOD1. In order to facilitate noise immission simulation, a specific CityGML Noise ADE was created which extends CityGML feature classes by noise data like the number of cars during daytime and nighttime or the type of roadbed and road material. Also a new feature type representing noise protection walls was modelled. The complete dataset comprises buildings, roads, railways, DTM, and noise barriers with most of them including noise specific extensions. It is forwarded to the noise simulation application as one CityGML file, accessed via a Web Feature Service over the Internet [29].

Municipalities nowadays create and exchange their 3D city models according to the CityGML standard. They use their models mainly for urban planning, city business development, and tourism. For example, the cities of Berlin [30], Stuttgart, Cologne, Munich, Düsseldorf, Recklinghausen, Frankfurt, and many more are providing CityGML compliant 3D city models. The VEPS 3D project has investigated different ways to employ 3D city models (and CityGML) in urban planning and public participation [31].

Within the OGC Open Web Services Testbed no. 4 (OWS-4), CityGML was employed together with IFC in a homeland security scenario. Here it was shown how an emergency operator can find a suitable location and building for a field hospital after the explosion of a dirty bomb [25]. Generally, CityGML can provide important information for disaster management [32]. This includes emergency route planning and indoor navigation [15, 33] as well as the extension of CityGML by dynamic water surfaces within a specific ADE for flooding scenarios [34].

CityGML may also become an important data source for training simulators. In [35] the requirements on 3D city models for training simulators are discussed in general, and in [36] concerning emergency car-driver training in particular.

The field of 3D cartography is not well developed yet. One important reason has been the lack of semantic 3D models in the past, because class information and thematic attributes are prerequisites for cartographic styling, symbolisation and generalisation. CityGML provides rich data that can be exploited for non-photorealistic rendering [37]. User defined styling may be specified using an appropriate 3D extension of the OGC Styled Layer Descriptors and Symbology Encoding (SLD/SE) standards [16].

2.5 Conclusions

CityGML is both a semantic model (specified by a formal data model) and an exchange format for virtual 3D city and landscape models. Rules for the acquisition and structuring of urban objects follow implicitly from this semantic modelling. The long discussions with the big group of people during the development of CityGML is (of course) not a guarantee, but a reasonable basis for the definition of these structures. It is now a topic of future work to bring CityGML to a wider adoption and discuss and learn from the experiences of the much broader user base.

The data model of CityGML balances between strictness and genericity. For this purpose it consists of three main parts: 1) the core thematic model with well-defined LODs, classes, spatial and thematic attributes, and relations; 2) *GenericCityObjects* and generic attributes allow the extension of CityGML data ‘on-the-fly’; and 3) ADEs facilitate the systematic extension for specific application domains.

CityGML also balances between simple objects and objects with complex thematic and spatial structures. Data is given high flexibility to grow with respect to their spatial and semantic structuring as well as their topological correctness in different stages of data acquisition or along a city model processing chain (see [26]).

Finally, CityGML is complementary to visualization standards like X3D or KML. While these address presentation, behaviour, and interaction of 3D models, CityGML is focused on the exchange of the underlying urban information behind the 3D objects.

It also complements building information models (BIM) and the IFC standard on a smaller scale and by topographic feature classes and georeferencing.

2.6 Acknowledgements

CityGML has been developed by the Special Interest Group 3D of the initiative Geodata Infrastructure North-Rhine Westphalia, Germany (GDI NRW). The international standardization was prepared by the CityGML Standard Working Group of the OGC and was supported by EuroSDR. I thank the members of these groups and especially my co-editors of the CityGML specification document for their contributions to this joint work. Special thanks go to Claus Nagel for his help in the preparation of the illustrations.

References

1. Teller J. , Keita A. K. , Roussey C. , Laurini R., 2005. Urban Ontologies for an improved communication in urban civil engineering projects. In: Proc. of the Int. Conference on Spatial Analysis and GEOmatics, Research & Developments, SAGEO 2005 Avignon, France, June, 20th-23rd.
2. Hamilton, A., Wang, H., Tanyer, A. M., Arayici, Y., Zhang, X., Song, Y., 2005. Urban information model for city planning, ITcon Vol. 10, Special Issue From 3D to nD modelling
3. Gröger, G., Kolbe, T.H., Czerwinski, A., Nagel, C., 2008. OpenGIS City Geography Markup Language (CityGML) Encoding Standard, Version 1.0.0, OGC Doc. No. 08-007r1, Open Geospatial Consortium
4. CityGML Homepage: http://www.citygml.org
5. Cox, S., Daisy, P., Lake, R., Portele, C., Whiteside, A., 2004. OpenGIS Geography Markup Language (GML3), Implementation Specification Version 3.1.0, OGC Doc. No. 03-105r1.
6. Köninger, A., Bartel, S., 1998. 3D-GIS for Urban Purposes. Geoinformatica, 2(1), March 1998
7. Coors, V., Flick, S., 1998. Integrating Levels of Detail in a Web-based 3D-GIS, In: Proc. of ACM GIS '98 in Washington D.C., USA, ACM Press
8. Schilcher, M., Guo, Z., Klaus, M., Roschlaub, R., 1999. Aufbau von 3D-Stadtmodellen auf der Basis von 2D-GIS (in German only). Zeitschrift für Photogrammetrie und Fernerkundung (ZPF), Vol. 67, No. 3
9. Döllner, J., Buchholz, H., 2005. Continuous level-of-detail modeling of buildings in 3D city models. In: Proc. of ACM GIS 2005, ACM Press
10. ISO 19109, 2005. Geographic information — Rules for application schema.
11. Booch, G., Rumbaugh., J. & Jacobson, I. 1997. Unified Modeling Language User Guide. Addison-Wesley.
12. Gamma, E., Helm, R., Johnson, R.E., 1995. Design Patterns. Elements of Reusable Object-Oriented Software, Addison-Wesley Longman
13. Herring, J., 2001. The OpenGIS Abstract Specification, Topic 1: Feature Geometry (ISO 19107 Spatial Schema). OGC Document Number 01-101.

14. Foley, J., van Dam, A,. Feiner, S. & Hughes, J. 1995. Computer Graphics: Principles and Practice. Addison Wesley, 2nd Ed.
15. Lee, J., Zlatanova, S., 2008. A 3D data model and topological analyses for emergency response in urban areas. In: Zlatanova, Li (eds.), Geospatial Information Technology for Emergency Response, Taylor & Francis
16. Neubauer, S., Zipf, A., 2007. Suggestions for Extending the OGC Styled Layer Descriptor (SLD) Specification into 3D – Towards Visualization Rules for 3D City Models. In: Proc. of 26th Urban and Regional Data Management, October 2007 in Stuttgart, Taylor & Francis
17. Molenaar, M., 1992. A topology for 3D vector maps. ITC Journal 1992-1.
18. Zlatanova, S. 2000. 3D GIS for Urban Development. PhD Thesis, ITC Dissertation Series No. 69, The International Institute for Aerospace Survey and Earth Sciences, The Netherlands
19. Oosterom, P., Stoter, J., Quak, W. & Zlatanova, S. 2002. The balance between geometry and topology. In: Proc. of 10th Int. Symp. SDH 2002, Springer
20. ISO 19775, 2005. X3D International Specification Standards, Web 3D Consortium.
21. Quadt, U., Kolbe, T. H., 2005. Web 3D Service Implementation Specification, OGC Discussion Paper, Doc. No. 05-019
22. Stadler, A., Kolbe, T.H., 2007. Spatio-semantic Coherence in the Integration of 3D City Models. In: Proceedings of the 5th International Symposium on Spatial Data Quality, Enschede 2007.
23. Liebich, T., Adachi, Y., Forester, J., Hyvarinen, J., Karstila, K., Reed, K., Richter, S., Wix, J., 2007. Industry Foundation Classes IFC2x Edition 3 – Technical Corrigendum 1, International Alliance for Interoperability, http://www.iai-international.org.
24. Benner, J., Geiger, A., Leinemann, K., 2005. Flexible generation of semantic 3D building models. In: Gröger, Kolbe (eds.), Intern. ISPRS / EuroSDR / DGPF-Workshop on Next Generation 3D City Models. Bonn, Germany, EuroSDR Publication No. 49.
25. Lapierre, A., Cote, P., 2007. Using Open Web Services for urban data management: A testbed resulting from an OGC initiative for offering standard CAD/GIS/BIM services. In: Coors, V., Rumor, M., Fendel, E. M., Zlatanova S. (eds): Urban and Regional Data Management. Proceedings of the 26th UDMS, October 10-12, 2007, Stuttgart, Taylor & Francis
26. Kolbe, T. H., Nagel, C., Stadler, A., 2008. CityGML – A Framework for the Representation of 3D City Models from Geometry Acquisition to full Semantic Qualification. In: Proc. of ISPRS Congress 2008 in Bejing, China
27. Wilson, T., 2008. OGC KML. OGC Encoding Standard, Version 2.2.0, OGC Doc. No. 07-147r2, Open Geospatial Consortium
28. Pittarello, F., De Faveri, A., 2006. Semantic Description of 3D Environments: a Proposal Based on Web Standards. In: Proceedings of the Web 3D Symposium 2006 in Columbia, Maryland, 18-21 April 2006, ACM Press.
29. Czerwinski, A., Kolbe, T. H., Plümer, L., Stöcker-Meier, E., 2006. Spatial data infrastructure techniques for flexible noise mapping strategies. In: Tochtermann, Scharl (eds.), Proc. of the 20th International Conference on Environmental Informatics - Managing Environmental Knowledge. Graz 2006.
30. Döllner, J., Kolbe, T. H., Liecke, F., Sgouros, T., Teichmann, K., 2006. The Virtual 3D City Model of Berlin - Managing, Integrating, and Communicating Com-

plex Urban Information. In: Proceedings of the 25th Urban Data Management Symposium UDMS 2006 in Aalborg, DK, May 15-17. 2006.
31. VEPS 3D Homepage: http://www.veps3d.org
32. Kolbe, T. H., Gröger, G., Plümer, L., 2008. CityGML – 3D City Models and their Potential for Emergency Response. In: Zlatanova, Li (eds.), Geospatial Information Technology for Emergency Response, Taylor & Francis
33. Mäs, S., Reinhardt, W., Wang, F., 2006. Conception of a 3D Geodata Web Service for the Support of Indoor Navigation with GNSS. In: Proc. of 3D GeoInfo 2006 in Kuala Lumpur, LNG&C, Springer
34. Schulte, C., Coors, V., 2008. Development of a CityGML ADE for dynamic 3D flood information. Joint ISCRAM-CHINA and GI4DM Conference on Information Systems for Crisis Management 2008 in Harbin, China
35. Bildstein, F., 2005. 3D City Models for Simulation and Training – Requirements on Next Generation 3D City Models. In: Proc. of the Int. ISPRS Workshop on Next Generation 3D City Models in Bonn, Germany. EuroSDR publication no. 49
36. Randt, B., Bildstein, F., Kolbe, T. H., 2007. Use of Virtual 3D Landscapes for Emergency Driver Training. In: Proc. of the Int. Conference on Visual Simulation IMAGE 2007 in Scottsdale, Arizona 2007
37. Döllner, J., Walther, M., 2003. Real-Time Expressive Rendering of City Models. In: Proc. of the 7th International Conference on Information Visualization in London, IEEE Press

Chapter 3
Research and Development Program for the Third National Geographic Information System Project in Korea: Korean Land Spatialization Program

Byung-Guk Kim, Hoon-Sung Choi and Yeun J. Jung

Abstract. Under control of the Ministry of Land, Transport and Maritime Affairs (MLTMA) and Korea Institute of Construction & Transportation Technology Evaluation and Planning (KICTTEP), Inha University, Incheon, Korea, has conducted Korean Land Spatialization Program (KLSP). KLSP aims at providing not only geospatial information technology, but also service and business models based on the technology for the future ubiquitous society. KLSP is one of the Value Creator (VC) 10 projects, which was planned by the Korean Government, for the future. The first program of KLSP, from 2006 to 2012, initiated with $132 million (US dollars) of National Fund and $42 million of Private Fund. The second program will be conducted from 2012 to 2016. This paper introduces KLSP and attracts international collaboration to this great program.

Keywords: National Geographic Information System (NGIS), Research and Development, Geospatial Information, Ubiquitous Society

3.1 Introduction

3.1.1 History

Ubiquitous technology has become one of main issues in Korean society. The technology has changed the current society of e-Land and e-Government to the future society of u-Land, u-Government, and u-City. The primary component of the ubiquitous technology is geospatial information based on geographic information. Geospatial information has been utilized for various fields such as making maps, environment,

Korean Land Spatialization Group
byungkim@inha.ac.kr

national defense, prevention against disaster, and resource management. Recently, major companies including Microsoft and Google plan for a new future industry in commercializing geospatial information.

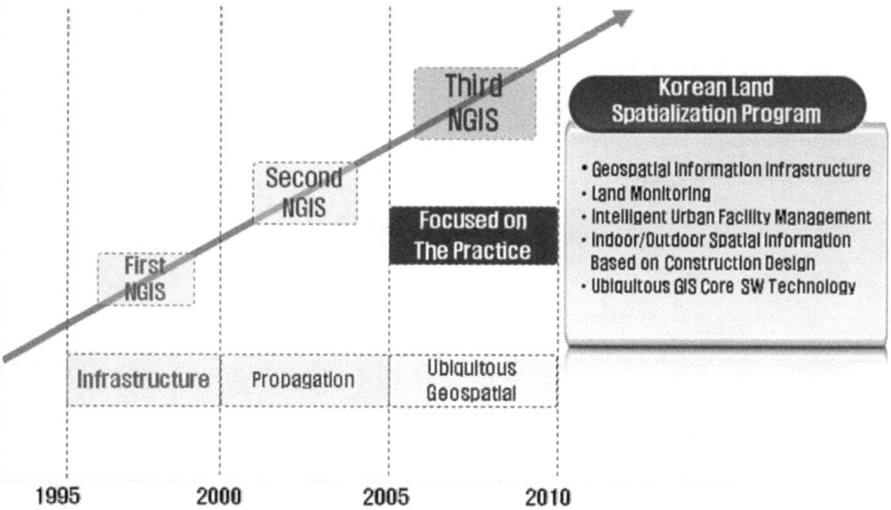

Fig. 3.1. Conceptual Diagram of NGIS Programs in Korea

From the recognition of the value of geospatial information by a gas explosion in Daegu in 1995, the Korean government carried out the first (1996-2000) and the second (2001-2005) NGIS projects, and has been pushing ahead with the third NGIS project (2006-2010). Fig. 3.1 shows conceptual diagram of the NGIS Programs in Korea. The first and second NGIS projects included development of GIS digital data and GIS technology, which may be the backbone of digital map, thematic map, geographic information, underground facility map, and cadastral map. However, existing GIS digital data only came from digitalizing existing paper maps, therefore the data may be inadequate in preparing a fast transition of geographic features, and potentially expensive to apply to various fields. In addition, GIS digital data that are not updated have a limit to be used as reliable data.

Table 3.1. History of KLSP

Date	Contents
Dec 29, 2005	Approved as the R&D part of "the Third NGIS Program"
May 22, 2006	Selected as VC-10 program in "the Construction and Transportation R&D Innovation Roadmap"
Nov 01, 2006	Elected Byung-Guk Kim as center director of KLSP
Feb 27, 2007	KLSG Office Opening
May 15, 2007	Select principal research institution for the third core project
Sep 11, 2007	Select principal research institutions for the first, second, fourth, and fifth core projects and initiate the first and second year's programs
Jul 16, 2008	Conduct self-evaluation by KSLG
Aug 26, 2008	Finish the first and second year's programs
Aug 27, 2008	Initiate the third year's program

To solve abovementioned problems, Ministry of Land, Transport and Maritime Affairs (MLTMA) planned the Korean Land Spatialization Program (KLSP), a R&D Project, as the third NGIS project. The primary objective of KLSP is to take a lead in technological developments for the realization of "Ubiquitous Infrastructures for Digital Korea." As shown in Table 3.1, KSLP was actually initiated with selecting center director, Byung-Guk Kim, who established center administration office, Korean Land Spatialization Group [1].

3.1.2 Construction of KLSP

KLSP aims at leading the world market through the connection between Korean Information Technology (IT) and GIS technology. As shown in Fig. 3.2, KLSP consists of five core research projects and one research coordination project to practically utilize and commercialize the results of core research. The five core projects include

(1) geospatial information infrastructure, (2) land monitoring, (3) intelligent urban management, (4) indoor/outdoor spatial information based on construction design, and (5) ubiquitous GIS core software technology.

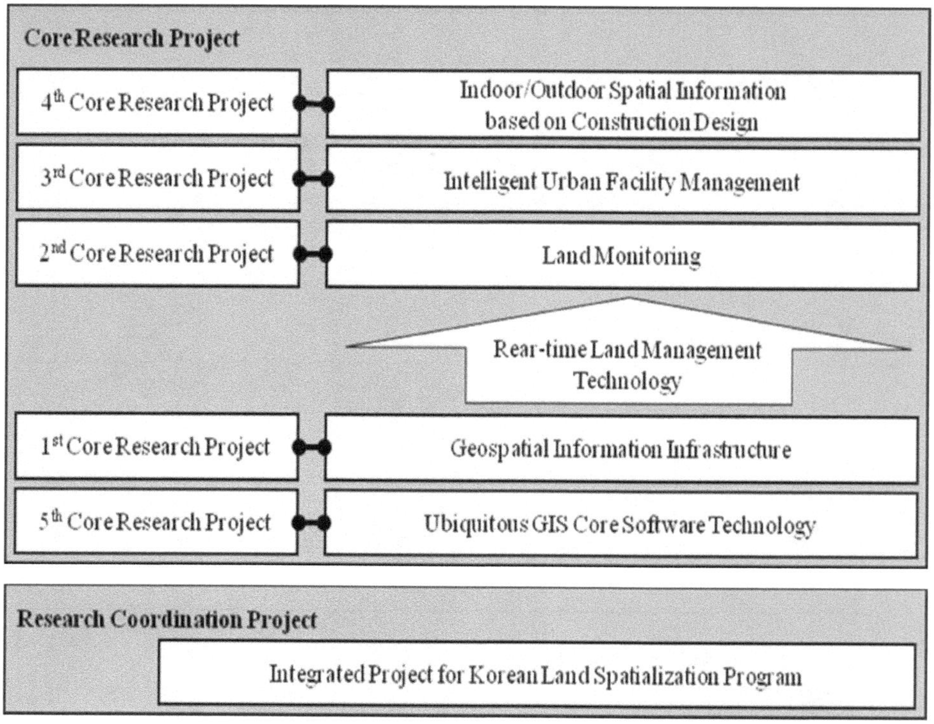

Fig. 3.2. Projects of KLSP

3.2 Core Research Projects and Research Coordination Project

3.2.1 The First Core Research Project: Geospatial Information Infrastructure

Geospatial information infrastructure is the first requirement to embody intelligent land nationwide, and includes development of base technology for developing advanced geodetic reference frameworks, high-tech mapping instruments and equipment, and real-time updates of geospatial information. This project is divided into two unit research projects.

The objective of the 1st unit research project is to develop technology and management system for advanced geodetic reference frameworks with three main subprojects; (1) development of advanced geodetic reference frameworks; (2) development of high-precision geoid model; (3) development of application systems based on integrated space geodesy. These projects enable the community to obtain more accurate geospatial information with horizontal error tolerance of mm unit and vertical error tolerance of cm unit. Fig. 3.3 shows the configuration of the first core research project.

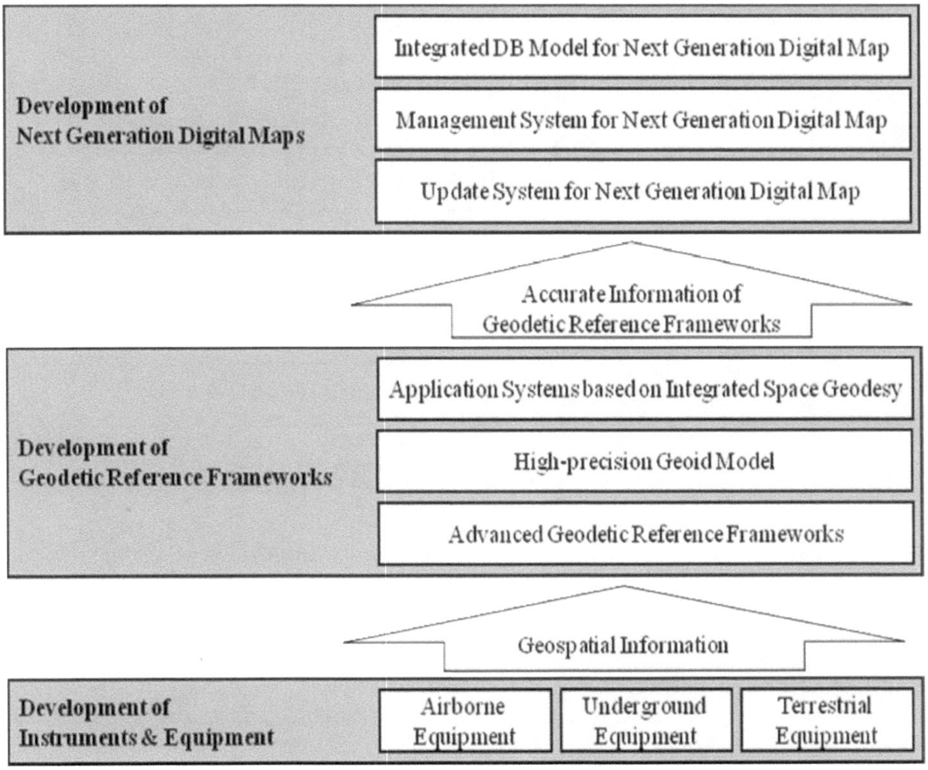

Fig. 3.3. Configuration of the First Core Research Project

The 2nd unit research project develops instruments and equipments for obtaining airborne, underground, and terrestrial geospatial information as well as digital mapping technology for the next generation. The main issue of this project is Unique Feature Identifier (UFID), which could be assigned to every geospatial features. Eventually, continuous and real-time geospatial information would be obtained by UFID, and utilized in various fields.

3.2.2 The Second Core Research Project: Land Monitoring

As shown in Fig. 3.4, land monitoring is also composed of two unit research projects, which include: (1) data acquisition of land monitoring and (2) data processing and applying of land monitoring. The objective of the second core research project is to integrate individual monitoring systems, and set a solid foundation for real-time integrated monitoring system. Using the results of this research, the decision making process would become more efficient in various fields such as resource management, prevention of disasters, environment, and land change.

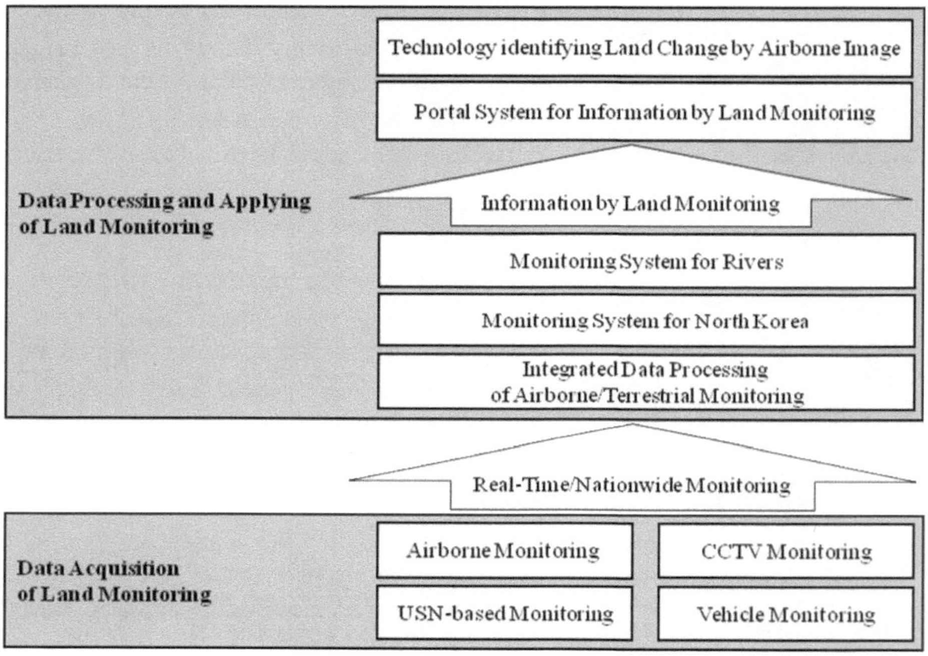

Fig. 3.4. Configuration of the Second Core Research Project

The 1st unit research project is conducted to know how to get land monitoring information. The project includes development of technology for airborne monitoring, CCTV monitoring, Ubiquitous Sensor Network (USN)-based monitoring, and vehicle monitoring. One of the main issues in the project is USN. USN consists of spatially distributed autonomous devices for sensing physical or environmental conditions, and these devices are collectively connected through a wireless communication network in order to support real-time measurements. Real-time and periodical geospatial information would be obtained by using the USN.

Data processing and applying of land monitoring is carried out in the 2nd unit research project. Using land monitoring technology developed in the 1st unit research project, this project seeks practical applications such as integrated data processing, monitoring system for North Korea, monitoring system for rivers, and technology identifying land change by airborne image. Likewise, portal systems for land monitoring information are also developed for the public.

This project for land monitoring will contribute to take the lead in airborne monitoring by preparing for Global Earth Observation System of Systems (GEOSS), and satisfy government agencies to efficiently monitor by means of developing a real-time and nationwide monitoring system. In addition, integration of conventional and emerging technology for land monitoring is expected to make it possible to accomplish the safe and comfortable society, with the support of application areas such as natural disaster monitoring and mitigation as well as natural resource/environment monitoring.

3.2.3 The Third Core Research Project: Intelligent Urban Facility Management

Development of intelligent urban facility management requires prevention and prompt action against accidents of urban facilities such as water and sewer systems, electricity and gas lines, traffic signal and utility poles, communication cables, regional heating facilities, and many others. This core research project also has two unit research projects.

The 1st unit research project is conducted to develop USN-based monitoring systems for urban facility management. Underground and ground facilities should be managed differently because each requires different monitoring technology and, particularly, because failure of underground facilities is often associated with urban emergencies. Projects for underground facilities try to develop Underground Facility Sensor Network (UFSN) and advanced monitoring and management systems. For ground facilities, USN-based technology and related monitoring and management systems are developed. Real-time monitoring and efficient lifecycle management for urban facilities would be feasible by means of abovementioned developments.

Urban facility management also requires an integrated and centralized platform to support the decision making process in an effective and efficient manner. In the 2nd unit research project, centralized urban facility management platform is developed with related sub-projects; (1) research and development of UFID for urban space; (2) awareness-based facility management system; (3) SOA/GRID-based simulation. Eventually, this project enables future urban cities to obtain any information and service whenever and wherever they are needed. Fig. 3.5 shows the configuration of the third core research project.

3.2.4 The Fourth Core Research Project: Indoor/Outdoor Spatial Information based on Construction Design

Based on construction design, the fourth core research project aims at developments of digital maps and related technologies for indoor/outdoor spatial information. CAD and GIS integration would be beneficial for building next-generation geospatial information because GIS data can be updated with the construction of CAD data having richer spatial contents. In this sense, the 1st unit research project seeks the renewal system of geospatial DB based on construction design. The project that utilizes

construction design develops GIS DB renewal system and indoor/outdoor location-awareness technology. These developments enable efficient renewal of building information by digital map and highly organized resource and labor management on the construction site using their location information.
In the 2nd unit research project, acquisition and application of indoor spatial DB, the current GPS technology for outdoor spatial information is utilized for the construction of indoor spatial information that includes DB, data management systems, sever systems, and many others. The results of this project can reduce the renewal cost of geospatial information, achieve up-to-date information, and make optimized construction management as well as enhanced construction workers' safety. Fig. 3.6 shows the configuration of the fourth core research project.

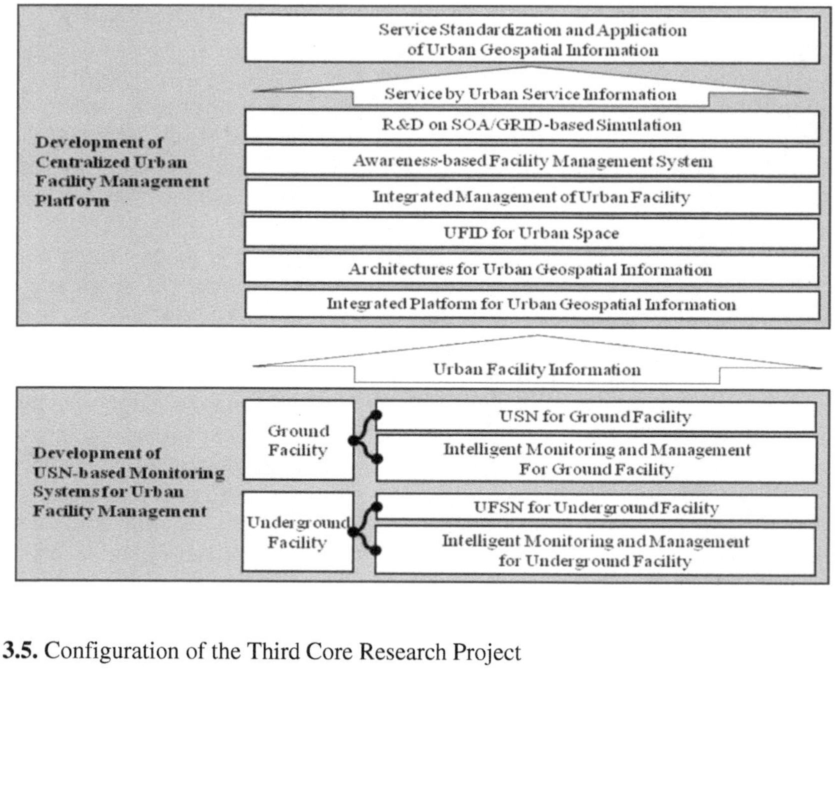

Fig. 3.5. Configuration of the Third Core Research Project

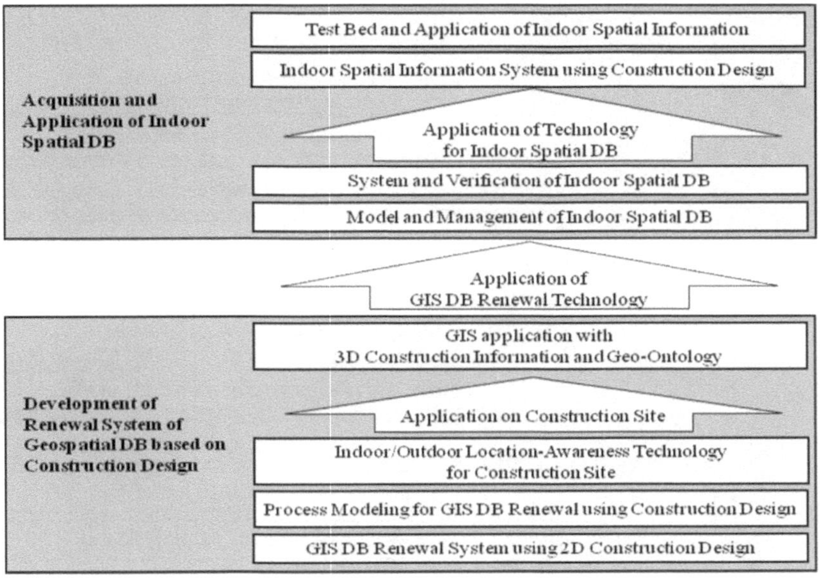

Fig. 3.6. Configuration of the Fourth Core Research Project

3.2.5 The Fifth Core Research Project: Ubiquitous GIS Core SW Technology

Ubiquitous Geographic Information System (u-GIS) is a part of the major ubiquitous computing technology trend. Not only are the ubiquitous computing infrastructures required, but also advanced positioning and mapping technology are required to realize u-GIS. The research of the fifth core project, ubiquitous GIS core software technology, is divided into two unit research projects; (1) processing and management of u-GIS geospatial information and (2) customized land information. The 1st unit research project develops storage and management systems for u-GIS spatial data, GeoSensor data, and mobile u-GIS data. In addition, research study for convergence and analysis of u-GIS data is also conducted in this project. Main sub-projects of the 2nd unit research project include developments of platform, mobile application technology, and visualization technology for customized land information. Particularly, for the visualization technology, augmented reality, which displays real space and virtual space together, is developed to support the efficient decision making process. Through this project, existing service systems from the standpoint of providers can be changed into customized service systems from the standpoint of end-users. The technology developed in the fifth core research project would contribute to the maximum use of u-GIS data, activate related businesses, and provide customized land information with the public through direct participation of end-users. Fig. 3.7 shows the configuration of the fifth core research project.

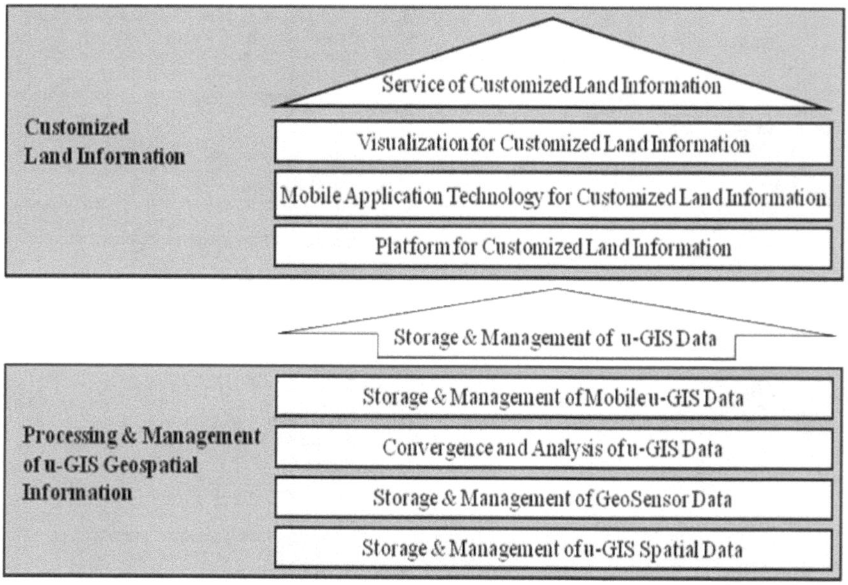

Fig. 3.7. Configuration of the Fourth Core Research Project

3.2.6 Research Coordination Project

The key objective of the research coordination project is to efficiently control and support five core projects. The project consists of three unit research projects; (1) construction of test-bed; (2) service modeling and standardization; (3) R&D portfolio and business modeling. Fig. 3.8 shows the configuration of the research coordination project.

Results of five core research projects will be operated and tested on a test-bed selected in the 1st unit research project. It is very important to operate the test-bed for verification of their applicability before commercializing and using the results. After finishing KLSP, the test-bed will be operated and maintained by the integrated operation center, and it will be used for testing future GIS technology, which can be applied to various fields.

The 2nd unit research project develops service models for the results of the core projects and supports to establish standardization. The integrated service model developed by connection analysis among results will be a basis for building business models and business plans. For supporting standardization, it is recommended to follow international standardization in developing technology of core research projects. The new technology of KLSP will be used as a leading technology through international standardization organization.

The 3rd unit research project includes R&D portfolio and business modeling. The R&D portfolio is conducted for efficient project management to enable organic integration of results, prevent research overlap, and increase feasibility of KLSP. In business modeling, overall commercial roadmap is established in the first place, and individual business models are developed to make a profit.

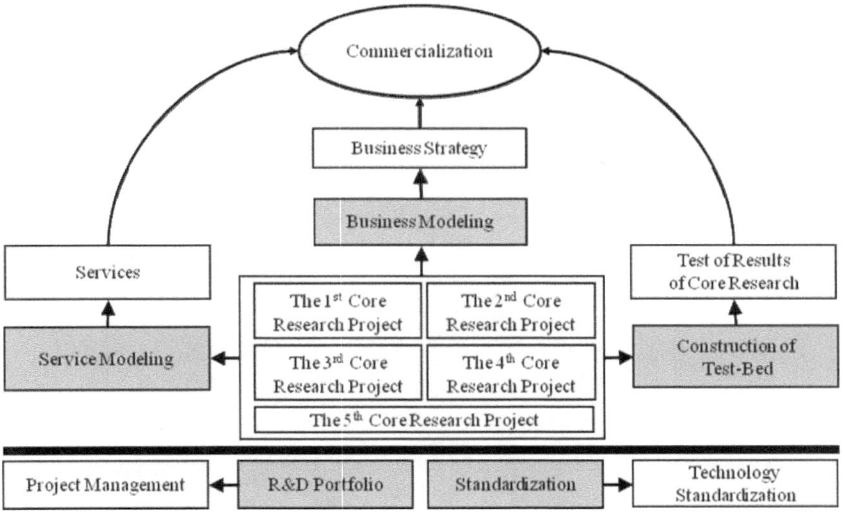

Fig. 3.8. Configuration of Research Coordination Project

3.3 Vision of KLSP

As shown in Fig. 3.9, KLSP seeks future land information not only for public sector but also for private sector by means of well-established IT infrastructure, support of the government, and advanced base technology. Technology developed in KLSP would be the fountainhead technology of u-City projects, u-Eco projects, and other projects in Korea, and they would build the nation's competitiveness in the world market. Particularly, it is expected to achieve the Korean GIS market of $1.5 billion ($300 million, 2008) and the 5th place in the rank of the World's GIS technology (10th place, 2008) in 2011, after KLSP is carried out successfully.

3.4 Summary

The Korean government, in recognizing the importance of geospatial information technology to the future ubiquitous society, assigned an R&D grant of over $130 million to the Korean Land Spatialization Program, which is composed of 5 core research projects and one research coordination project. Dr. Byung-Guk Kim of Inha University was appointed as center director of the program in a proposal competition in 2006. The world's largest research program for geospatial information technology,

KLSP, aims at technological developments for the realization of "Ubiquitous Infrastructures for Digital Korea." Particularly, this program targets practical implementation of the results of projects and commercialization in the market. KLSP will be an important milestone in the Korean GIS field and will provide the nation's growth engine for the next thirty years. In order to achieve the missions, authors put great emphasis on international collaboration and hope that this paper would be helpful to introduce KLSP to the world.

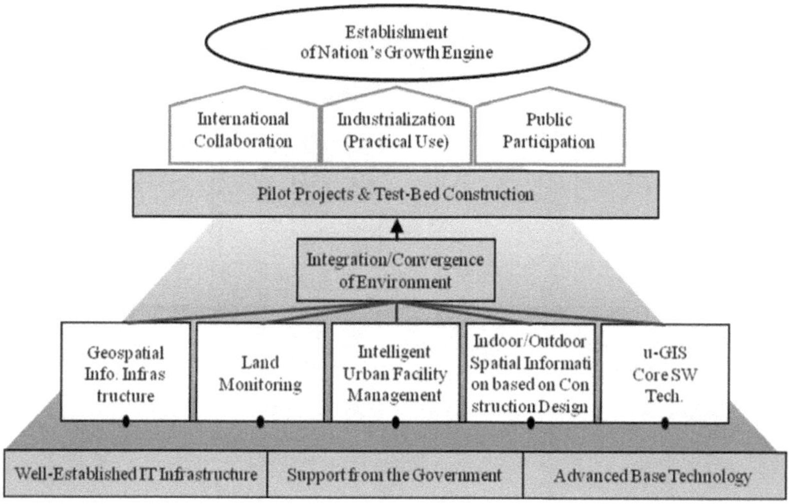

Fig. 3.9. Vision of KLSP

3.5 Acknowledgments

The authors wish to thank the directors and researchers in the five core research projects and research coordination project. Sincere thanks should go to the Ministry of Land, Transport and Maritime Affairs and to the Korea Institute of Construction and Transportation Technology Evaluation and Planning, both of which initiated the program and have provided the government's full support.

References

1. Korean Land Spatialization Group: Research Plan for Korean Land Spatialization Program. Technical Report. Incheon, Korea (2006)
2. Korean Land Spatialization Group: The First and Second Year's Report for Korean Land Spatialization Program. Technical Report. Incheon, Korea (2008)

Part II: Papers

Chapter 4
Construction Operators for Modelling 3D Objects and Dual Navigation Structures

Pawel Boguslawski and Christopher Gold

Abstract. This work presents new operators for construction of 3D cell complexes. Each cell in a complex is represented with the Augmented Quad-Edge (AQE) data structure. Cells are linked together by the dual structure and form a mesh. This structure stores information about the geometry and topology of a modelled object. Navigation in the mesh is possible using standard AQE operators. A new set of atomic operators was developed to simultaneously construct the Primal and the Dual structures. This allows 3D tetrahedralization as well as the construction of different types of objects, such as buildings composed of multiple rooms and floors.

4.1 Introduction

3D modelling is getting very important in many fields of science and industry. Current methods are often insufficient because many do not support complex connected 3D structures, and it is impossible or very difficult to integrate different models to analyze more complex influences. An example is geology where 3D models are essential and very often other, lower dimensional ones are not adequate. Very important is spatial connection and the influences between layers - for example between different rock formations. Another field is GIS and in particular modelling of buildings. Disaster management systems, simulation of terrorist attacks and looking for escape routes from buildings and other multi-level structures in urban areas are vital especially after 9/11. Many emergency management systems use 2D models and do visualization in 3D. Others use 2.5D models to navigate inside buildings [1], but 3D models are essential to this issue. Results presented by Kwan and Lee [2] shows that extending 2D building models to 3D can significantly improve the speed of rescue operations. The next improvement that gives better calculations in a model is the combination of geometry and logical models [3, 4]. The logical model would be used

Department of Computing and Mathematics, University of Glamorgan, Wales, UK
e-mail: pbogusla@glam.ac.uk, cmgold@glam.ac.uk

for computing escape routes and the geometry model for visualization. Lee in his work [5] described current 3D methods of modelling with particular reference to emergency management. He presents a 3D Navigable Data Model (NDM) to support building evacuation. This is a digital geographic database of a transportation system that can support emergency guidance operations at a high level. An emergency situation in a real building was simulated.

The most essential item in each case is topology - adjacency and connectivity between objects. In a general topological model a dual navigation graph is derived from the geometric one (primal outline graph). For real-time updating we need automatic dual graph construction during dynamic (insertion and deletion) primal graph maintenance. New structures which will store both of them simultaneously will improve the efficiency of calculations significantly in the above issues.

4.2 Related works

3D data models can be classified into: Constructive Solid Geometry (CSG), boundary-representations (b-rep), regular decomposition, irregular decomposition and non-manifold structures [6]. For our research b-reps and irregular decomposition models are the most relevant. The well known b-reps are: Half-Edge [7], DCEL [8], Winged-Edge [9] and Quad-Edge [10]. Irregular decomposition models (e.g. for constructing a 3D Delaunay tetrahedralization) can be constructed with a Half-Faces [11], G-maps [12] and Facet-Edges [13]. The most important for us are the Quad-Edge (QE) and (its extension) the Augmented Quad-Edge (AQE) [6] data structures. These structures are suitable to construct models and their duals at the same time. We use dual space to connect cells in cell complexes and to navigate between them. Navigation and data structures are the same in both spaces. Eventually both spaces are connected together and we don't need any additional pointers for this connection. Other data structures like Half-Edge or Winged-Edge used widely in CAD systems don't provide for management of the duality. The only similar structure which preserves primal and dual 3D subdivisions is the Facet-Edge. In a cell complex modelled with this structure common faces are not stored twice – faces are shared by two cells. This is less memory consuming than the AQE structure [6]. However we decided to use AQEs in our project as we need to keep common faces separate because of the nature of the modelled objects. We believe that our solution is more suitable for real-time simulations and computation.

4.2.1 Quad-Edge

The Quad-Edge (QE) was introduced by Guibas and Stolfi [10]. Each QE consists of 3 pointers: R, N and V. 4 QEs connected together in a loop by the R pointer to create an edge. They are connected in an anticlockwise (CCW) direction. The next pointer N points to the next edge (with the same shared vertex or face). All edges connected by this pointer form a loop. This is a CCW connection as well. The pointers R and N are directly used in Rot, Sym and Next standard navigation operators. Rot uses R and returns the next quad from a loop of four. Sym calls Rot twice. Next uses N and return the next edge from the loop (Fig. 4.1). The V pointer is used to point to coordinates of vertices in the structure.

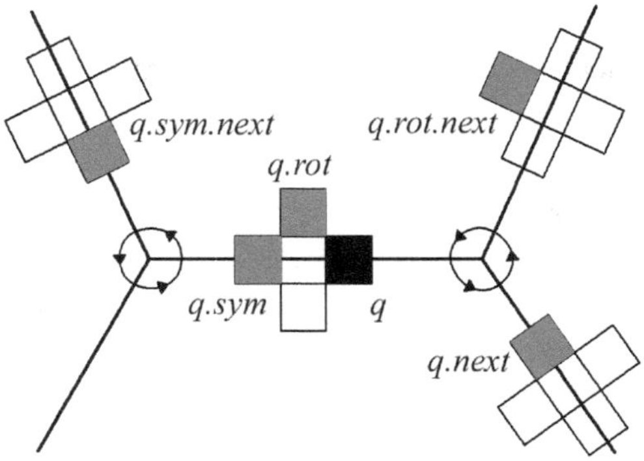

Fig. 4.1. Navigation in a structure with Quad-Edges; q represents origin quad.

To construct and modify graphs only two operations are needed – MakeEdge to create a new primal/dual edge pair, and Splice to connect/disconnect it to/from the graph under construction. Tse and Gold showed that Euler Operators could be easily developed on the basis of QEs rather than half-edges or winged-edges [14].

The QE structure was originally used to describe both the primal and dual graphs simultaneously. The particular example was triangulation modelling, showing both the primal Delaunay triangulation and the dual Voronoi tessellation. Either graph may be navigated by simple pointer-following, using Rot and Next pointers, with "Vertex" pointers to primal and dual graph nodes.

Duality of structures is very important for many purposes. As stated in the previous section, many applications demand of 3D models to include geometry and a graph of connections between elements at the same time. It may be carried out using the dual. Having dual graphs at the same time are also advantageous for 2 and 3 dimensional electromagnetic simulation [15].

4.2.2 Augmented Quad-Edge

In 3D primal and dual graphs may also be defined: the dual of a face is an edge, the dual of an edge is a face, the dual of a node is a volume and the dual of a volume is a node. Ledoux and Gold defined an extension of the Quad-Edge - the Augmented Quad-Edge (AQE) [6]. In this approach the "Vertex" pointer to a face (usually unused) was assigned to the dual edge of that face. This was called "Through" and allowed the navigation between cells, via the dual edge. "Adjacent" is the next operator that uses the dual structure. It is a combination of Through and Next operators. It is used to reach the adjacent cell (e.g. the next room) in the structure (Fig. 4.2). If we are in one room/cell and want to go to the next one we have to find a wall/face between those two rooms/cells (this is quad q_f in Fig. 4.2) and use the Adjacent operator (q_f.adjacent). In result we have an adjacent wall/face from the next room/cell. The original QE operators were restricted only to a single cell.

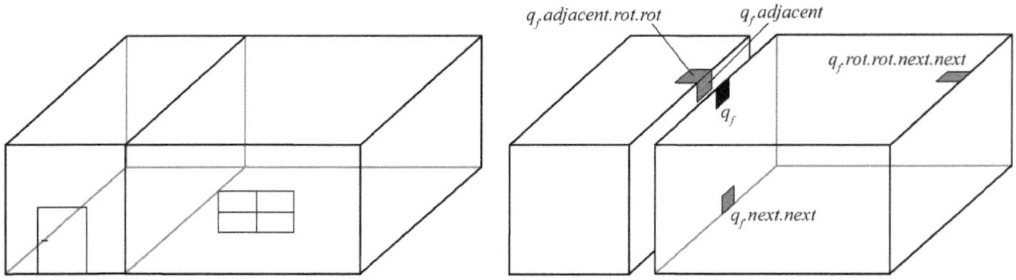

Fig. 4.2. The AQE structure is suitable structure for modelling 3D objects – e.g. building interiors. The set of operators allow for navigation inside one cell as well as between cells; q_f represents origin quad. [6]

The target application was the 3D Voronoi/Delaunay structure. The difficulty with this model was that the construction operators were complex, in particular the implementation of the through pointer maintenance.

4.2.3 GML/ISO19107

In GIS systems the ISO 19107 standard is used widely. It uses the GML schema to describe objects. For example, a simple building composed of rooms consists of a collection of Solid components, each bounded by an ExteriorSurface composed of SurfacePatches which have one exteriorRing each. This LinearRing has a sequence of coordinate positions, or a sequence of Point components each with a coordinate position. There is no navigation mechanism between Solids, or between Surfaces/Patches on the same Solid. This must be done by matching coordinates, or Point IDs, for the LinearRing defining the SurfacePatch.

Our data structure has a "Next" pointer permitting navigation within the ExteriorSurface ("Shell") of any solid. By using the Next pointer of the dual we may navigate between adjacent Solids. Our Solid is a volumetric component distinct from its Shell. Our structure handles lower-dimensional components, e.g. partly connected walls or edges. Our structure can easily be exported to GML by navigating the graph structure and outputting the various components as they are encountered.

GML currently assumes matching coordinates or Point IDs. To import from GML we must make these matches between GML components to reconstruct the adjacency structure. Where coordinates are only approximate and matching points have approximately similar coordinates some tolerances must be used. This is ongoing research.

4.3 Current work

The work reported here is the continuation of a larger project. Ledoux and Gold [6] saw the need for a data structure for the simultaneous maintenance of both the primal and dual spatial subdivisions. Although the dual can always be obtained from the primal, there are many applications where it is needed directly: an obvious example is for the navigation from primal cell to primal cell. Especially for disaster management, where repeated traversals are needed for real-time shortest-path planning through a changing network, such processes are basic. They demonstrated that, with a space-filling set of primal cells and another space-filling set of dual cells (where each cell is a simple 2-manifold shell), it was possible to provide a mechanism for navigation throughout. It was based on the observation that the dual-entity of a face in one space is an edge in the other. Thus transforming the pointer to a face within a cell in the first space into a pointer to the corresponding edge in the second space gave a link between the two – and hence a way to navigate. This Augmented Quad-Edge was a direct modification of Guibas and Stolfi's [10] Quad-Edge structure.

We were already familiar with the Voronoi-Delaunay duality in 3D, and so this was our test-bed to validate the principle and develop functional code. It was entirely successful and, for example, drawing a single Voronoi cell or a single Delaunay tetrahedron was greatly simplified. The algorithms used were the standard incremental insertion and flipping methods of Shewchuk [16] and others. There was a great deal of local searching and sewing needed to maintain the primal and dual structures simultaneously. An effect of a simple tetrahedralization is shown in the Fig. 4.3.

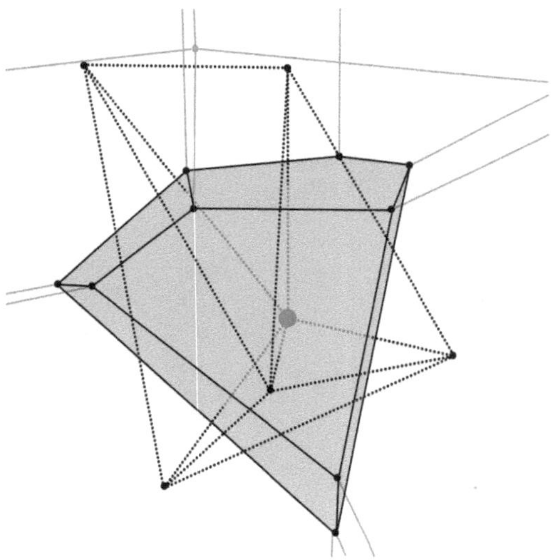

Fig. 4.3. Tetrahedralization. Gray cell with black edges is a Voronoi cell. This cell enclosed one Delaunay point (gray big vertex). Dotted black edges form Delaunay tetrahedral cells.

However, rooms in buildings, or subsurface catacombs, are not easily made directly from Voronoi cells – so the point-insertion and flipping algorithms were inappropriate. The current work attempts to produce a set of local, and general, construction operators that allow the maintenance of the dual structure as a by-product of the (manual) construction of buildings from floor-plans, stairways, etc.

This is in principle very similar to the Euler Operators developed for CAD systems [7, 17] –, but with the ability to create and link multiple simple shells (as in non-manifold CAD systems) – and with the simultaneous maintenance of the dual. We are still working on the construction of exact equivalents of Euler Operators, but we have developed an alternative approach, based on double-face elements, that can achieve the same results.

In the Quad-Edge structure there are four individual pieces ("Quads") associated with each edge of a simple oriented 2-manifold. These are linked (by pointers in our implementation) so that the "first" has a pointer to a vertex, the "third" points to the other vertex forming the edge, the "second" (anticlockwise) points to the right-hand face and the "fourth" points to the left-hand face. An additional "Next" pointer in each Quad points to the next Quad-Edge in anticlockwise order around the associated vertex or face. The (2D) dual is represented, and one may navigate directly around e.g. Voronoi cells or Delaunay triangles. Frequently the face pointers remain unused, unless an explicit face object is desired, e.g. to assign an attribute such as colour.

We observed above that in 3D each face of a cell is penetrated by a dual edge. Indeed, it is penetrated by one edge for each of the vertices of the face – each of these is surrounded by its own dual shell. Thus if the face pointer is re-assigned to point to a dual edge, then this need never be changed. (Inversely, the related face pointer of the dual edge points back and also remains unchanged.) Construction is thus simplified if these are created together, already linked. However, an edge has two sides, with associated faces, and during construction low-level elements need to be linked to form the final structure. Initially our atomic elements were complete edges – but this necessitated the updating of the re-assigned face pointers, which is undesirable.

Our final choice was to create an atomic element from a "half edge" (essentially two of our four Quads) together with the linked half-edge in the dual. This reduced our construction process to the snapping-together of the appropriate half edges, to form complete edges. This atomic element is shown in Fig. 4.4.

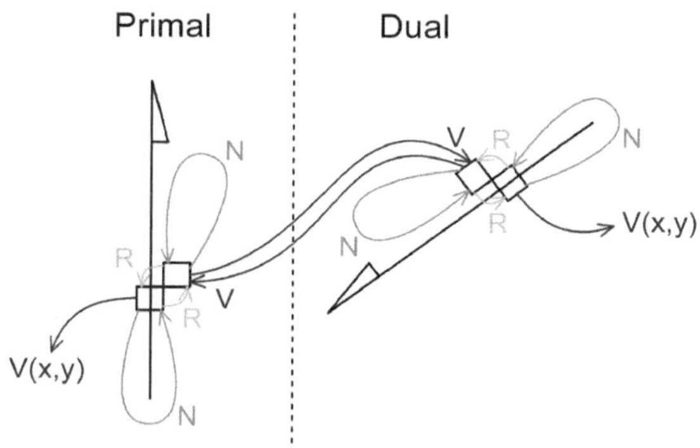

Fig. 4.4. Half Edges are created in the primal and dual. They are linked permanently.

Our current construction method starts by building one side of a complete "double face" – a "half-face", then the other, and then joining them to form a "sandwich". Fig. 4.5 shows the process to connect half-edges together in primal space, and to make the necessary connections in the dual. The result is a half-face based on the building design. This is repeated for the matching half-face.

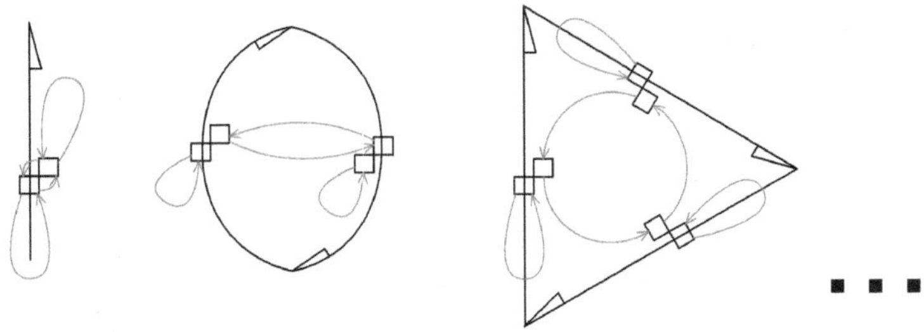

Fig. 4.5. Construction of a half face in the primal. Half edges are added one by one to the loop. Dual is not shown.

These two are combined to form the completed "double face" by snapping the matching half-faces together (Fig. 4.6). The result is a completely defined shell – although of zero volume.

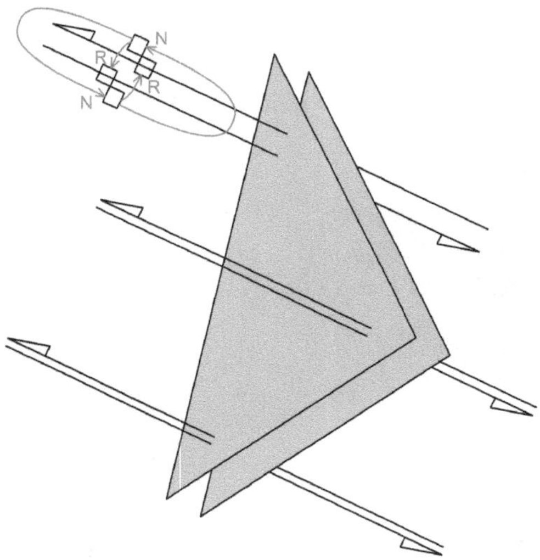

Fig. 4.6. Construction of a double face. Half edges in the primal and dual are linked together and now they form full edges.

These "walls" (or "floors", etc.) are then joined together as desired to give our final building structure. As all operations are defined to be reversible, one edge at a time of the double-face is un-snapped, the same is done for the receiving face, and then the half-faces are snapped back together in the new configuration (Fig. 4.7). This directly mimics the construction of a cardboard model using sticky tape!

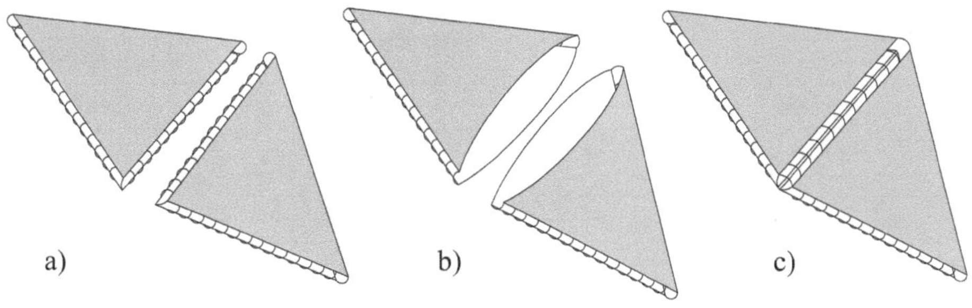

Fig. 4.7. Snapping two faces: a) two separate faces; b) unsnapping edges; c) snapping edges into new configuration.

Visualization of a model is very easy. To get all points, edges or faces of a cell we use breadth –first traversal of the graph. All cells are connected by a dual structure which is a graph too.

The process of a construction was first validated using the previous Voronoi-Delaunay model (see Fig. 4.3). It satisfactorily replaces the complex stitching process previously used. It was then used to construct both simple and more complex building structures (see Fig. 4.8). Note that the sequence of "manual" operations is relatively straightforward.

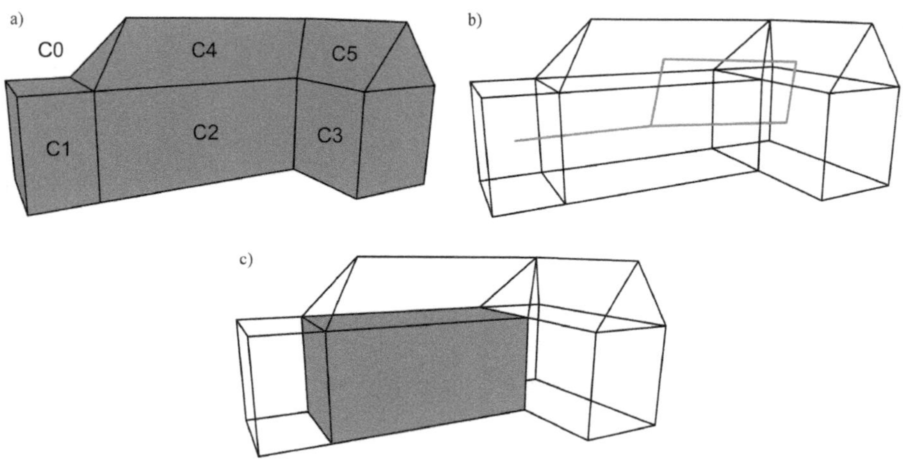

Fig. 4.8. Objects constructed using AQEs and the new construction operators; a) model with faces shown; C1-C5 – cells of a model, C0 – outside cell, b) mesh of edges; black – primal and grey – dual structure and c) one volume/room selected.

There are 98 edges used in this model of a building grouped in six cells of one cell complex. Five cells C1-C5 form the building and one C0 is its outside. Outside is an environment of a building. It can be air, ground etc. We need this outside cell for navigation and to have properly built dual graphs. Each edge in the model consists of four quads and each edge in the primal has been assigned two half edges in the dual. This gives us 784 quads. To obtain the number of pointers we have to multiply the number of quads by three. In total we have 2352 pointers in the model. There are also 26 vertices (21 in the primal and 5 in the dual).

You can see that number of elements in this model is too high to show the process of construction. We decided to present only a middle part of the building to make this example clear (Fig. 4.9). All steps in Table 4.1 are high level operators used to construct a model. The MakeFace operator takes a list of points (Pxx) to create a face. The SnapFaces operator takes two quads (Qxxx) from faces to snap them together. Finally we have three cells C2, C4 and C0. C0 is outside of the model.

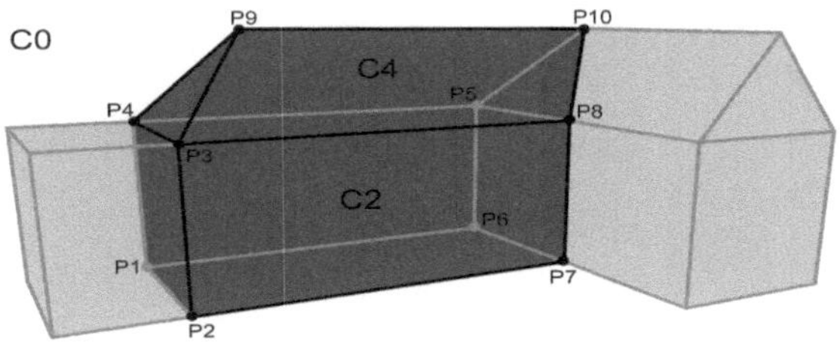

Fig. 4.9. Process of construction is presented only for a middle part of a building.

Table 4.1. Process of the model construction. Only high level operators are shown.

Cell 2	Cell 4
MakeFace(P1,P2,P3,P4) – left face	MakeFace(P4,P3,P9) – left face
MakeFace(P1,P4,P5,P6) – back face	SnapFaces(Q193,Q153)
SnapFaces(Q33,Q1)	MakeFace(P3,P8,P10,P9) – front face
MakeFace(P2,P1,P6,P7) – bottom face	SnapFaces(Q217,Q149)
SnapFaces(Q65,Q5)	SnapFaces(Q217,Q201)
SnapFaces(Q65,Q33)	MakeFace(P8,P5,P10) – right face
MakeFace(P3,P2,P7,P8) – front face	SnapFaces(Q249,Q145)
SnapFaces(Q97,Q9)	SnapFaces(Q249,Q225)
SnapFaces(Q97,Q65)	MakeFace(P5,P4,P9,P10) – back face
MakeFace(P4,P3,P8,P5) – top face	SnapFaces(Q273,Q157)
SnapFaces(Q129,Q13)	SnapFaces(Q273,Q193)
SnapFaces(Q129,Q97)	SnapFaces(Q273,Q229)
SnapFaces(Q129,Q41)	SnapFaces(Q273,Q257)
MakeFace(P6,P5,P8,P7) right face	
SnapFaces(Q161,Q45)	
SnapFaces(Q161, Q141)	
SnapFaces(Q161, Q109)	
SnapFaces(Q161, Q77)	

We do not have to construct bottom face of C4 explicitly. The faces of C4 are being connected to C2. The top, two-sided face from C2 is shared with C4 and is a bottom face of C4.

In all cases the dual graph – a necessity for building navigation and real-time escape route planning – is constructed on-the-fly.

We believe that this construction process is both satisfactory and intuitively clear when using the higher-level commands. It is not presented in this paper but each operator has an inverse version or is self-reversible, although no formal proofs have yet been developed. We are exploring the relationship with conventional CAD Euler Operators, and we hope to provide equivalent operations in the near future – with automatic dual construction as the major requirement. Overall, the ready availability of the dual graph should provide new insight into spatial relationships – for navigation, sound propagation, the solutions of Maxwell's Equations [15] and many other applications.

4.4 Future work

4.4.1 Euler operators

The next issue we want to undertake is to develop set of Euler operators. This will open the door to CAD system development. Many CAD applications implement a boundary representation using Winged-Edge and Half-Edge structures [17, 18]. Those structures don't include a dual graph which is essential in emergency management systems. We think that it is possible to implement duality in Euler operators for cell complexes. It works for single cells. Future work will show if more complex operations on many connected cells are possible.

4.4.2 Storage

Many GIS 3D applications are based on tetrahedral mesh stored in databases such as Oracle [19]. Modern big database engines support 3D geometry storage and operations. Our approach comes from graph theory rather than database design. Also in [19] only tetrahedral shapes are allowed – ours is more general. Our future task is to develop methods for saving our graph structures in a database or on disk.

We can assign unique IDs to points and quads in a structure. Each cell in one space encloses exactly one point in dual space. Thus solids can be identified by ID of the dual point and its boundary is associated with the dual point. In addition faces have dual penetrating edges and thus take the edge ID.

4.4.3 Overlapping

Currently coordinates of the vertices of adjacent cells in the same cell complex must be identical. In many models objects permeate each other due to data errors. We would like to include mechanisms in the future that can manage this issue.

4.5 Applications

To prove the correctness of our algorithms we developed a simple application. We used Delphi to write it and an in-house graphics engine based on OpenGL for visualization. This engine was written for previous projects. To be sure that the presented methods are suitable for modelling building interiors we reconstructed the simple project (Fig. 4.10a) originally presented in [5].

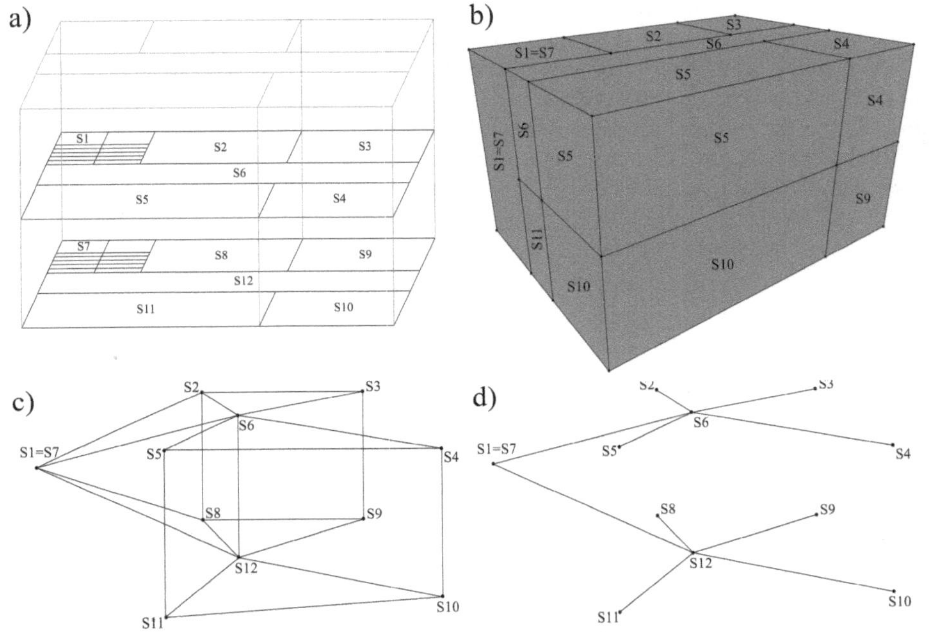

Fig. 4.10. Example structure representing a 3D model of a building interior. S1, S7 – staircase, S2-S5, S8-S11 rooms, S6, S12 - corridor; a) spatial schema – original version [5], b) volumetric model of rooms, c) complete graph of connections between rooms, d) graph of connections between rooms – only passages.

In our approach the geometry of the building is in the primal (Fig. 4.10b) and topology in the dual space - connection graph (Fig. 4.10c). Each point S1-S12 in this graph represents a separate room, corridor or stair cage. Connection between these objects is represented by edges. Edges have weights as attributes. The decision if there is a passage or not depends on this weight. A strong connection means that there is a passage like door, window, etc. A weak connection means no passage, but the possibility to go to the next room still exists. This model predicts such situations as making holes in walls and the creation of new passages. When weak connections are removed from the graph, we have the same logical network as in case of [5] (Fig. 4.10d). The only difference between the models is that we connected the staircase into one room (originally the stair case was divided into two levels).

4.6 Conclusions

The presented 3D structure and construction operators create a solid base for 3D models where both geometry and topology are essential. It seems to be valid for modelling and navigation in spatial objects like multilevel buildings in urban areas. The structure consists of two graphs. The first one represents the geometry i.e. rooms in a building and the second one indicates the connectivity between objects (rooms). We do not need to manipulate the dual space directly. It should be noted that we have achieved this by creating and joining individual faces only in the primal space. We achieved a straightforward mechanism for 3D volumetric modelling which preserves the dual space for navigation, useful for example in disaster management, with no additional effort. Other approaches are also possible – e. g. standard CAD Euler Operators. We are exploring these other possibilities.

Real-time systems for emergency response are very important nowadays. They can save many human lives and a lot of money in hazardous situations. The lack of a good 3D model has been a problem. Perhaps our structures and methods will be useful in real-time applications in the future and they will improve existing solutions.

4.7 Acknowledgments

This research is supported by the Ordnance Survey and EPSRC funding of a New CASE award.

References

1. Raper, J., Slingsby, A.: Navigable Space in 3D City Models for Pedestrians. Advances in 3D Geoinformation Systems. Peter van Oosterom, Sisi Zlatanova, Friso Penninga, Elfriede Fendel (Eds.) Springer, pp. 49-64 (2007)
2. Kwan, M-P., Lee, J.: Emergency response after 9/11: the potential of real-time 3D GIS for quick emergency response in micro-spatial environments. Computers, Environment and Urban Systems, vol. 29, pp. 93-113 (2005)
3. Pu, S., Zlatanova, S.: Evacuation Route Calculation of Inner Buildings. In: Geo-Information for disaster management. Eds P. van Oosterom, S. Zlatanova and E. M. Fendel, pp. 1143-116. Springer Verlag, Heidelberg (2005)
4. Meijers, M., Zlatanova, S., Pfeifer, N.: 3D geo-information indoors: structuring for evacuation. In: The First International Workshop on Next Generation 3D City Models, G. Groeger & T. H. Kolbe (Eds.), pp. 11-16 (2005)
5. Lee, J.: A Three-Dimensional Navigable Data Model to Support Emergency Response in Microspatial Built-Environments. Annals of the Association of American Geographers, Blackwell Publishing, vol. 97, pp. 512-529 (2007)
6. Ledoux, H., Gold, C. M.: Simultaneous storage of primal and dual three-dimensional subdivisions. Computers, Environment and Urban Systems 31 (4), pp. 393-408 (2007)
7. Mantyla, M.: An introduction to solid modelling. Computer Science Press, New York, USA (1998)
8. Muller, D. E., Preparata, F. P.: Finding the intersection of two convex polyhedra. Theoretical Computer Science, vol. 7, pp. 217-236 (1978)
9. Baumgart, B. G.: A polyhedron representation for computer vision. In: National Computer Conference, AFIPS (1975)

10. Guibas, L. J., Stolfi, J.: Primitives for the manipulation of general subdivisions and the computation of Voronoi diagrams, ACM Transactions on Graphics, vol. 4, pp. 74-123 (1985)
11. Lopes, H., Tavares, G.: Structural operators for modelling 3-manifolds. In: Preceedings 4th ACM Symposium on Solid Modeling and Applications, Atlanta, Georgia, USA, pp. 10-18 (1997)
12. Lienhardt, P.: N-dimensional generalized combinatorial maps and cellular quasi-manifolds. International Journal of Computational Geometry and Applications, vol. 4 (3), pp. 275-324 (1994)
13. Dobkin, D. P., Laszlo, M. J.: Primitives for the manipulation of three-dimensional subdivisions. Algorithmica, vol. 4, pp. 3-32 (1989)
14. Tse, R. O. C., Gold, C. M.: TIN Meets CAD - Extending the TIN Concept in GIS. Future Generation Computer systems (Geocomputation), vol. 20(7), pp. 1171-1184 (2004)
15. Sazanov, I., Hassan, O., Morgan, K., Weatherill, N. P.:Generating the Voronoi – Delaunay Dual Diagram for Co-Volume Integration Schemes. In: The 4th International Symposium on Voronoi Diagrams in Science and Engineering 2007 (ISVD 2007), pp. 199-204 (2007)
16. Shewchuk, J. R.: Delaunay Refinement Mesh Generation. Ph.D. thesis, School of Computer Science, Carnegie Mellon University, Pittsburg, USA (1997)
17. Lee, K.: Principles of CAD/CAM/CAE system, Addison-Wesley/Longman, Reading (1999)
18. Stroud, I.: Boundary Representation Modelling Techniques, Springer (2006)
19. Penninga, F., van Oosterom, P.: First implementation results and open issues on the Poincare-TEN data structure, Advances in 3D Geoinformation Systems. Peter van Oosterom, Sisi Zlatanova, Friso Penninga, Elfriede Fendel (Eds.) Springer, pp. 177-197 (2007)

Chapter 5
A Multilayered Space-Event Model for Navigation in Indoor Spaces

Thomas Becker, Claus Nagel and Thomas H. Kolbe

Abstract. In this paper a new conceptual framework for indoor navigation is proposed. While route planning requires models which reflect the internal structure of a building, localization techniques require complementary models reflecting the characteristics of sensors and transmitters. Since the partitioning of building space differs in both cases, a conceptual separation of different space models into a multilayer representation is proposed. Concrete space models for topographic space and sensor space are introduced. Both are systematically subdivided into primal and dual space on the one hand and (Euclidean) geometry and topology on the other hand. While topographic space describes 3D models of buildings and their semantically subdivisions into storey's and rooms, sensor space describes the positions and ranges of transmitters and sensors like Wi-Fi access points or RFID sensors. It is shown how the connection of the different layers of the space models describe a joint state of a moving subject or object and reduces uncertainty about its current position.

5.1 Introduction

Over the last decade, personal navigation systems (PNS) became an established tool for route planning and guidance of individual traffic using cars and other vehicles. From the technical aspect this was made possible mainly due to the availability of global localization techniques like GPS on the one side and the acquisition and provision of the road network for large parts of the world on the other side.

Pedestrian navigation systems are not as successful as car navigation systems yet. The crucial point is that pedestrians can freely move inside and outside of buildings, but satellite localization still only works outdoor and information about navigation space is only available for the outdoor environment so far. However, indoor navigation would be helpful for finding locations like shops, police, rest rooms, or (check-in)

Institute for Geodesy and Geoinformation Science
Technische Universität Berlin
{becker, nagel, kolbe}@igg.tu-berlin.de

counters inside of airports, shopping malls, or public buildings. In an emergency response situation indoor navigation could provide escape routes from buildings and fastest routes for rescue personnel to a disaster area [7].

In general, navigation comprises 1) the determination of the location of a subject or object, 2) the determination of the best path (often the fastest, the shortest, or the cheapest) from a start to an end location, and 3) guidance along the path which includes monitoring of the difference between the current position and the path and enforcement of appropriate actions to minimize the difference. Thus, in order to facilitate indoor navigation the problem of data availability on the indoor navigable space has to be solved and appropriate localization techniques and methods need to be developed.

In the past, different models for structuring indoor space and localization methods have been proposed. As we will discuss in sections 5.2 and 5.3, in most cases space is partitioned due to route planning and addressing criteria on the one hand and localization technology and sensor characteristics on the other hand. Often they are mixed within one model, which has the disadvantage that changes to the building structure or sensor configurations may affect the entire model. For example, changes to room topology (e.g., a door will be closed permanently or a new door will be installed within a wall) does not necessarily have an impact on the localization infrastructure like Wi-Fi access points or RFID sensors and vice-versa.

In section 5.4 we present a new framework in which different concepts of space are combined to a multilayer space-event model. One concrete concept of space deals with the 3D topographical representation of buildings and another with the 3D representation of sensor and transmitter placements and ranges. While the first will facilitate route planning, the second will facilitate localization; both together then facilitate navigation. Each space concept is separated into primal and dual space on the one side and geometry representation and topology on the other side. The different space concepts are then linked by an n-partite graph where nodes represent spaces and the states of a guided subject or object at the same time. Edges represent events like leaving or entering a room (in topographic space) or change of signal strength within the range of a transmitter (in sensor space). The actual position is given by the so-called joint state, which helps to reduce uncertainty about the true absolute position in real space.

Finally, in section 5.5 we draw some conclusions and point to future work.

5.2 Related work

Substantial work has already been done in the area of indoor navigation. In the following, we give a brief overview of current developments on specific systems and underlying information structures needed in order to support location services and route planning in indoor environments.

The OntoNav system [3] describes a semantic indoor navigation system. It proposes an indoor navigation ontology which provides semantic descriptions of the constituent elements of navigation paths such as obstacles, exits, and passages. Furthermore, specific user capabilities/limitations are modeled allowing for a user-centric navigation paradigm and the application of reasoning functionality.

In the field of mobile robot navigation, Kulyukin et al. [4] present an indoor navigation system for assisting the visually impaired. The system is designed as a tool to help visually impaired customers navigate a typical grocery store using a robot shopping cart. For localization, the system relies on RFID tags deployed at various locations in the store.

In order to simplify complex spatial relationships between 3D objects in built environment, Lee [6] introduces a topological data model, the Node-Relation-Structure (NRS). The NRS is a dual graph representing the connectivity relationships between 3D entities by Poincaré Duality. In the context of emergency response, Lee et al. [7] show the use of this simplified NRS representation of the indoor environment for routing purposes.

Within the REWERSE project, Lorenz et al. [2] provide an approach for the automated partitioning of the building interior not only into rooms, but also into smaller parts, so called cells. The internal structure of the building is hence represented as a hierarchical graph enabling localization and route planning on different levels of detail. The partitioning is based on the automatic cell-and-portal decomposition of polygonal scenes proposed by Lefebvre and Hornus [5].

Liao et al. [13] propose an approach to track moving objects and their identity in indoor environments. Based on a Voronoi graph providing a natural discretization of the environment, the locations of people are estimated using noisy, sparse information collected by id-sensors such as infrared and ultrasound badge systems.

Kolodziej [9] provides a comprehensive overview and discussion of existing technologies and systems in the context of indoor navigation. Various approaches and algorithms for indoor localization using different kinds of sensor systems are described, which form the basis for Location Based Services (LBS).

5.3 Detailed analysis of previous approaches

In this section, some of the approaches mentioned in section 5.2 are revisited and analyzed in more detail with focus on their geometric and topological representation of indoor space. It is assumed, that for the realization of localization and navigation systems the built environment such as a building is represented geometrically in Euclidean space, particularly in IR3. Therefore, a building can be described using geometric and topological representations defined in ISO 19107 [1].

5.3.1 The Node-Relation-Structure (NRS)

The Node-Relation-Structure (NRS) proposed by Lee [6, 7] is a data model which allows for a simplified representation of the complex topological relationships between 3D spatial objects in indoor environments, such as rooms within a building. These relationships, i.e., adjacency and connectivity relations, are directly derived from 3D geometry and topology of the interior entities. They are transformed into a dual graph structure in topology space using the Poincaré Duality. The dual graph enables the efficient implementation of complex computational problems within indoor navigation and routing systems.

5.3.2 Poincaré Duality

The NRS utilizes the Poincaré Duality in order to simplify the complex spatial relationships between 3D objects by a combinatorial topological network model. Solid 3D objects in primal space, e.g., rooms within a building, are mapped to vertices (0D) in dual space. The common 2D face shared by two solid objects is transformed into an edge (1D) linking two vertices in dual space. Thus, edges of the dual graph represent adjacency and connectivity relationships which may correspond to doors, windows, or hatches between rooms in primal space. Fig. 5.1 illustrates this duality transformation. A formal definition of the Poincaré Duality is given by Munkres [8].

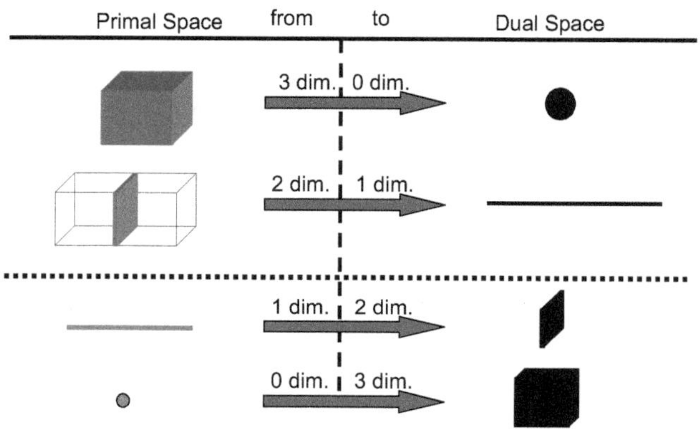

Fig. 5.1. Principles of Poincaré duality as shown by Lee [6]; for further information on the mathematical definition of Poincaré duality, see [8].

Since the resulting combinatorial model only represents topological relations, it does not contain metric information. However, metric information is needed in order to implement 3D spatial queries in the NRS such as shortest path operations. For this purpose, a complementary geometric network model is derived in Euclidean space by applying mathematical skeletonization algorithms and centroid calculations to the 3D spatial objects. By relating both graph representations, a geometric-topological network model can be established applicable to complex 3D spatial queries.

Fig. 5.2 illustrates the approach of Lee in a way that allows for the distinct separation of primal space from dual space on the one hand, and geometry and topology on the other hand. This structure forms the basis for the framework proposed in the next section. The NRS data model supports the implementation of indoor navigation systems, e.g., in the context of emergency response, since the complete indoor environment of a building is described by a graph with an embedding in IR3. This graph represents topological adjacency and connectivity relationships between spatial objects as well as metric information. Accordingly, methods for indoor routing can be efficiently applied.

Generally, the dual representation of the indoor environment can be understood as a room-to-room connectivity graph. However, indoor navigation approaches like those proposed by OntoNav [3] and Lorenz [2, 17] rely on a further spatial decomposition of rooms according to the modus of navigation, e.g., to represent navigable and non-navigable areas with respect to the capabilities and limitations of moving persons. Moreover, the partitioning of indoor space into smaller units may also be induced by limited propagation areas of sensor-based positioning systems, e.g., systems based on RFID tags, which do not cover the spatial extent of an entire room.

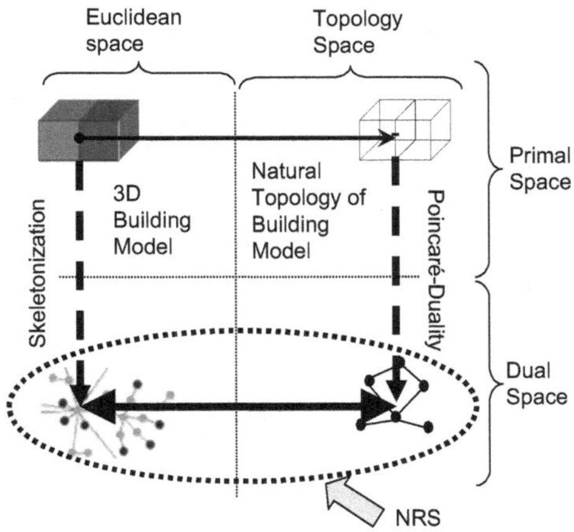

Fig. 5.2. Graphical summary of Lee's approach.

5.3.3 The need for the decomposition of rooms into smaller units

In semantic 3D building models, the free space within buildings is modeled by non-overlapping room objects (see [10, 11]). Whereas this representation of indoor environment is suitable for the derivation of a room-to-room connectivity graph, Lorenz [2] and Lefebvre [5] propose a more differentiated decomposition of the semantic room entities. The room itself is geometrically fragmented into so-called cells, which again represent non-overlapping parts of the room. Based on the topological relationships of the resulting cells, a cell-to-cell connectivity graph can be derived by applying the duality transformation proposed by Lee [7].

The importance of a fine-grained subdivision of space and its dual cell-to-cell representation is exemplified within a fire escape scenario illustrated in Fig. 5.3. The figure shows several rooms connected by doors to a corridor. Whereas in Fig. 5.3a) no further partitioning is applied to the topographic room objects, the corridor in Fig 5.3b) is subdivided into disjoint cells representing partially accessible passages of the corridor with respect to adjacent doors. The corresponding dual graph representations are also shown in Fig. 5.3. The task within this fire scenario is to find an evacuation route from the upper left room to the staircase. As a constraint for the modus of navigation, rooms affected by fire, i.e., the left part of the corridor, are marked as non-navigable.

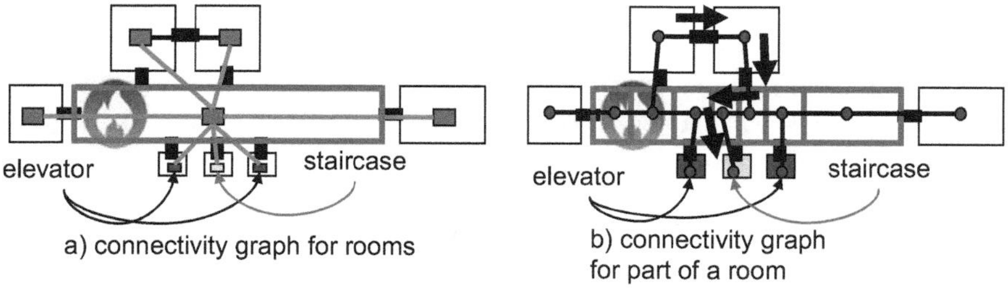

Fig. 5.3. The effect of spatial decomposition of rooms along escape routes.

Based on the room-to-room connectivity graph, this task cannot be performed since the corridor is only represented by a single vertex in the dual graph and is completely marked as non-navigable. However, the semantic decomposition of the corridor into single cells allows for its dual representation by several vertices. Since only two cells are affected by fire and thus marked as non-navigable, a valid escape route can be computed based on the cell-to-cell connectivity graph (denoted using black arrows in Fig. 5.3b).

Smaller partitions of topographic space and the corresponding semantic decomposition of room objects provide the necessary means for a more precise indoor route planning. Although the approach of Lee [7] introduces a multi-scale representation of spatial objects within the geometric network model, this representation is the result of skeletonization processes of 3D spatial objects in Euclidean space (see Fig. 5.2), and thus does not follow semantic decompositions as proposed by Lorenz et al. [2, 17]. As shown in the previous example, these decompositions of room space allow for a more detailed planning of escape routes.

Furthermore, the single partitions can be individually addressed by sensor-based positioning and tracking systems to provide a more accurate location of moving subjects or objects. Lorenz et al. [2] describe such a system by integrating a Wi-Fi sensor model using so-called fingerprints. Fingerprints represent measurements of the signal strength of Wi-Fi transmitters at discrete locations within a room (see Fig. 5.4). The cell decomposition of the room is performed based on different fingerprint measurements which are modeled as attributes of room cells. This approach allows for localization within rooms. However, the illustrated modeling approach also faces substantial disadvantages. Since the partitioning of topographic Euclidean space follows the characteristics of sensor space, there is no separation of the different space concepts any more. Instead of a spatial partitioning of topographic space according to geometrical, semantic or rule-based aspects, the decomposition is decisively influenced by the sensor model, e.g., by the received signal strength of the transmitter or signal source.

Accordingly, both space representations cannot be modeled individually. Changes to building topology or sensor configuration would both affect the entire structure. Furthermore, the integration of another kind of sensors or transmitters, e.g., RFID tags within a Wi-Fi based system, induces further modeling complexities, since the same room cell in topographic space could be covered by various overlapping sensor propagation areas, e.g., based on Wi-Fi signal strength and RFID signal strength.

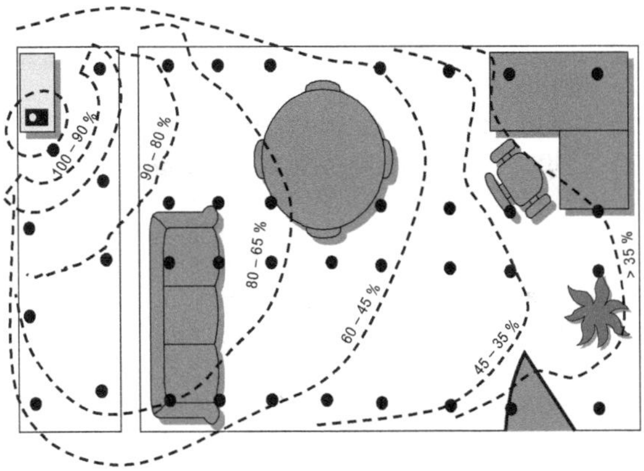

Fig. 5.4. Signal propagation area of a Wi-Fi transmitter including discrete areas of different signal strength and measurement points.

5.4 The proposed model

Due to limitations of existing modeling approaches discussed in the previous sections, we propose a novel framework for a multilayer space-event representation. A crucial aspect of this framework is the clear separation of different space models, e.g., topographic space and sensor space.

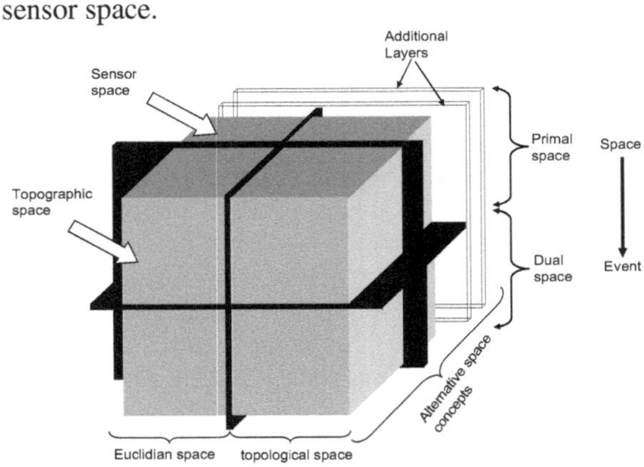

Fig. 5.5. Multilayer combination of alternative space concepts.

This approach allows for the decomposition of a specific space into smaller units according to respective semantics, without influencing other space representations.. Furthermore, we show how to connect the layers, i.e., space models, in a well-defined way and to derive a valid and unique joint state embracing all linked layers at a given point in time. Based on joint states, e.g., between topographic space and sensor space, the proposed multilayer modeling approach can be utilized to enable localization and route planning strategies. Fig. 5.5 illustrates the proposed modeling framework.

Within the framework, alternative space models are represented as separate layers. In Fig. 5.5, the layer to the front exemplarily represents topographic space, whereas the sensor space is depicted by the layer in the back. Each layer can further be divided into four segments (indicated by black cutting planes). The vertical division corresponds to space representations within Euclidean space respective topology space on the one hand. The horizontal partitioning indicates primal and dual space on the other hand. Consequently, each space model is given by four distinct space representations.

The separation of layers results from different space models with different partitioning schemas. For example, in topographic space geo-objects such as buildings may be represented using semantic 3D building models (see [10, 12]). Further semantic decompositions into, e.g., rooms, walls, doors, etc. can be applied within these model. However, the notion of sensor space substantially differs from topographic space. The sensor space is rather decomposed according to signal characteristics such as propagation and signal coverage areas. Besides topographic and sensor space, further alternative concepts of space can be incorporated into the framework by adding additional layers. The number of layers is unbounded. For example, in the area of philosophy different definitions for space (e.g., movement space, activity space, visual space etc.) can be encountered which can also be used to describe a built environment. However, the notion of space and its semantic decomposition again differs from topographic or sensor space. Since each layer provides a valid and consistent representation of space, the common framework itself is to be seen as a valid multi-layered space representation, which can be used as a whole to describe, for example, the indoor environment of buildings.

For each layer, topological relationships such as connectivity and adjacency relations between 3D spatial objects are represented within topology space (i.e., the right side of Fig. 5.5). In primal space, topology is induced by the corresponding 3D geometry in Euclidean space. By applying a duality transformation based on Poincaré duality, the 3D cells in primal topology space are mapped to nodes (0D) in dual space. The topological adjacency relationships between 3D cells are transformed to edges (1D) linking pairs of nodes in dual space. The resulting dual graph represents a Node-Relation-Structure as proposed by Lee [7]. Furthermore, the dual graph can also be seen as a state transition diagram. The active state is represented by a node within the dual graph and denotes the spatial area the guided subject or object is currently in. Once the subject or object moves into a topologically connected area, another node within the dual graph and thus a new active state is reached. The edge connecting both nodes represents the event of this state transition. Therefore, events are related to the movement of subjects or objects through the explicit topological representation of space. Accordingly, our modeling approach is a space-event model. Under the assumption that the space is subdivided into disjoint areas, exactly one node within the NRS respectively the state transition diagram can be active.

5.4.1 Topographic Space / Layer

The topographic layer is illustrated in Fig. 5.6. For indoor navigation, the topographic space represents the interior environment of buildings and its semantic decomposition into building elements like rooms and doors in order to enable route planning. Semantic building models for the representation of topographic 3D objects nowadays become increasingly available in the context of Building Information Modeling (BIM), such as the Industry Foundation Classes (IFC) [12] and in the field of 3D city modeling. The City Geography Markup Language (CityGML) [10, 11] defines a geospatial information model for the representation of 3D topographic urban objects including buildings.

According to the general space concept of layers, the topographic space can be described by four distinct representations. The upper left element of Fig. 5.6 illustrates the non-overlapping 3D geometry representation of built environment in Euclidean space. This geometry information can be directly derived from IFC and CityGML building models. The upper right element represents the induced natural topology of the 3D spatial objects according to ISO 19107. Since disjoint partitioning of Euclidean space is assumed, the relation between both upper elements can be expressed with the "Realization" association between geometric and topological objects defined by ISO 19107. Accordingly, associated objects in either space must share a common dimension and are related by 1:1.

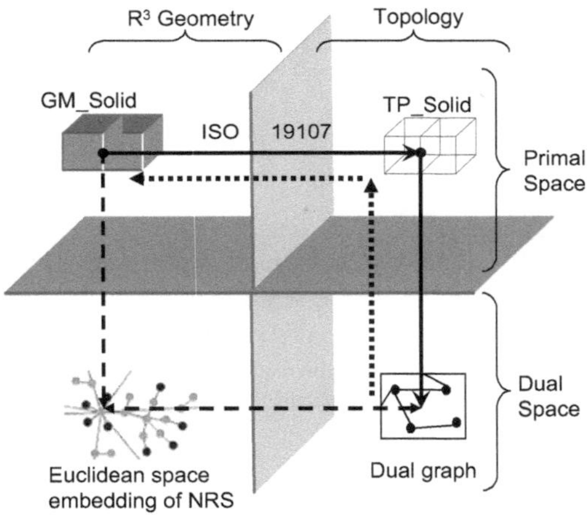

Fig. 5.6. The topographic space.

Whereas the upper part of Fig. 5.6 represents the primal Euclidean respectively topology space, their dual representations are depicted by both lower elements. For the lower right part, topology is represented as dual graph based on the NRS model and is derived from topology in primal space by Poincaré duality transformation. As mentioned in section 5.3, the NRS does not contain metric information which is, however, necessary in terms of spatial 3D queries such as shortest path calculation. In order to integrate metrics, one possible solution could be the usage of the methods "representativePoint()" and "centroid()" defined for GM_Objects in ISO 19107. For 3D solids, these methods return a point geometry representing the centroid of the volumetric object. This point representation could be stored attributively within the NRS. Since nodes of the NRS are directly related to TP_Solids in primal topology space, which, in turn, are directly related to GM_Solids in primal Euclidean space (depicted by dotted arrows in Fig. 5.6), this metric information can be uniquely derived. Furthermore, weights representing, for example, distances between rooms can be assign to the edges of the NRS. These weights could be derived from primal Euclidean space accordingly.

The lower left element of the topographic layer finally represents the Euclidean space embedding of the NRS. The dual transformation of Euclidean space results in a geometric network model [7]. This dual graph representation is derived by mathematical functions such as skeletonization processes.

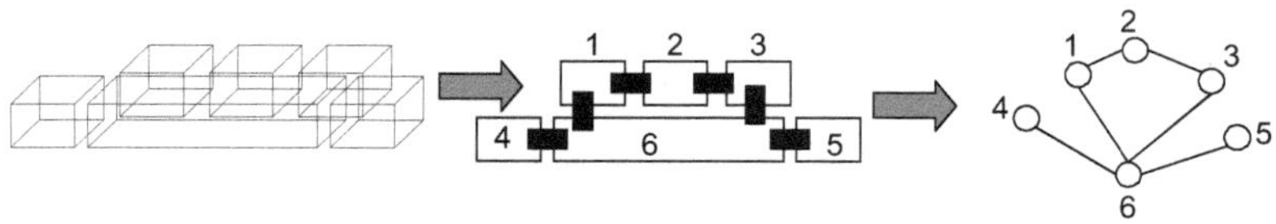

Fig. 5.7. Example for the partitioning of building interior into rooms and its representation in dual space.

5.4.2 Sensor Space / Layer

The concept of space-event modeling allows for consistent specification and interpretation of various space concepts. This ensures equivalent interpretations of sensor space and topographic space. When arranging sensors within a building (e.g., Wi-Fi), transmission ranges may overlap, which requires their decomposition into disjoint regions in order to define unambiguous states. As a state one can define the range or different signal strength areas. The event can be understood as an entry into a sensor area or as the crossing of a certain threshold value.

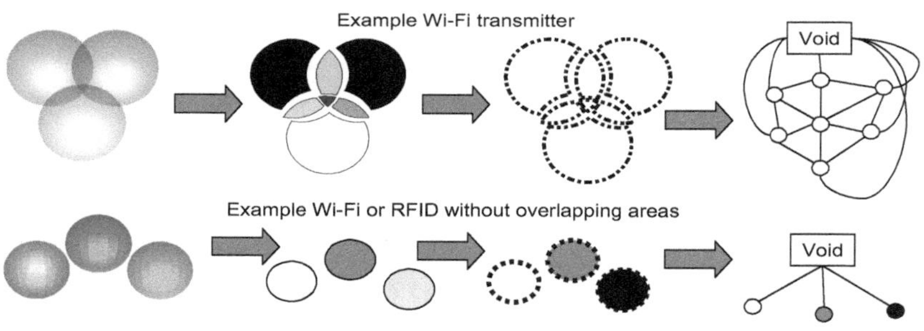

Fig. 8. Example for partitioning into cells and their representation in dual space.

Like in the topographic layer, the accuracy of positioning correlates to the granularity of partitioning. Hence with smaller cells, navigation gains in precision. To describe areas with no sensor coverage, an additional state called "void" is defined for every sensor system. This state is needed when the navigating subject or object leaves the range of a sensor without other sensors around, e.g., when leaving the building. For sensor systems covering the whole interior building area, the state "void" only represents the outside building environment. Fig. 5.8 illustrates the modeling of sensor space in the case of overlapping transmitter/sensor ranges. Fig. 5.9 further specifies different geometric and topological representations of sensor space.

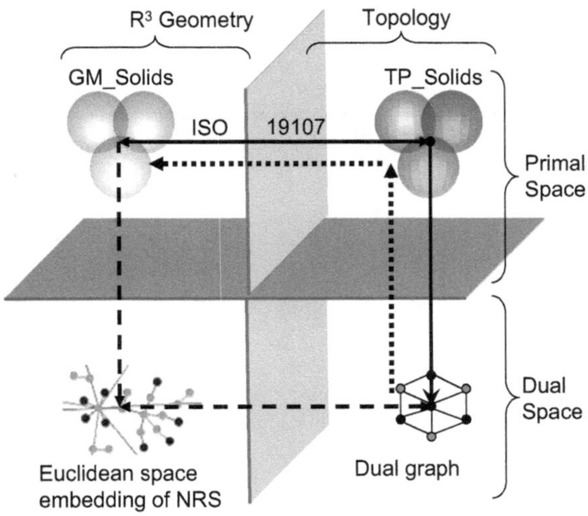

Fig. 5.9. Sensor space.

In IR^3, the partitioned sensor areas are represented as GM_Solids (upper left part) and their topological representation as TP_Solids (upper right part). The two representations are linked by the "Realization" association defined in ISO 19107. The Poincaré Duality defines the mapping from the topological representation to a dual graph structure (lower right part), representing a state transition diagram. To allow for quantitative evaluation of state distances, a metric is needed within the graph structure (like in the topographic layer). This metric is defined by explicit linking of nodes and corresponding GM_Solid objects. The distances between GM_Solids are then assigned attributively to the graph edges, resulting in a geometrical network of sensors in IR^3 (lower left part). The link between GM_Solids and the sensor network (both defined in Euclidean space) embodies potential mathematical algorithms for network derivation, e.g., Delaunay Triangulation, Voronoi Diagram, etc.

5.4.3 Decomposition of buildings

As shown in Fig. 5.10, a building can be partitioned within Euclidean space into smaller units both on the layer of sensor space and the layer of topographic space. The coexistent spaces of a building have their respective correspondence in dual space. Hence, a 3D cell in primal space has its own representation as a node in dual space. The cellular space (outlined with a dashed rectangle) is given by the set of smallest units (cells) of these partitions. The smallest unit of a partition may be a part of a room but also a complete room. As indicated in Fig. 5.10, the decomposition can be hierarchical. However, only the smallest units (cells) are considered in the following.

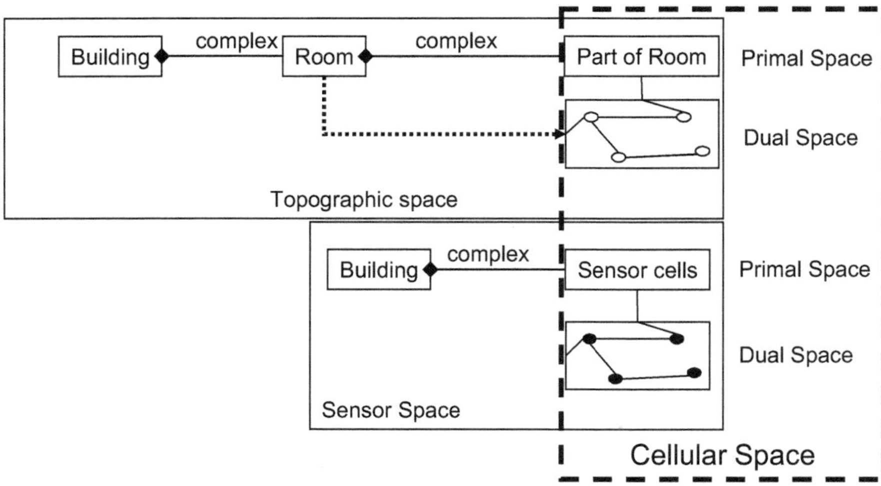

Fig. 5.10. The decomposition approach of a building.

5.4.4 Connecting all N space layers by a N-partite graph

As was illustrated before, the node-relation-structure NRS (bottom right in Fig. 5.5, Fig. 5.6, Fig. 5.8) within each layer constitutes a graph. The nodes represent the possible states of a navigating subject or object and correspond to cells with volumetric extent in primal space while the edges represent state transitions, i.e., events caused by the movement of a subject or object. They correspond to adjacency relations between the cells in primal space within the same space model (e.g., neighbored rooms in topographic space).

If we assume that each space model is based upon a disjoint partitioning of (Euclidean) space, a navigating subject or object can only belong to one cell at a time and thus always only one state may be active. Since we have different space layers with different partitioning, each layer contains such a state transition graph with exactly one active state. The overall state is then given by the joint state of all space models, i.e., all layers.

However, only certain combinations of states between different layers are valid. These combinations are expressed by additional edges between the nodes of different layers. These edges are called joint-state edges. The overall structure then constitutes an N-partite graph, where all the nodes from all N layers are included but are separated into N partitions which are connected by the joint-state edges. Furthermore, the graph also contains the state transition (or cell adjacency) edges. This is illustrated in Fig. 5.11 which shows an example with three space models / layers. The dashed lines represent state transitions / cell adjacencies within the layers and the continuous lines joint-state edges between different layers.

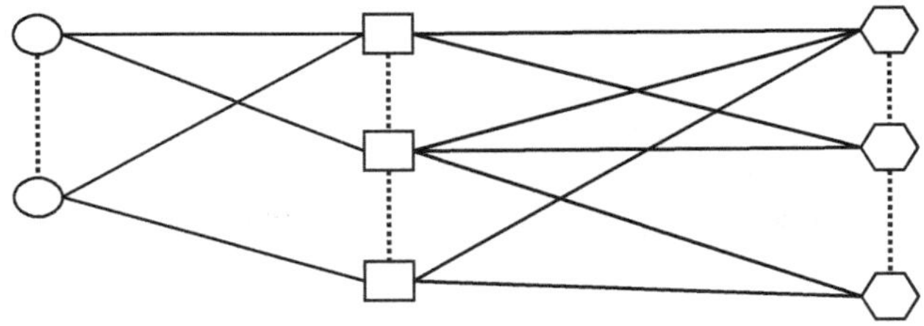

Fig. 5.11. Example for a tripartite graph containing nodes from three layers. Nodes of different layers are connected by joint-state edges. Only one state in each partition can be active and active states must be connected to each other by joint-state edges. The dashed edges represent cell adjacencies within each layer.

The joint-state edges can be automatically derived by pair wise intersection of the respective geometries between different layers. If the intersection of the interior of a cell from one space model (layer) with the interior of a cell from another space model is non-empty, a joint-state edge exists between the corresponding nodes of the respective NRS. In other words, if two cells from different space models do not overlap or are contained within each other there will be no valid joint-state in which these nodes are active at the same time.

5.4.4.1 Connections between both layers

Not only the nodes of the NRS between different layers can be combined, but also connections between layers of the other three quadrants (cf. Fig. 5.5) can be useful. For example, the connection of the geometries in primal space (see connection of upper left parts in Fig. 5.12) would allow for a common 3D visualization within Euclidean space. If geometry is represented according to ISO 19107 in $\mathbb{R}^3$ the spaces are represented as GM_Solid objects which can be visualized together in one 3D scene. This is illustrated in Fig. 5.13 where both the building topography and the position and range of transmitters and sensors are shown.

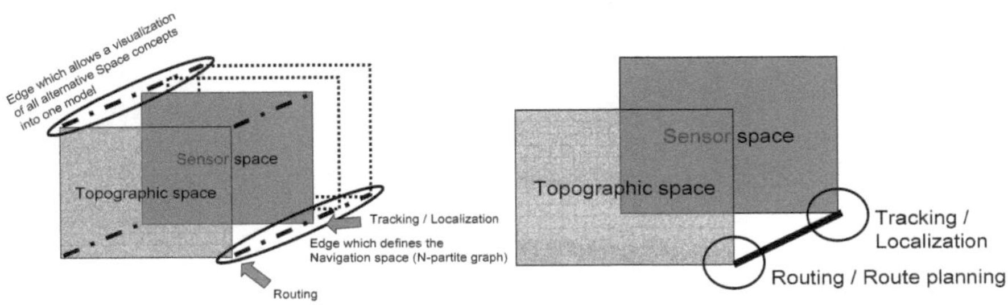

Fig. 5.12. Connection between layers (left); especially for topographic and sensor space (right).

The dashed edges between the different layers in Fig. 5.12 comprise not only the possibility of a common visualization, but generally define additional constraints to the model. They enforce an identical spatial reference system and the possibility of determining the absolute position in 3D space within a building.

Fig. 5.13. Examples for combined visualizations (left: RFID; right: Wi-Fi).

The NRS of topographic space, marked by the lower circle on the right side of Fig. 5.12, facilitates route planning within the building. Therefore, it is already useful for itself for emergency planners in order to calculate escape routes without the need for an additional sensor model. On the other side the NRS of sensor space, marked by the upper circle on the right side of Fig. 5.12, can be used in a decoupled way for tracking and localization without knowing the actual position in topographic space. The edge between the two NRS denotes the joint-state connection combining both graphs to the N-partite graph (in this example a bipartite graph) which defines the valid states of the entire model. The existence of this joint-state connection not only allows the determination of relative positions with respect to a sensor, but also the absolute position determination within the sensor and topographic space. The uncertainty about the absolute position in Euclidean space can be restricted to the intersection volume of all 3D cell geometries associated with the active nodes in the joint-state.

In addition, the N-partite graph allows also for assessment of localization infrastructure and estimation of location uncertainty with a given building decomposition in topographic space and a given sensor / transmitter configuration in sensor space.

5.4.5 Example for Modeling Proposal

The following example illustrates the representation of a building floor both in topographic space and sensor space. It demonstrates how geometrical-topological representations are mapped to dual space and which joint-state edges are established within the N-partite graph.

In Fig. 5.14 a building floor consisting of 6 rooms is shown. The entire floor is covered by signals of three Wi-Fi access points. Since the ranges of the access points A, B, and C overlap, the range geometries are partitioned accordingly into A, AB, B, BC, and C. From the dual space transformation the bipartite graph at the bottom right is derived. The joint-state edges are drawn by dashed lines. They indicate for example, that node 4 can be jointly active only with node A. The other type of edges marks adjacencies (and possible state transitions in each space model). Now, if one moves from cell 4 to cell 3 in topographic layer, one must pass the cells 1 and 2 or 6 in this layer. In the sensor layer one passes the cells A, AB, B, BC, and C. In Fig. 5.11 a joint-state for a given location is highlighted.

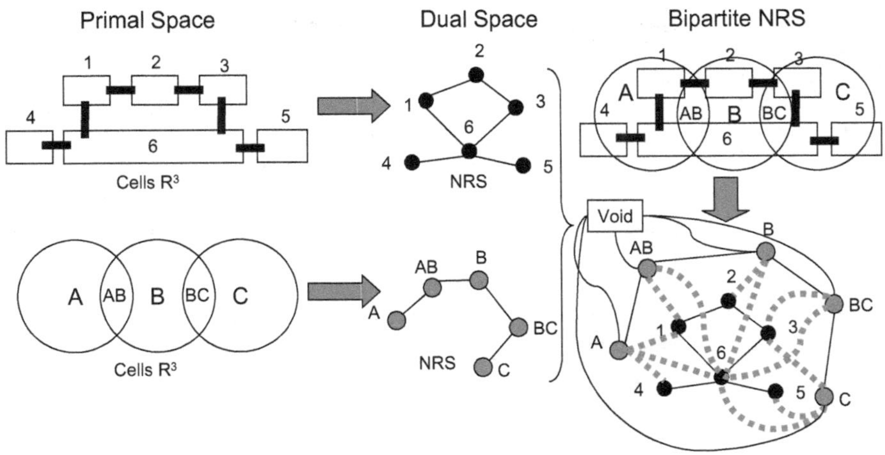

Fig. 5.14. Simple example for modeling building space by using a bipartite graph.

Sensor events indicate movement and will lead to respective state transitions. By using the joint-state edges the possible locations can then be restricted to those cells in topographic space which are connected to the new state in sensor space by an edge. The investigation of further properties of and constraints implied by the N-partite graph will be the subject of future work.

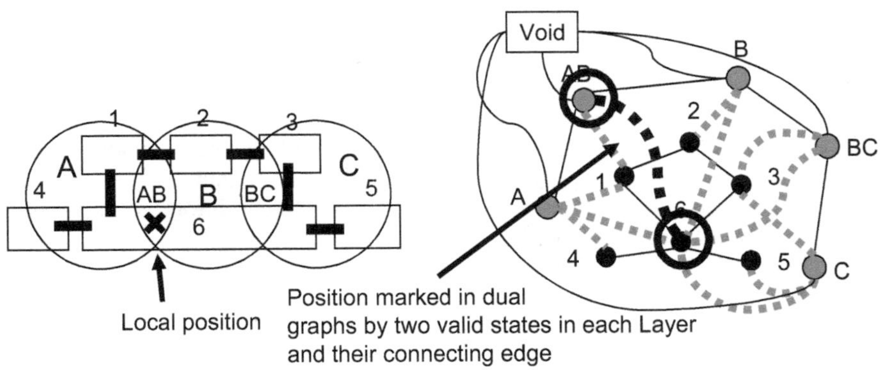

Fig. 5.15. Example for a joint state; for each partition of the bipartite graph only one state is active and the active states are connected by a joint-state edge.

5.5 Conclusion and Outlook

We have presented a novel concept for the modeling of indoor spaces to be used for route planning and localization/tracking within indoor navigation systems. The concept extends previous work from Lee [6, 7] and others [2, 3] to a multilayer representation of specific decompositions of buildings according to different semantic criteria.

As an example the decomposition of 3D building models within topographic space and the representation of transmitters and sensors within a distinct sensor space were introduced and discussed. The model reflects the duality of space and events by means of the Poincaré duality of topological models. Nodes in dual space represent possible states of a navigating subject or object. Joint-states between the different space models mutually constrain possible locations in either space model. The advantage of the multilayered representation is that space models for different sensors and topography can be represented independently from each other and that changes to one of the models do not affect the structure of the other models.

In the future, we intend to further examine the properties of the N-partite graph. First, different "void" nodes may be distinguished denoting different disconnected areas which do not provide sensor coverage but which may only be reached by crossing certain sensor / transmitter ranges. Second, it should be investigated how the structure can be used to plan sensor / transmitter deployment. By both means uncertainty of localization can be minimized.

5.6 Acknowledgements

The presented work was mainly carried out within the collaboration project "Indoor Spatial Awareness" funded by the Ministry of Transportation and Construction of South Korea. We thank Ki-Joune Li and Jiyeong Lee for fruitful discussions.

References

1. Herring, J., 2001: The OpenGIS Abstract Specification, Topic 1: Feature Geometry (ISO 19107 Spatial Schema), Version 5. OGC Document Number 01-101. / prepared by Technical Committee ISO/TC211 Geographic information — spatial schema, available at: http://www.iso.org/iso/iso_catalogue/catalogue_tc/catalogue_tc_browse.htm
2. Lorenz, B., Ohlbach, H. J., Stoffel, E.-P., Rosner, M., September 2006: NL Navigation Commands from Indoor WLAN fingerprinting position data, Technical Report of REWERSE-Project, Munich, Germany http://www.pms.ifi.lmu.de/publikationen/idefixStatic/rewerse-publications.html#REWERSE-DEL-2006-A1-D7, (Accessed June 2008),
3. Anagnostopoulus, C., Testsos, V., Kikiras, P., Hadjiefthymiades, S. P., 2003: OntoNav: A Semantic Indoor Navigation System. In: First International Workshop on Managing Context Information in Mobile and Pervasive Envi-ronments, Vol. 165, Ayia Napa, Zypern
4. Kulyukin, V., Gharpure, C., Nicholson, J., 2005: RoboCart: Toward Robot-Assisted Navigation of Crocery Stores by the Visually Impaired. In: Proceed-ings of the international Conference on Intelligent Robots and Systems, IROS 2005, IEEE/RSJ
5. Lefebvre S, Hornus S: Automatic cell-and-portal decomposition. Technical Report 4898, INRIA, 2003. http://artis.imag.fr/Publications/2003/LH03/ (Ac-cessed June 2008).

6. Lee, J., 2004: 3D GIS for Geocoding Human Activity in Microscale Urban Environments. In: M.J. Egenhofer, C. Freksa, and H.J. Miller (Eds.): GIS-cience 2004 , Springer, Berlin, Germany
7. Lee, J., Zlatanova, S., 2008: A 3D data model and topological analyses for emergency response in urban areas. Geospatial Information Technology for Emergency Response – Zlatanova & Li (eds), Taylor & Francis Group, London, UK
8. Munkres, J. R., 1984.: Elements of Algebraic Topology. Addison-Wesley, Menlo Park, CA
9. Kolodziej, K. W., Hjelm, J., 2006: Local Positioning Systems – LBS Applications and Services, Taylor & Francis Group, London, UK
10. Gröger, G., Kolbe, T.H., Czerwinski, A., 2007: OpenGIS City Geography Markup Language (CityGML), Version 0.4.0, OGC Best Practices Paper Doc. No. 07-062
11. Kolbe, T.H., Gröger, G. & Plümer, L. 2005: CityGML – Interoperable Access to 3D City Models. In P. van Oosterom, S. Zlatanova & E.M. Fendel (eds), Geo-information for Disaster Management; Proc. of the 1st International Symposium on Geo-information for Disaster Management', Delft, The Neth-erlands, March 21–23, 2005. Springer.
12. Adachi, Y., Forester, J., Hyvarinen, J., Karstila, K., Liebich, T., Wix, J. 2003: Industry Foundation Classes IFC2x Edition 3, International Alliance for Interoperability, http://www.iai-international.org.
13. Liao, L., Fox, D., Hightower, J. Kautz, H., Schulz, D., 2003: Voronoi Tracking: Location Estimation Using Sparse and Noisy Sensor Data. In: Proc. of the International Conference on Intelligent Robots and Systems, IROS 2003, IEEE/RSJ
14. Lewin, B., Cassimeris, L., Lingappa, V. R., Plopper, G., 2006: Cells. Jones & Bartlett, USA
15. Mishra, A. R., 2004. Fundamentals of Cellular Network Planning and Optimisation. John Wiley & Sons, Chichester, UK
16. LaValle, S. M, 2006: Planning Algorithms. Cambridge University Press, USA
17. Lorenz, B., Ohlbach, H. J., Stoffel, E.-P., Rosner, M., September 2007: Towards a Semantic Spatial Model for Pedestrian Indoor Navigation. In: Lecture Notes in Computer Science - Advances in Conceptual Modeling – Foundations and Applications, Volume 4802/2007, Springer, Berlin, Germany

Chapter 6
Towards Defining a Framework for Automatic Generation of Buildings in CityGML Using Building Information Models

Umit Isikdag and Sisi Zlatanova

Abstract. Increased demand for tools that allow merging of Building Information Models with GIS models is observed in the last several years. Professionals from both domains are looking for solutions to seamlessly integrate such models for various purposes such as, building and construction analysis, urban planning, tourism, cadastre, homeland security, etc. Researchers suggested that the best approach for such integration is harmonised semantics, which will allow formal mappings between the design (BIM) and real world (GIS) models. Although many geometric models have been developed in both domains, the number of semantic models is relatively few. Two most prominent semantic models in the design and real worlds are currently IFC and CityGML. Several studies demonstrate the transfer of information from IFC models into the CityGML but the literature is lacking a formal and descriptive framework for automatic generation of buildings in CityGML using the IFC models. This paper presents preliminary ideas for defining a semantic mapping, which will allow automatic transformations between the two models.

6.1 Introduction

Several ways and methods exist to acquire geometrical and semantic information from building models and to represent this information within the 3D Data Models. Detailed geometric information about buildings can be obtained from digital building models (i.e. CAD drawings), by measuring existing buildings using laser scanning methods, surveying or photogrammetric techniques. Semantic information can be acquired by, querying various databases where information about a building is stored or using the semantic information stored in Building Information Models (BIM). This - acquired information- can later be transformed into the geospatial environment and

Delft University of Technology
egetera@superonline.com, S.Zlatanova@tudelft.nl

stored in form of a geometrical and/or topological geospatial model, in a physical file or in a geospatial database. In this paper we concentrate on transformation of information from BIM (digital models) to the 3D geospatial models (and specifically to CityGML).

Since many years, the AEC industry is benefiting from Computer Aided Design (CAD) systems for creating digital building models. CAD systems are developed with the aim of modelling objects for assisting production/construction process. Models defined with CAD systems usually exist before the final product (or building), and are designed for representing the maximum level of detail in terms of geometry of the model. In CAD systems one or a group of building elements can be modelled in two or three dimensions, within a complex geometrical representation. The element geometries can be modelled by using CSG, Sweeping or BRep methods and geometries can contain curves, splines and surface patches [22]. Until recent years, most of the CAD models created were in two dimensions, were not object oriented and rich in semantic information, and spatial relations between the buildings elements were not preserved within the models. Storing semantic information was not the focus of CAD systems, but with the advent of Product Lifecycle Management (PLM) in production management and Building Information Modelling (BIM) in the AEC industry this situation is now radically changing. Several Building Information Models have been reported in the literature, most commonly known ones are CIS/2 [2] and IFC models [12], which offer means to define objects with geometry and semantics.

On the other hand, geospatial information systems are developed to represent objects that already exist around us. They are defined for representing the objects in a simplified but efficient way (specifically in terms of geometry). Geospatial models represent large number of objects mostly in 2D/2,5D with more simple geometric representations and building geometries are represented by BRep and Sweeping methods. In addition the geometries are mostly created with straight lines (i.e. consisting of polygons, polyhedrons). Attribute information is an important aspect of geospatial information models and usually stored in databases. As [24, 29, 1] indicated that CAD software supports a broad range of 3D primitives and free-form curves, while these primitives and free-form curves are not present in the GIS world. Recently 3D Information models were developed for representing the real world objects, i.e. CityGML [6, 19] in fact the geometry is still limited to simple representations.

Many researchers have investigated the differences and the similarities between CAD and GIS and suggested approaches for transforming information from one to other. As mentioned by [27] the lack of object definitions in the CAD files, different scale representations, transformation of the local (CAD) coordinates into a geospatial coordinate system, the existence of parametric shapes in CAD files that can not be converted into GIS objects, and different levels of detail between CAD models and their representation in the geospatial environment appeared as main barriers that prevent CAD-GIS data transformation. [27] draws an attention to the need of integrated geometric models and harmonised semantics between two domains in order to tackle with the interoperability problems between AEC and geospatial information domains. [29] pointed out that developing uniform data types for both CAD and geospatial information models would eliminate the need for conversion between different formats.

Several attempts have been made to simplify BIM models and integrate them with GIS. For example in a recent effort, [16] demonstrated the transfer of information from an industry standard BIM (IFC) to the (ESRI) Shapefiles and Geodatabases. In parallel, commercial software for conversion from IFC to CityGML and vice versa is in development [13, 26]. OGC has completed tests on the integration of CityGML and IFC models in OWS-4 testbed [21]. However a formal framework for strict (semantic and geometry) conversion is not available yet.

This paper suggests that information from (semantic) design and real-world models can automatically be exchanged if a formal framework can be made available. Following the background sections, we present our preliminary ideas for establishing framework for transformation of information between IFC and CityGML. Section 6.2 presents an introduction to Building Information Modelling, the following (third) section presents the background on the representation of buildings within 3D geospatial models. The 4th section presents an overview on information mapping needs which will serve as a basis when establishing the framework for transformation of information between IFC and CityGML models.

6.2 Building Information Modeling

The fragment nature of the Architecture Engineering Construction (AEC) industry has resulted in significant barriers to communication between the various stakeholders, which in turn has significantly affected the efficiency and performance of the industry. [20] indicated that, US$15.8B is lost annually in the U.S Capital Facilities Industry due to the lack of interoperability. In recent years, Building Information Modelling has become an active research area in order to tackle the problems related to information integration and interoperability. Today, Building Information Models (BIMs) are promising to be the facilitators of integration, interoperability and collaboration in the future of the construction industry. According to [23] a BIM is a computable representation of all the physical and functional characteristics of a building and its related project/life-cycle information, which is intended to be a repository of information for the building owner/operator to use and maintain throughout the life-cycle of a building.

6.2.1 Industry Foundation Classes (IFC)

Today, the current key efforts in the area of BIM are the Industry Foundation Classes (IFC) and the CIMSteel Integration Standards 2(CIS/2). The distinctive characteristics of the BIMs can be identified as follows:

1. *Object Oriented:* most of the BIMs are defined in an object-oriented nature.
2. *Data-rich / Comprehensive:* BIMs are data rich and comprehensive as they cover all physical and functional characteristics of the building.
3. *Three dimensional:* BIMs always represent the geometry of the building in three dimensions.
4. *Spatially-related:* Spatial relationships between building elements are maintained in the BIMs in a hierarchical manner.
5. *Rich in semantics:* BIMs maintain a high amount of semantic (functional) information about the building elements.

6. *Supports view generation:* The model views are subsets or snapshots of the model that can be generated from the base information model. BIMs therefore support view generation.

Today IFC is seen as being the strongest player of the BIM world. IFC is the effort of IAI/buildingSMART whose goal is to specify a common language for technology to improve the communication, productivity, delivery time, cost, and quality throughout the design, construction and maintenance life cycle of buildings. Each specification (called a 'class') is used to describe a range of things that have common characteristics. These IFC-based objects aim to allow AEC/FM professionals to share a project model, while allowing each profession to define its own view of the objects contained within the model. In 2005, IFC became an ISO Publicly Available Specification (as [17]).

In IFC model geometries of the building elements are always represented in 3D using one of/or combination of CSG, Sweeping and BRep methods. IFC model usually represents a building element by multiple geometric representations (i.e. a Sweeping Representation and a BRep). The spatial relations between the building elements are preserved within the models' spatial structure. The information about the geographic location of the building is stored as an attribute of *IfcSite* object of the model. The main structure of the building is represented within IFCSHAREDBUILDINGELEMENTS data model. This data model defines the entities for representing the basic components of the building. Each entity in this data model is a subtype of *IfcBuildingElement* entity. The EXPRESS-G Representation of IFCSHAREDBUILDING ELEMENTS data model is shown in Fig.6.1.

6.2.2 Model Views

In order to support several phases and the stakeholders of the construction life-cycle several views of the BIM needs to be generated [8]. The BIM model views are generated by a declaration of a model that is a subset of another model, or by declaration of a model that is derivable from a BIM (in this case the view should not be a superset of a the BIM).These views can be generated from files or databases by using application interfaces and web interfaces. These views can either be transient or persistent depending on the need.

If the model view is persistent, it will be stored in a physical file or a database, otherwise if it is transient the physical storage of the view is not necessary. The model views are updated using EXPRESS-X [8] or XSL [14] languages. The Model Views can be used for eliminating semantic differences between the BIM and another domain model, and can also help in the model simplification process, i.e. [16] demonstrated the use of BIM model views while mapping information from the IFC model to an application specific view, which is developed to interact with a GIS.

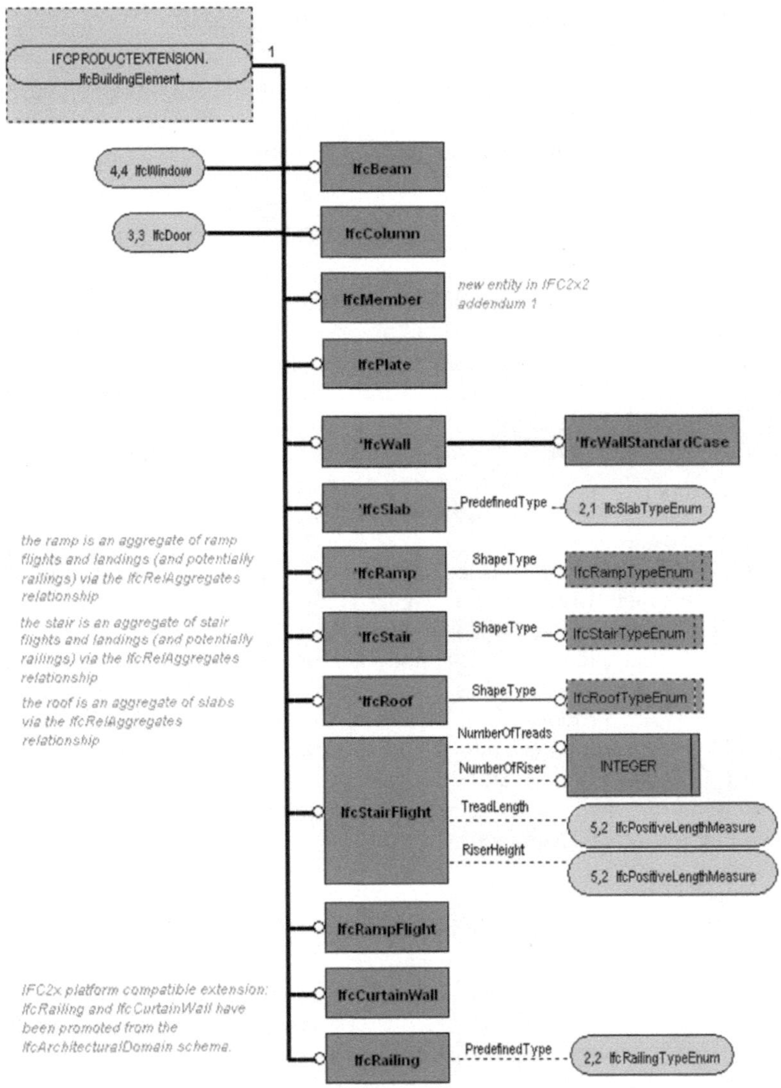

Fig. 6.1. EXPRESS-G Representation of IFCSHAREDBUILDING ELEMENTS data model

6.3 Buildings within 3D Geospatial Models

In geospatial model a 3D geospatial object can be defined by its geometric or topologic representation or by using both representations [28, 1]. As explained in [20] topologic objects can be mapped into geometric objects.

6.3.1 3D Geometrical Representations

[25] summarised six main approaches for representing 3D objects by computers. These approaches are primitive instancing, spatial occupancy enumeration, cell decomposition, constructive solid geometry, sweeping and representing the object using boundaries (Boundary Representation, BRep). Two of these geometrical representations sweeping and BRep are used commonly to represent 3D objects in 2,5D vector models. Swept or extruded 3D geometries are created by GIS applications on demand. Research in the area demonstrated the use of cell decomposition and 3D spatial occupancy enumeration (voxels), to represent 3D objects and also outlined the need for using CSG representations.

In ISO 19107 (Standard for Geographic Information: Spatial Schema, compliant with OGC Abstract Specifications Topic 1: Spatial Schema) geometric model, a primitive solid object (GM_Solid) is constructed by its boundaries (bounding surfaces) thus its 3D representation method is BRep. Other solid types such as Cone, Cylinder, and Sphere are defined as sub-classes of Surface Patch (GM_SurfacePatch), Parametric Curve Surface (GM_ParametricCurveSurface) and Gridded Surface (GM_GriddedSurface) and are interpreted as surfaces. According to [18] a composite solid (GM_CompositeSolid) can be is generated by using a set of solids that join in pairs on common boundary surfaces to form a single solid. The resulting model will also be in the form of a BRep model.

6.3.2 Buildings within CityGML

[7] defines CityGML as a common semantic information model for the representation of 3D urban objects that can be shared over different applications. The current version of CityGML refers to only earth surface objects and above earth surface objects, but research and first tests has been reported on incorporating underground objects [10, 9]. CityGML is designed as an open data model and XML-based format for the storage and exchange of virtual 3D city models. It is implemented as an application schema of the Geography Markup Language 3 (GML3). CityGML defines the classes and relations for the most relevant topographic objects in cities and regional models with respect to their geometrical, topological, semantic and appearance properties. CityGML is applicable for large areas and small regions and can represent the terrain and 3D objects in different levels of detail simultaneously. In CityGML 5 levels of detail (LOD) were defined in order to represent city objects. In terms of representing buildings 4 out of 5 LODs are used. As explained by [7] as follows:

1. LOD0 is essentially a two and a half dimensional Digital Terrain Model, over which an aerial image or a map may be draped.

2. LOD1 is the well-known blocks model comprising prismatic buildings with flat roofs.
3. A building in LOD2 has differentiated roof structures and thematically differentiated surfaces.
4. LOD3 denotes architectural models with detailed wall and roof structures, balconies, bays and projections. High-resolution textures can be mapped onto these structures.
5. LOD4 completes a LOD3 model by adding interior structures for 3D objects. For example, buildings are composed of corridors, rooms, interior doors, stairs, and furniture.

In LODs 2-4 of CityGML the building façade is defined in form of Boundary Surfaces i.e. WallSurface, Roof Surface, GroundSurface or ClosureSurface. The BuildingInstallation class is used for representing building elements like balconies, chimneys, dormers or outer stairs, which affect the outer appearance of a building. In LODs 3-4 the openings in the BoundarySurfaces can be represented as Doors and Windows. In LOD4, the Room class is used to represent the interior of the building. A Room can be defined by its Boundary Surfaces i.e., Interior Wall Surface or Floor Surface. The movable objects i.e. lamps, table and chairs are represented with Building Furniture class. Building Installation class is used to model other immovable building elements such as pillars and stairs.

6.4 The Framework

The IFC model consists of 4 conceptual layers as Resource, Core, Interoperability and Domain. The Interoperability layer defines concepts (or classes) common to two or more domain models (i.e. Architecture, Construction management) defined in IFC. As explained in [15] it is through the schemata defined at the interoperability layer that, multiple domain models can be plugged into the common IFC core. The entities in the Interoperability Layer and specifically in the 'Shared Building Elements' part of IFC specification can form an interesting starting point for the development of the transformation framework between IFC and CityGML models. The 'Shared Building Elements' part of the IFC specification, contains entities that represent the basic components of a building such as a beam, column, wall or slab (Fig. 6.1).Each instance of these entities corresponds to an element of a building in a BIM. The '*Representation*' attribute of each entity in this set (i.e. IfcColumn, IfcBeam) will refer to an 'IfcProductRepresentation' entity which defines the geometry of that building element. Similarly the orientation and absolute location of a building element can be acquired from the 'IfcObjectPlacement' entity referred from the 'ObjectPlacement' attribute. In summary, a building elements' geometry is referred from a BIM object that contain semantic information about that element.

Transforming information from IFC to CityGML requires a two-step approach: transforming semantic information and transforming geometries. Since the objects (classes) in these two models are very diverse, the two steps cannot be performed separately. An object in one of the models might be mapped to a group of objects (and vice versa), which requires a careful consideration on the order or converting geometries and semantics.

In order to perform a successful transformation operation;

- A set of rules (a rule base) needs to be clearly defined in the first stage, in order to define the semantic mappings between the classes of two models for each LOD of CityGML.
- The second stage will be building up the rules/algorithms for geometric model simplification. A BIM model view can facilitate model simplification in this stage.
- The final stage will be defining the information that will be transformed to form the attributes of the CityGML objects for each LOD.

The following sections present a general overview of (semantic and geometric) information transformation from IFC to form CityGML models in different LODs of CityGML, mainly concentrating on the transfer of geometric information.

6.4.1 Transfer of Semantic Information from IFC to *_AbstractBuilding* class

The information required for populating the attributes of the *_AbstractBuilding* class in CityGML model can be acquired from several different IFC entities. For example an object count of the *IfcStorey* entity will provide the number of stories of the building, and the *IfcBuilding* entity will help in determining the number of stories above the groud. On the other hand the information for determining the CityGML *RoofType*, can be acquired from *IfcRoof* entity.The year of construction of the building can be acquired from *IfcWorkSchedule* entity.

6.4.2 Generation of Buildings in CityGML LOD 1

In the LOD 1 of the CityGML model, the geometry of a building is represented as a prismatic object with a flat top (Fig. 6.2). The geometry of the building can either be represented with *gml:SolidType* as a volumetric object or the exterior surface (façade) of the building is represented with *gml:MultiSurfaceType*. In fact, in LOD 1 every wall is represented with a single face (flat surface).

In order to generate a building model in LOD 1, the façade of the building can be acquired from the *IfcWall* entities (representing the façade walls) and *IfcSlab* entities (representing the building roof). In this situation, the façade needs to be simplified using a BIM model view, in such a way that each wall (and roof slab) will only be represented with a single face (flat surface). An alternative way of getting a simplified geometry is using bounding box representation of the elements. The extensions to building such as balconies also need to be disregarded when constructing the LOD 1 model. In addition, if there is any curvature in the geometry of walls and roof slab, these should also be eliminated.

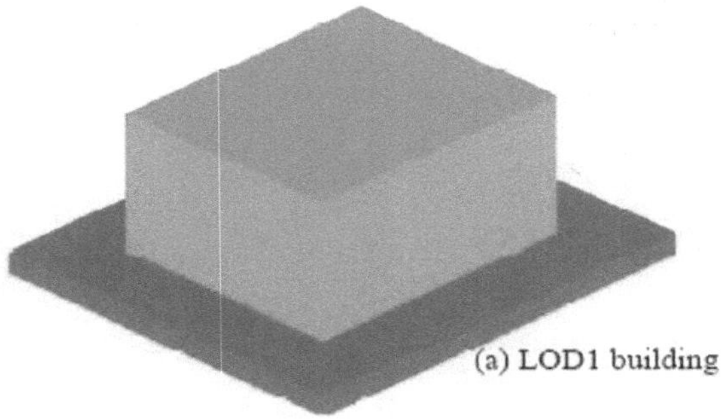

Fig. 6.2. Geometrical Representation of a building in CityGML LOD 1 [7]

On the other hand, as the model in LOD 1 represents the building in a simple form, another approach for generating the building representation would be acquiring the footprint of the building from the *IfcSlab* entity(representing the ground or the top floor) and the height of the building from the *IfcSlab* entity that are referenced from the *IfcRoof* entity (in other words the Roof Slabs). In this case, the BRep model of the building can easily be constructed using the coordinates of the building footprint and the height of the building. The Terrain Intersection Curve (TIC) can be generated either using the building footprint acquired from the IfcSlab entity or using *IfcWall* entities that form the building façade.

6.4.3 Generation of Buildings in CityGML LOD 2

In LOD 2 of the CityGML model, the outer façade of the building can be represented in a greater detail. The biggest differences between the LOD 1 and LOD 2 (in terms of geometrical representation) are, i.) the outer walls of the building can be represented with multiple faces and ii.) the curve geometries of the building façade can be represented within the model structure (Fig. 6.3). On the other hand LOD 2 enables the representation of outer building installations such as chimneys, balconies, dormers, outer stairs etc. These installations can be represented within the *BuildingInstallation* class as *gml:Geometry* (as an aggregate of multiple geometrical types). Starting from LOD 2 (in order to represent the semantic differentiation between the building elements) different parts of the outer façade of the building can be denoted within different classes of CityGML. These different classes are aggregated under *_BoundarySurface* class and can be used to explicitly differentiate Roof Surface, Wall Surface, Ground Surface and Closure Surfaces. Each of these surfaces is represented with *gml:MultiSurfaceType.* (i.e as an aggregate of multiple surfaces).It should be noted here that the openings on the façade are only represented in LOD 3 and LOD 4, thus the façade representation in LOD 2 will not include the representations of openings such as door and windows (but there is no obligation against the representation of openings as a part of a face, i.e. wall surface, in the façade.).

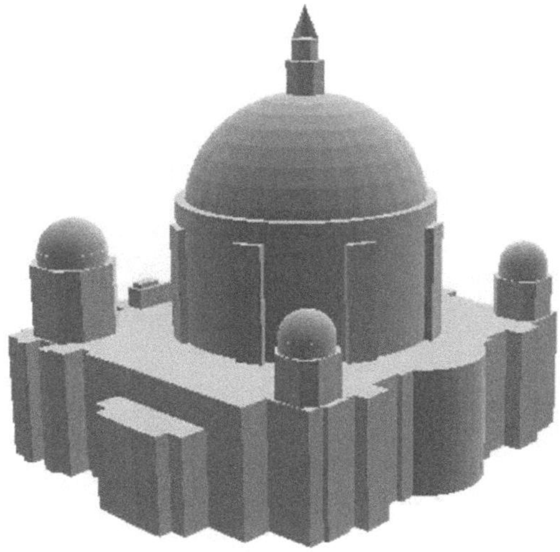

Fig. 6.3. Building façade with curved geometries represented in CityGML LOD 2 [3]

The façade of the buildings in LOD 2 can be wrapped by raster-based 2D textures (Fig. 6.4).

In CityGML LOD 2, the façade of the building can be represented as a result of an IFC model simplification. In order to generate a building model in LOD 2, *IfcWall* and *IfcSlab* entities need to be used as a primary resource. Entities such as *IfcColumn*, *IfcBeam*, *IfcCurtainWall*, can also help in forming the building façade at LOD 2. In order to preserve the semantic information about the building elements it can be advised to map the information from the IFC entities to the classes aggregated under the *_BoundarySurface* class.

In such a case, the *WallSurface* objects in the CityGML model will be generated by acquiring the outer wall face from the *IfcWall* enitiy. The geometric structure of the *IfcWall* entity is usually defined in form of a Sweeping and CSG Solid, thus a Sweeping and CSG to BRep conversion will usually be needed before the simplification process. After this conversion process, the model simplification will serve for eliminating the sides of the wall that will not be transferred into the CityGML model. If there are curtain walls in the building, geometric information from *IfcCurtainWall* entity can also be mapped to *WallSurface* class of CityGML. In this mapping process the openings in the *IfcWall* and *IfcCurtainWall* classes can be omitted, as they can not be represented in this LOD, or the surface geometry of the Door and Windows can be represented within the *Wall Surface* class, but in this case semantic information will be lost (i.e. a Door/Window will be interpreted as a face of the wall).The curved surfaces in the *IfcWall* can be represented by the *WallSurface* class, but the BRep model will contain more objects to represent the curvature of the wall surface.

The *RoofSurface* objects can be generated by simplifying the geometry of *IfcSlab* entities that are referenced from the *IfcRoof* entity. Similar to the walls, the BRep structure of the roof slab first needs to be constructed (by Sweeping to BRep conversion) before the simplification process.

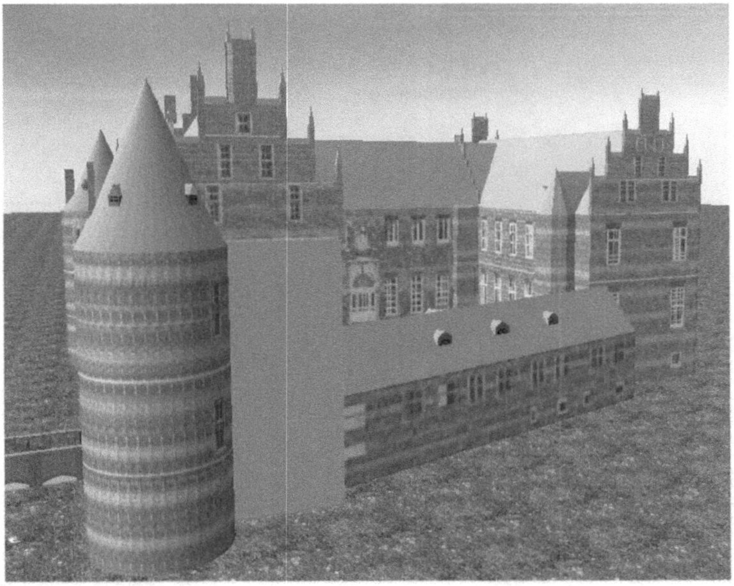

Fig. 6.4. CityGML LOD 2 model showing a building façade with raster texture [4]

The *GroundSurface* object can be generated either by acquiring the geometric information from the *IfcSlab* entity that is used to represent the ground floor of the building, or by joining the centrelines of the façade walls (represented by *IfcWall* entities) at the ground level, for forming a closed polygon, but some line clipping operations can be required in this latter approach if the centrelines of the walls cross each other at some points.

Columns and Beams that are visible from the outside of a building can be represented within *BuildingInstallation* class (as a *gml:Solid* or *gml:MultiSurface*). Geometric information regarding these building elements can be acquired from *IfcColumn* and *IfcBeam* entities. Similarly the stairs of the building can be represented in a LOD 2 model, by the *BuildingInstallation* class. In such a case, the geometry of the stairs needs to be acquired from the *IfcStair* entity. The geometrical representation of building elements, which are represented without a semantic definition in IFC, can be obtained from *IfcBuildingElementProxy* entities.

6.4.4 Generation of Buildings in CityGML LOD 3

In a LOD 3 CityGML model, the façade of the building can include openings such as doors and windows (Fig. 6.5). The openings in the building are represented with *Door* and *Window* classes which are defined as the sub-classes of the (abstract) class *Opening*. Each opening is represented with *gml:MultiSurface* geometry. The *Window* class is used to represent the windows inside and outer façade of the building. Similarly the *Door* class is used for modelling the doors that are between the adjacent -inside-spaces and located at the outer façade of the building.

Fig. 6.5. Geometrical Representation of a building in CityGML LOD 3 (CityGML Dataset, Four Buildings in LOD 3)

Similarly the door and window classes in the IFC model (*IfcDoor* and *IfcWindow*) are referred from the (IFC) opening element (*IfcOpeningElement*) but, in contrast to the CityGML object model, in IFC model, neither *IfcOpeningElement* is an abstract class, nor the *IfcDoor* and *IfcWindow* are the subclasses of the *IfcOpeningElement.* The *IfcOpeningElement* is an element that is used the describe the geometry and semantics of an opening which can contain multiple door and windows, thus an *IfcOpeningElement* can refer to multiple *IfcDoor* and *IfcWindow* elements.

The doors and windows (in the IFC model) are represented with the help of two different relationships. First, the opening is referred from its spatial container (i.e. a wall) by

IfcWall→IfcRelVoidsElement →IfcOpeningElement **(1)**

relationship (*IfcRelVoidsElement* is the element that enables the semantic relationship between a building element and an opening). As the next step, the door or window is referred from the opening by

IfcOpeningElement→ IfcRelFillsElement→IfcWindow/IfcDoor **(2)**

relationship (in this case *IfcRelFillsElement* is used for establishing the semantic relationship between the opening and the element that fills that opening)

The *Door* and *Window* objects in CityGML can be generated by the information acquired from the *IfcDoor* and *IfcWindow* classes. In order to acquire geometric information from these IFC classes (i.e *IfcWindow*) using its related container (i.e. *IFcWall*), the following path of relationships needs to be followed:

IfcWall→IfcRelVoidsElement→IfcOpeningElement→
IfcRelFillsElement→IfcWindow **(3)**

The coarse geometric representation of doors and windows in IFC, is very similar to the wall/slab representation (i.e. they are represented as Sweeping and CSG models), but a (finer) geometric representation of these elements is usually presented in form of a BRep. If the coarse geometric representation would be used when acquiring information from the IFC model, a Sweeping /CSG to BRep conversion will become a need. In IFC model, additional semantic information such as the panel type/number and the opening direction is also provided for the door and window objects in *IfcDoor* and *IfcWindow* classes, but CityGML *Door* and *Window* classes do not contain any specific (non-generic) attributes to store such information.

6.4.5 Generation of Buildings in CityGML LOD 4

The LOD 4 of CityGML provides the option to represent the interior structure of a building. The main CityGML classes that enable this representation are *Room* and *IntBuildingInstallation.* The *Room* is described as the semantic object for modelling the free space inside the building. According to CityGML specification the *Room* should be closed topologically (i.e. by using a *ClosureSurface* when necessary). The geometry of the room is generally represented by *gml:Solid* , but if the topologic correctness of the boundary can not be guarantied its geometry can also be represented by *gml:MultiSurface* . In order to preserve the semantic information related to a room (in a LOD 4 model) different parts of the *Room* can be represented within the classes aggregated under the *_BoundarySurface* class. These classes can be used to semantically differentiate the *CeilingSurface*, *InteriorWallSurface*, and *FloorSurface* (Fig. 6.6).

In CityGML LOD 4 model, the *Room* (*gml:Solid*) objects are topologically connected by surfaces at the doorways. The *Room*s are defined as being adjacent if the share an *Opening* or a *ClosureSurface*. The *Room*s can contain, movable objects such as furniture which are represented with *BuildingFurniture* class. A building can also contain other installations such as interior stairs, railings and pipes. These are represented by *IntBuildingInstallation* class in CityGML LOD 4 model. The geometry of both *BuildingFurniture* and *IntBuildingInstallation* objects are represented by *gml:Geometry* type.

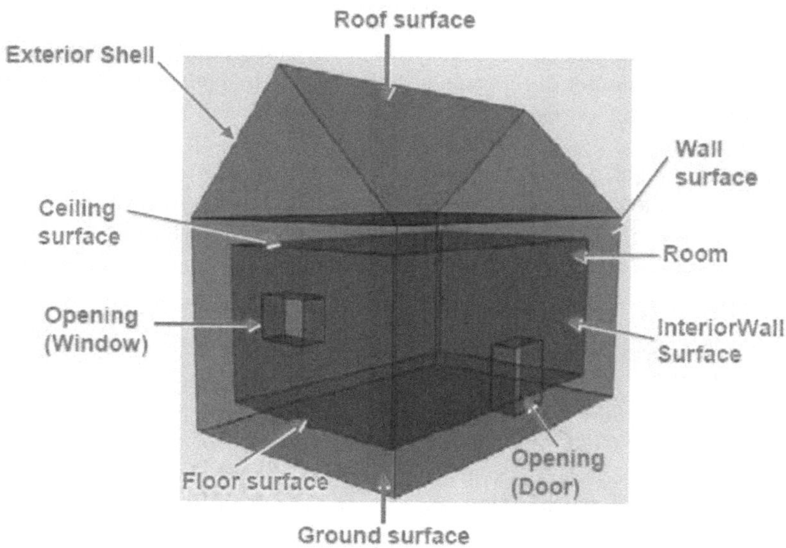

Fig. 6.6. Indoor Representation of a building in CityGML LOD [7]

The IFC model does not have a specific class for denoting room spaces; in fact, the model has a generic class (*IfcSpace*) for defining the spaces in the building. *IfcSpace* class represents an area or volume bounded actually or theoretically. A space (*IfcSpace*) is generally associated to a building storey (*IfcBuildingStorey*) or in case of exterior spaces to a site (*IfcSite*). As explained in [14], a space may span over several connected spaces, therefore a space group provides for a collection of spaces included in a storey. A space can also be decomposed in parts, where each part defines a partial space. *IfcSpace* is represented within multiple geometric representations (usually by a combination of Sweeping and CSG and rarely with a BRep). As mentioned earlier, a Sweeping/CSG to BRep conversion can also become a need (in this case) to transfer information from the *IfcSpace* entity for generating the *Room* object in CityGML model. Other approach for mapping room geometries can be acquiring information from *IfcWall* and *IfcSlab* entities that form room boundaries. In this approach the *FloorSurface* and *CeilingSurface* objects for a *Room* object at storey N of a building, will be generated by the information acquired from *IfcSlab* entities located in storey N and N+1. The *IfcSlab* entity at storey N will be used for generating the *FloorSurface* and *IfcSlab* entity at storey N+1 will be used for generating the *CeilingSurface.* Information acquired from *IfcWall* entities that are representing the interior walls will help in generating the *InteriorWallSurface* objects of CityGML. On the other hand, information acquired from the *IfcWall* entities that are representing the exterior (façade) walls of the building can be used to generate both the *Wall Surface* objects (for representing the building façade) and *InteriorWallSurface* objects (for representing the interior walls of the exterior rooms of the building). The interior *Door* and *Window* objects can be generated by the information acquired from *IfcDoor* and *IfcWindow* entities by using the approach explained in the previous section.

Structural elements such as beams and columns and other installations such as stairs, ramps and railings are semantically differentiated and represented in separate classes (as *IfcBeam*, *IfcColumn*, *IfcStair*, *IfcRailing*, *IfcRamp*) in the IFC model. The information acquired from these elements and installations can only be represented as *IntBuildingInstallation* in CityGML LOD 4. A Sweeping/CSG to BRep conversion also becomes a need specifically for representing the beams and columns in CityGML, as IfcBeam and *IfcColumn* entities are usually represented as Swept or CSG solids.

Furnitures in the building is represented with IfcFurnishingElement entity, and the geometry of the furniture is usually represented by BRep in the IFC model. Information acquired from the IfcFurnishingElement can be used for generating the BuildingFurniture in CityGML LOD 4 model.

6.5 Conclusion

Information from design and real-world models can automatically be exchanged if a formal framework can be made available. Following a background literature review, the paper presents a general overview of (semantic and geometric) information transformation from BIM (specifically from IFC models) into CityGML models. The information provided in the paper can contribute to the efforts towards such a building a formal framework. The focus of this work was on transforming the building information as buildings are the key elements of the city fabric. On the other hand in terms of information modelling, the building elements explicitly described in different LODs and the building representation in higher LODs of CityGML, provides the possibility for building up rules for semantic matching between BIMs and CityGML.

Although there are explicit semantic classifications on both models a transformation needs to tackle problems due to semantic mismatches between the classes in two models. For example, an opening on a slab is represented with the *IfcOpening* entity in the IFC model, in contrast the CityGML model does not provide a class for representing an opening that does not contain a door or window. Similarly, CityGML model does not provide a class for representing the storey of a building. On the other hand, an entity of the IFC model can correspond to different objects of CityGML also due to the semantic mismatches between two models. For example, the CityGML model provides two different objects for representing the *FloorSurface* and *CeilingSurface* of a room, but these surfaces are represented with a single entity in the IFC model (i.e. *IfcSlab*).

Another problem in the transformation process might exist due to differences in granularity of both models. A BIM (i.e IFC model) is a finer-grained model, in terms of representation of building elements, when compared with the CityGML model. This difference in levels of granularity can cause problems when exchanging information between two models, i.e a *Window* object in CityGML might correspond to 3 or 4 different '*IfcWindow*' entities in the IFC model. In this case 1-to-1 object matching will cause the creation of redundant CityGML objects when transforming information from IFC to CityGML model, on the other hand if a need for transforming the information vice versa appears, the resultant IFC model might become very coarse grained, i.e. 3 or 4 windows might be represented with a single *IfcWindow* entity.

The overview presented in this paper, only concentrated on unidirectional information transformation (i.e. from a BIM to the CityGML models) as the need for such a transformation appears more eminent today, in fact bidirectional transformation might also be required to support renovation related tasks where an information model of a building usually does not exist.

The findings of the study indicate that:

- IFC model contains all necessary information for representation of buildings in different LODs of CityGML model.
- It is possible to define rules for transforming geometrical information from IFC entities to CityGML objects.
- It is possible to define rules for facilitating semantic matching between two models.

The results of the previous research illuminate that the model views (BIM sub-models) can play role in the semantic matching process in addition to simplifying the model before the geometrical transformation.

The ongoing research will focus of formally defining the stages of the transformation framework, constructing the rule base for semantic matching, developing model views and algorithms for achieving geometric/semantic model simplification, and testing the developed models views and algorithms. The future research will also investigate the possible benefits and investigate methods of bidirectional transformation.

References

1. Breuning, M. and S. Zlatanova, 2006, in: Zlatanova&Prosperi (Eds.) 3D Geo-DBMS, in 3D large scale data integration: challenges and opportunities, CRC Press, Taylor&Francis Group, pp. 87-115
2. CIS2, 2007, Model documentation webpage. CIMSteel Integration Standards Relase 2, 2007, http://www.cis2.org [last accessed 10-2007]
3. CityGML Dataset, City of Berlin, http://www.citygml.org/ (last accessed 06-2008)
4. CityGML Dataset, Castle Herten, http://www.citygml.org/ (last accessed 06-2008)
5. CityGML Dataset, Four Buildings in LOD 3, http://www.citygml.org/ (last accessed 06-2008)
6. CityGML, Exchange and storage of Virtual 3D city Models 2008, available at http://www.citygml.org/ [last accessed 06-2008]
7. CityGML Implementation Specification, 2007, Candidate OpenGIS Implementation Specification (City Geography Markup Language) https://portal.opengeospatial.org/files/?artifact_id=16675 [last accessed 12-2007]
8. Eastman, C and Jeng, T. S. 1999, A database supporting evolutionary product model development for design, Automation in Construction, 8(3):305-323
9. Emgard, L. & S. Zlatanova, 2008, Implementation alternatives for an integrated 3D information model, in: Van Oosterom, Zlatanova, Penninga and Fendel (eds.), 2008, Advances in 3D Geoinformation Systems, Lecture Notes in Geoinformation and Cartography, Springer-Verlag, Heidelberg, pp. 313-329

10. Emgard, L. and S. Zlatanova, 2008, Design of an integrated 3D information model. In Coors, Rumor, Fendel & Zlatanova (eds.), Urban and regional data management: UDMS annual 2007 (pp. 143-156), Taylor & Francis Group, London, UK
11. IAI/BuildingSmart, Web Site for International Alliance for Operability http://www.iai-international.org
12. IFC, 2008, Model documentation webpage. International Alliance for Operability, 2008, http://www.iai-international.org/Model/R2x3_final/index.htm [last accessed 06-2008]
13. IFC Explorer, 2008, Tool for viewing and conversion of IFC models http://www.iai.fzk.de/www-extern/index.php?id=1040&L=1[last accessed 06-2008]
14. IFC Model View Definition, 2007, IFC Model View Definition Format, http://www.iai-international.org/software/mvd.shtml [last accessed 05-2007]
15. IFC Technical Guide, 2000,The Techincal Guide for Industry Foundation Classes http://www.iai-international.org/Model/documentation/IFC_2x_Technical_Guide.pdf [last accessed 12-2000]
16. Isikdag,U., 2006, Towards the Implementation of Building Information Models in Geospatial Context, PhD Thesis, University of Salford, UK.
17. ISO 16739 Industry Foundation Classes, Release 2x, Platform Specification http://www.iso.org/iso/iso_catalogue/catalogue_tc/catalogue_detail.htm?csnumber=38056
18. ISO 19107 Geographic information -- Spatial schema http://www.iso.org/iso/iso_catalogue/catalogue_tc/catalogue_detail.htm?csnumber=26012
19. Kolbe, T. and G. Gröger, 2003, Towards unified 3D city models. Proceedings of the ISPRS Comm. IV Joint Workshop on Challenges in Geospatial Analysis Integration and Visualization II2, September 8-9, 2003 in Stuttgart, 8p.
20. Lake, R., Burggraf, D., Trninic, M., Rae, L., 2004, Geography Mark-up Language: Foundation for the Geo-Web, John Wiley and Sons
21. Lapierre, A. and P. Cote, 2008, Using Open Web Services for urban data management: a testbed resulting from an OGC initiative offering standard CAD/GIS/BIM services, in Coors, Rumor, Fendel & Zlatanova (eds.): Urban and Regional Data Management; UDMS Annual 2007, Taylor and Francis, London, pp. 381-393
22. Lattuada, R., 2006, Three-dimensional representations and data structures in GIS and AEC. In: Zlatanova& Prosperi (eds.): Large-scale 3D data integration - Challenges and Opportunities, Taylor&Francis, pp.57-86
23. NBIMS, 2006 National BIM Standard Purpose, US National Institute of Building Sciences Facilities Information Council, BIM Committee, http://www.nibs.org/BIM/NBIMS_Purpose.pdf [last accessed 02-2008]
24. Pu, S. and S. Zlatanova, 2006, Integration of GIS and CAD at DBMS level, </span>In: Fendel&Rumor (Eds); Proceedings of UDMS'06 Aalborg, Denmark May 15-17, 2006, TU Delft, 2006, pp. 9.61-9.71
25. Requicha, A.A.G., 1980, Representations for Rigid Solids: Theory, Methods and Systems, ACM Computing Surveys,12(4):434-464
26. Safe Software, 2008, FME Desktop Translator/Converter Software http://www.safe.com/products/desktop/formats.php [last accessed 06-2008]

27. Van Oosterom, P. van, J. Stotter and e. Janssen, 2006, Bridging the worlds of CAD and GIS, In: Zlatanova& Prosperi (eds.): Large-scale 3D data integration - Challenges and Opportunities, Taylor&Francis, pp.9-36
28. Van Oosterom, P, J. Stoter, W. Quak and S. Zlatanova, 2002, The balance between geometry and topology, in: Advances in Spatial Data Handling, 10th International Symposium on Spatial Data Handling, D.Richardson and P.van Oosterom (Eds.), Springer-Verlag, Berlin, pp. 209-224
29. Zlatanova, S., S. Pu and W.F. Bronsvoort, 2006, Freeform curves and surfaces in DBMS- a step forward in spatial data integration, in Nayak, Pathan1&Garg (Eds.), Proceedings of the ISPRS Commission IV Symposium on 'Geospatial Databases for Sustainable Development', 27-30 September, 2006, Goa, India; Archives of ISPRS Vol. 36, Part 4A, pp. 407-412

Chapter 7
Managed Objects for Infrastructure Data

Erik Kjems, Lars Bodum and Jan Kolar

Abstract. Using data objects to describe features in the real world is a new idea and several approaches have already been shown to match scientific paradigms exceedingly well [1, 2, 3]. Depending on the required level of abstraction, it is possible to represent the world more or less closely to reality. In the realm of 3D Geoinformation research, this realism is often related to the way the spatial world is represented. By contrast, the 2D GIS community focuses on attribute data that describes additional states or characteristics of a feature. The main focus in 3D Geoinformation has always been on the representation of spatial objects, on relations like topology, ontology, and on storing and presenting them with more or less detail. The Centre for 3D GeoInformation (3DGI) at Aalborg University is currently participating in a project that explores objects that will not only contain geometry and associated attributive data but also will contain behavioural information. Our goal is to communicate the design and handling of these enhanced objects by means of the concept introduced in Java whereby objects are created in bytecode and subsequently executed within a Java virtual machine. This concept has already been implemented in the GRIFINOR (http://www.grifinor.net) platform [4]. This article will present the core ideas of relevance to this concept as it relates to current understanding of objects. Our work also offers suggestions on how to implemented such algorithms using real-life infrastructure data. Furthermore, we elaborate on the possibilities and challenges associated with moving from mostly static objects to dynamic objects in the area of 3D geoinformation.

7.1 Introduction

It is not always easy to comprehend the concept of managed objects (MOs) and immediately see new possibilities. This article describes new and related technologies by means of examples rather than code, and in terms of possibilities instead of implementation and resulting constraints. It is important to explore this concept intellectually

Centre for 3D GeoInformation
Aalborg University
{kjems, lbo, kolda@3dgi.dk}

and reflect on the nature of objects as entities that are so much more than mere geometry.

MOs represent a completely new paradigm for creating and handling objects in general. The term "managed objects" is a variation on the more often used term managed code (or bytecode as it is often termed). Managed code originates from Microsoft .NET programming, where it refers to a piece of code that cannot run without a virtual machine.

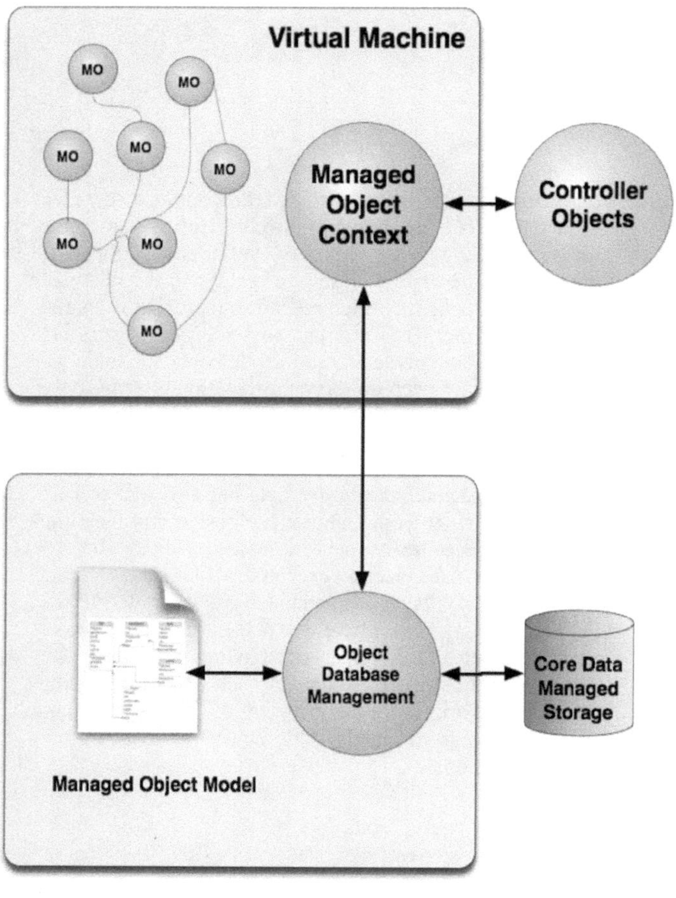

Fig. 7.1. The concept of communication/exchange of MOs. Object database management is connected to the Managed Object Model and to core data that is housed in a managed storage device (which could be any database). The virtual machine can execute the managed objects and the related managed object context. A group of controller objects exists to ensure the state of the objects.

The virtual machine makes the code independent of the hardware, eliminates reliance on external data sources and ensures that it is self-contained (see Fig. 7.1). It is also important to refer to very early work within this area. The first commercial application of this paradigm appeared in the original SmallWorld software in the late 1980s [1].

Existing systems, either commercial or open source, with traditional relational or object-relational data models will never be able to adopt these new objects, because they lack the underlying virtual machine that is needed to interpret the content of MOs.

Fig. 7.2. These four pictures present an overview of GRIFINOR from both a global and a local perspective. All objects are Java managed objects.

The biggest challenge in using MOs for 3D urban and landscape models is to design a whole new system from scratch. The system must use a virtual machine to handle its objects and to model MOs to achieve a virtual representation of real-world features. Such features must be dynamic because they can change shape, texture, position or behavior according to time, proximity or random chance. In this article, we focus on the challenges in creating and maintaining managed objects in relation to infrastructure projects. Our examples are chosen from selected cases in the context of a new research project called InfraWorld (http://www.infraworld.no). Since managed objects can contain so much more than geometry and attributes, it is important to first take a closer look at how such objects are created.

Strictly speaking, enhancing objects that are primarily geometric through the use of dynamic behavior is an obvious way of maximizing return on investment in a 3D model system [5]. Traditional maps focus on the esthetic dimensions of an object's structure. They can be visualized through virtual globes like Google Earth and Virtual Earth from Microsoft or through a dedicated model-map viewer. They can be used for planning purposes or even as a part of more specific tasks like disaster management. But the usage of model-maps can be vastly enhanced when the managed objects come alive and start interacting with one other, with the user, or even with a completely different computer on another continent.

Even though many technological obstacles have to be addressed, the biggest challenge will not be technological but instead will concern implementation at the level of individual organizations. At present it is not possible to meet this challenge because of a lack of knowledge surrounding the basic technological implications and means of creating advanced objects. Therefore, this issue will not be dealt with in this paper. However, it remains an important subject for future research.

The fundamentals of this field have been addressed by other researchers [6, 7, 8, 9]. Certain elements presented in this paper primarily indicate future development directions and are therefore of a more descriptive nature. For such elements, no detail on actual implementation is provided.

7.2 Managed Objects for Real-World Features

A rather simple object representing a building will neither change shape nor position very frequently but can still take advantage of managed objects. The information connected to a building will change continuously. Think of energy consumption, water supply, etc. But depending on the building and its size and function, MOs for buildings can connect or evaluate a significant volume of more important information that changes over time and is monitored by sensors and cameras. A guard at a construction site could benefit from a 3D model of the entire site where each of the buildings is visualized with a color code that corresponds to indoor temperature. If the temperature were within the desired range, the building would be colored green. Seeing the whole site at a glance gives an intuitive impression of whether all building air handling systems are operating nominally. Furthermore, color shifts could indicate a developing problem or the features/instances could be enhanced using clickable areas that could act as a user interface to provide additional information and context. Buildings in a model-map could change color to illustrate whether shops are open or closed. Parking lots could change color as a function of instantaneous occupancy. Apartments for sale can be colored according to price. The sign for a subway station could change color according to train arrival times, etc.

A building may not be the most obvious example to demonstrate the major advantages of MOs. Here is another: a traffic light is a very common object in the developed world. While the geometry of a traffic light is static, the instances of it are not. In principle the traffic light shifts between three different instances represented by three different colors. They shift according to timers or may even be controlled by local volumes of traffic. Implemented in a model, such an icon could be used for traffic simulation and traffic control.

The examples given here suggest that we wish to use a standard static 3D model map with simple feature geometry enhancements to represent all meaningful objects. Other objects in the city model, whether static or dynamic, would also be beneficial in the final visualization. Showing the position of buses in the road network would be an interesting feature. A change of color could give the user an indication of whether the service is running late.

These examples introduce managed objects as a practical concept. The most obvious use of MOs would be for objects that change over time or for objects that benefit from having analytical functions associated with them. Objects that could benefit from this type of dynamic behavior are mainly related to infrastructure. Infrastructure is a very broad term normally used for topological networks such as roads for traffic, power lines for electricity and pipes for water supply. Usually it takes a special solution to store and handle these spatial infrastructure databases. One of the most successful commercial applications has been the SmallWorld GIS. Several facility management companies around the world have used this product since it was introduced in the late 1980s. The SmallWorld application was developed around an object-oriented database technology similar to the one promoted in this article. Today most commercial GIS application suites offer solutions to handle objects either through object-relational principles or through APIs..

Roads generally have nodes that are perceived as crossroads, pipes have joints and power lines have connectors. But even so, it is not always clear when a feature becomes an identifiable object that is suitable for a MO. The power sector is part of the InfraWorld project and is therefore of great interest for MO research. MOs can be managed like all other objects using a hierarchical structure that makes it possible to identify larger items as a group of subordinated objects. Only the subordinated features of interest are used by the higher-level objects. Should it for some reason be necessary to add more subordinated features later on, one would most likely redefine the objects and regenerate the model. In a domain like power, every item in the design has to be identified clearly. Applying monitoring sensors at critical spots and visualizing them intuitively in a 3D model can help identifying upcoming power supply problems in short order.

The levels of detail that are available in the visualization or that are shown as identifiable features in the user-interface are controlling the geometrical extent of objects that can be generated. This issue is no different for objects handled in traditional model-map systems. By applying additional information to these features, a detail-oriented design can be arrived at. When excessive information is presented in the visualization it is likely that the user will not be able to intuitively access the data they need.

Many roads sense traffic directly through physical sensors placed just beneath the surface. They monitor the traffic and driving conditions along the road. The sensors register each passage and load of a vehicle and sometimes also detect speed. Other sensors monitor weather conditions along the road and report them to a control station. This kind of information could easily be visualized in a model-map. The greatest challenge for the inclusion of roads in the model will be the construction of 3D road objects. This is because roads are connected as geometrically challenging identifiable entities and therefore are difficult to handle as real objects.

Fig. 7.3. The power lines visualized here in 3D would be a perfect example for the use of Managed Objects and will be part of the InfraWorld project. Image courtesy "ViaNova Plan og Trafikk", Norway.

The point here is not only to create a model for visualization purposes as seen in many of the available applications, but also to make a design for the road object that complies with the rules for MOs as we have summarized them in this article. Then it will be possible to attach information of any kind, like the sensor data already mentioned - including traffic signals with information about instantaneous allowable road speeds or information about upcoming congestion. In-car navigation systems are about to become the first commercial systems to successfully incorporate model-maps. Navigation companies have succeeded in creating managed objects that can model roads. Most countries store data on roads in relational databases where the geographic reference is not given by a set of coordinates but by a road-id and distance along the road. To complicate the situation further, the distance is never provided with an exact distance unit specification but is rather a relative metric that can be used for locating features in the real world. This also means that there can be more or less than one distance unit between two iterations since roads routinely get redirected, reconstructed or changed. The database keeps track of these changes, but it is not always easy to integrate such information into a model-map.

7.3 Creating Managed Objects

First of all, creating managed objects (MOs) for model-maps is far from trivial. Since MOs can contain all kinds of information besides geometry, MOs must be understood as dynamic in nature and should be handled as such. An object has to be changeable even after it is created. Depending on the behavior programmed into the object, it may have to periodically demand attention either from the system, from other objects or from the user. It may also be necessary to recreate an object if it is required to handle new instances that get defined after the object is initially created.

Three different kinds of class definitions can be identified and used with GRIFINOR. They represent three different properties of class definitions and thus indirectly influence the ways of creating managed objects for the model-map:

- Instantiated
- Procedural
- Referenced

If objects carry all the data necessary to create the Java object, they will be of the type Instantiated and will need no additional data. This will probably be the typical case, and such objects will be fully standards-compliant. Such objects could for instance be an IFC building or a bridge, characterized by unique geometry.

If objects, on the other hand, are defined by parameters or other methods some computation has to be carried out by the Java objects before they can be interpreted in an application. These kinds of objects will be of the type Procedural. The geometry can for instance be derived through procedural algorithms that only need some main attributes for the feature to derive a complete geometry of it. This could be a road segment, which is derived from the centre alignment and information about the width of the road, the ditch construction and so forth.

Finally, objects can be dependent on external resources and be created due to influences coming from other external systems or other objects already created. These kinds of Java objects will be of the type Referenced. It could for instance be a road sign, which is defined almost by its number indicating exactly the kind of sign, size etc. Many countries have all signs stored in databases, which would make it rather natural to retrieve this kind of information from there and create referenced road sign objects for the model-map.

With these three different types of Java objects the most thinkable incidents of features can be created. It would also be possible to combine them and take advantage of both referencing and procedural methods.

7.4 Handling Managed Objects

Managed objects have to be treated differently than traditional objects. To be able to access MOs fast and to be able to take advantage of objects being stored in several distributed databases, the storage and indexing of the objects have to be especially designed. As in a pure object-oriented solution, all the contents in the system are stored and accessed in the form of objects, in the case of GRIFINOR as Java objects.

This means that features represented by objects may, when convenient, carry only attribute data, be associated with a certain functionality (behaviours), or merely trigger a certain functionality while carrying a minimum amount of data or even no content data at all.

It should also be stressed that the objects in GRIFINOR are organized or rather indexed for access through their position in geocentric coordinates. It is very essential for the system design to use a geocentric coordinate system, since this enables GRIFINOR to handle globally related data without any kind of map projections, which have spatial inconsistencies on a global level. Most model-maps are used for local areas like a city or minor landscape areas. For larger areas, the definition of datum and map projections can quickly become a challenge [5].

Objects that can change contents or even shape will need some kind of calibration. They will also need some kind of state, which can be identified as the state of origin. Working with dynamic objects is demanding and a whole new set of rules has to be set up. Even though the idea of one common model-map is reasonable, and the GRIFINOR system, which has implemented the concept of MOs, is designed with the goal to cover the whole world in one model, it might be difficult to apply one set of rules for one common model. But since each feature can be designed with its own set of rules, references and characteristics, it is not necessary to create one set of rules, which applies to all features but rather a few rules, which serve common functionality. Furthermore, features can even be the same in terms of real world objects and still comply with different rules. This is a breach of the traditional understanding of applying standards, where features are handled and described completely the same way as objects in the model.

One of many possible MOs behaviors is the ability to communicate (see Fig. 7.4). Internal communication between objects can be established and/or constrained in many different ways. The small example in Fig. 7.4 features 2 systems, and with 2+1 objects there are 10 different communication channels ((5*4)/2). This illustrates that many ways of communicating are possible with managed objects. And this example does not even include any user involvement. Most likely some kind of hierarchical communication structure will be implemented.

Being able to communicate with objects opens up a completely new way of using and interacting with model-maps. It is difficult to comprehend such possibilities at present, but try to imagine a system like Google Earth with such functionality.

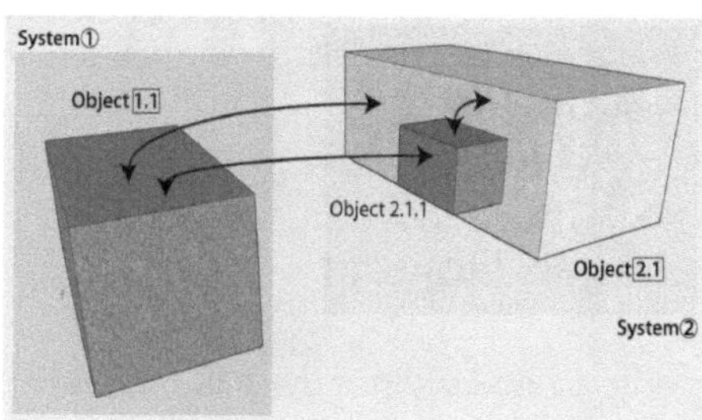

Fig. 7.4. This is an example of possible communication between managed objects. There are three different kinds of links between the two objects. First of all a link between the two main objects (1.1 and 2.1) is shown. Then there is a communication link between a main object and a subordinated object (1.1 and 2.1.1). Finally, a link between an object and its own subordinated object (2.1 and 2.1.1) is depicted.

The building industry would have great interest in identifying every element of a building during construction and tracking it back to its supplier and even manufacturer, because any kind of construction failure can be critical and life-threatening. The possibility to track back flawed equipment and thereby make suppliers accountable is worth a lot of money. It seems that the Industrial Foundation Class (IFC) standard and RFID technology may ultimately meet this need. The IFC standard helps identify the pieces and RFID technology will provide a reference between the virtual model and reality. Together, this offers the possibility of tracking the single pieces (or at least larger components) back to their origin. A system developed using the concept of MO will be able to support this kind of system and will even be able to handle additional dynamic information within the objects. At present, RFID technology is primarily used in logistics. This means that essential parts for a building must be tagged with a RFID chip and then tracked on the way to their destinations. This offers great advantages at the construction site because building assembly can be controlled and timed very precisely, thereby offering cost savings money. The advent of RFID chips on all kinds of production lines is only a first step toward tracking features in the real world. Knowing where things are is of increasing importance, and huge economic interests are connected to this type of information. In the real world, one major problem is the ability to search for this kind of information. That is where model-maps hold great promise.

7.5 Direct Comparison with Traditional Objects

One of the more surprising differences between MOs and traditional objects is that MOs will perform well without any type of standards needing to be implemented in advance. There exist certain elaborate standards like CityGML, KML or IFC. These standards give the ability to import geometrical data and immediately assemble and consistently name the resulting map elements. The system immediately understands how to deal with the different parts of the data. MOs do not depend on any kind of standard. As long as the objects can be identified by a Java virtual machine, it does not matter what the different parts are called. Of course it is important to identify data in the model, but this functionality is not dependent on any kind of standard, except for the standard format of Java bytecode. Its functionality is implicit in the concept. The system will reveal information required at the moment the information is requested. Prior to that instant, the viewer simply represents the object in a manner that is independent of the standard used to describe the data.

This means that it is possible to implement any kind of standard through the use of managed objects. Of course, standards can inform certain instances, which have to be taken into account while generating new objects into the model-map. But they are still not necessary for building up the model as long as the geometry is accessible and identifiable. A huge advantage is that MOs can always handle the newest standards, which can subsequently be used for importing data as long as Java bytecode is employed to model the corresponding MOs.

Several other issues concerning model-maps are influenced by the concept of virtual machines. For instance, can this concept handle topology in different ways depending on user requirements? Of course the geometry needs some kind of topology but in this case, or as implemented in GRIFINOR, topology is handled on the fly and does not necessarily need to be stored in a particular way. This allows for a flexible solution which suits the use of LOD very well. Since MOs are not stored in traditional RDBMSs, questions about topology do not have to be resolved here.

7.6 Perspective

Around 1990, a new concept of object-oriented technology for GIS was introduced, and one of the success stories from that decade was the product SmallWorld. It is still considered a very strong solution, especially within facility management. In this article we have proposed to make the product even stronger and more reliable than it was before. We wish to do this by introducing Managed Objects for 3D Geoinformation. The big challenge is to make the solution available as an Open Source software project and maintain development moving forward.

The main difference between this and other object oriented approaches is the concept of the virtual machine that can control each object separately and execute code within any object. This implies the need for a completely new way of handling objects for creation, storage, updating, communication, visualization and so forth. This can arguably be considered an all-new paradigm for 3D objects.

Unfortunately this also implies the need to develop systems from scratch since earlier solutions cannot realistically implement the concept of virtual machines. GRIFINOR on the other hand was developed with the virtual machine concept at its core and is available as an open source platform.

The advantages of MOs are tremendous since there is almost no limit to what a managed object can contain or do. The virtual machine concept for 3D objects has only been implemented recently in GRIFINOR because an understanding of its applicability and extensibility still has to be proven in applications. There is still a lot of work ahead but at present it seems that most limitations exist are on the side of system designers rather than in the MO concept itself.

References

1. CHANCE, A., NEWELL, R. & THERIAULT, D. (1990) An Object-Oriented GIS - Issues and Solutions. EGIS '90. Amsterdam, The Netherlands, EGIS Foundation.
2. VCKOVSKI, A. (1998) Interoperable and Distributed Processing in GIS, London, Taylor & Francis.
3 EGENHOFER, M. J. & FRANK, A. U. (1992) Object-oriented modeling for GIS. URISA Journal, 4, 3-19.
4. BODUM, L., KJEMS, E., KOLAR, J., ILSØE, P. M. & OVERBY, J. (2005) GRIFINOR: Integrated Object-Oriented Solution for Navigating Real-time 3D Virtual Environments. IN OOSTEROM, P. V., ZLATANOVA, S. & FENDEL, E. M. (Eds.) Geo-information for Disaster Management. Berlin, Springer Verlag.
5. KJEMS, E. & KOLAR, J. (2005) From mapping to virtual geography. IN BATTY, S. E. (Ed.) CUPUM '05: Computers in Urban Planning and Urban Management: Abstracts of the 9th International Conference. London, Centre for Advanced Spatial Analysis, University College London.
6. KOLAR, J. (2006) Global Indexing of Topographic Surface for 3D Visualization and Analysis. Ph.D. thesis, Centre for 3D GeoInformation - Department of Development and Planning. Aalborg, Aalborg University.

7. PILOUK, M. (1996) Integrated Modelling for 3D GIS. Wageningen Agricultural University and ITC, The Netherlands.
8. STOTER, J. & ZLATANOVA, S. (2003) Visualisation and editing of 3D objects organised in a DBMS. IN WOODFORD, P. (Ed.) EuroSDR comm. 5 workshop on Visualisation and Rendering. ITC, Netherlands, EuroSDR.
9. KOLAR, J. (2007) Managed Code Interoperability In Digital Earth Technology. Proceedings of the 5th International Symposium on Digital Earth, University of Berkeley, CA, USA.

Chapter 8
Integrating Terrain Surface and Street Network for 3D Routing

Arne Schilling, Sandra Lanig, Pascal Neis and Alexander Zipf

Abstract. For true 3D navigation applications that combine terrain data, city models, landmarks, and other 3D geo data, a real 3D street network is necessary for route calculations. In this paper we describe how such a network can be derived automatically using detailed digital terrain models. Furthermore a method is described for how the terrain model can be enhanced by integrating the roads directly into the triangulation system and by correcting the surface. All processes are encapsulated and made available as Web Processing Services in order to simplify their integration into Spatial Data Infrastructures.

Keywords: 3D Geo-visualization, Digital Terrain Models, Spatial Data Infrastructure, Outdoor Routing, Network Graph, Open Street Map, Mesh Manipulation, Triangulated Irregular Network, Road Flattening, SRTM.

8.1 Introduction

The development of 3D navigation systems for cars and pedestrians (on PDAs) has been a topic in tech newsletters and in applied science for some time now. Since route planning features have been introduced in popular virtual globe software, also makers of car navigation systems try to develop new presentation techniques based on their street network and to provide more clues for supporting the user's spatial cognition at e.g. decision points. A common technique is to perspectively deform a 2D map and to integrate a couple of landmarks. The development of a true 3D route planning and navigation system that can be used on the country road, in inner cities and by pedestrians requires a major effort in terms of data capturing, data processing, and maintenance. This involves generating an accurate digital terrain model (DTM), integration techniques for the display of already available 2D maps and the road network, detailed (i.e. recognizable) models of landmarks, generalized models of all other buildings

---

University of Bonn. Department of Geography
{schilling, lanig, neis, zipf}@geographie.uni-bonn.de

and structures, and possibly additional models of the street furniture. This has been realized so far only for smaller test areas.

Within the project GDI-3D (Geodaten Infrastruktur in 3D, http://www.gdi-3d.de), we have implemented a 3D GIS and information system based on standards of the Open Geospatial Consortium (OGC). The creation of a very detailed 3D city model of Heidelberg has been kindly supported by the local land surveying office. This model is used as a platform for testing new OGC standards and to see how they can become a part of a 3D spatial data infrastructure [1]. An OGC OpenLS Route Service for computing shortest or fastest routes for cars and pedestrians has been implemented and integrated into a range of applications already [9, 8, 4]. The route service could be quite easily converted into a true 3D route service (3DRS) by collecting height values from the DTM, without actually extending the OGC specification as suggested by [10]. However, the presentation together with a 3D landscape and city model turned out to be more problematic so that additional techniques for the preprocessing of the DTM and route network is necessary in order to produce acceptable results.

In the following sections, we show how open standards can be used in order to create a 3D route planning and navigation system that can also be used for close-up views and route animations. An important aspect is the geometrical integration of the road surface into the triangulation of the terrain model and a correction method in order to improve the surface quality.

8.2 Public Domain Street Map

OpenStreetMap (OSM) is a public project that aims at "creating and providing free geographic data such as street maps to anyone who wants them" [13] and puts all collected data under a public domain license. Setting off in early 2006, OSM is a typical Web 2.0 application to which everyone who likes can contribute and collect data using a GPS device. A big amount of the data comes also from importing other public domain data like TIGER (Topologically Integrated Geographic Encoding and Referencing system) for the US and AND (Automotive Navigation Data, donated) for the Netherlands. Although far from being complete, the coverage of OSM for central Europe and the US is already quite good and it's growing very quickly. It contains at least all the major streets and very detailed maps for major cities. It partly contains ways that are reserved for pedestrians and cyclists and it allows for additional attributes such as speed limits and other restrictions which make the design suitable for routing. Furthermore, it contains many objects like street lights, buildings, green areas, forest, and many others.

A first open routing application for Germany based on OSM has been implemented by us and is available at http://www.openrouteservice.org. The service is compatible to the OGC OpenLS (Open Location Service) specification. It already includes routing for pedestrians. Complementary OpenLS including Geocoder, Presentation Service, and Directory Service have also been added based on OSM data. Support for other regions will follow in the near future. It's the first web based routing solution that combines free data, open source and open OGC standards.

In the following section we show how OSM can be made available to 3D applications especially focusing on 3D routing and navigation.

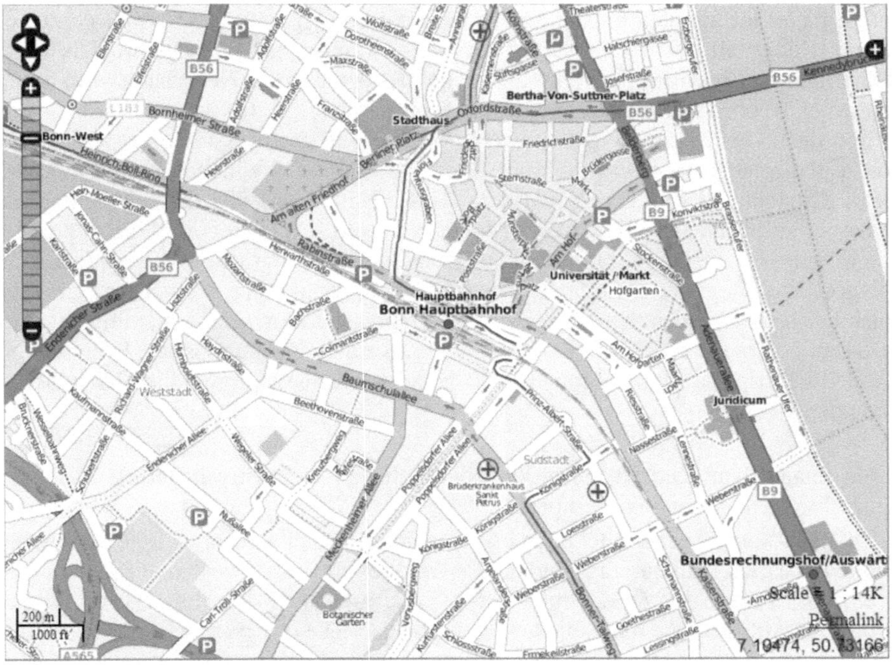

Fig. 8.1. Part of OpenStreetMap database showing Bonn (from http://www.openrouteservice.org).

8.3 Display of Roads in the DTM

The display of the road network in 3D differs from the usual maps produced by web map servers in several aspects. Within spatial data infrastructures (SDIs) geo data such as roads, transportation networks, land use areas, and point objects is usually provided by Web Feature Services (WFS). The question is how this data can be combined with a Digital Terrain Model (DTM). The idea of transforming existing 2D georeferenced vector features into 2.5D and 3D features and to combine them with terrain data is sometimes referred to as Smart Terrain (ST) [2].

Current car navigation systems feature a very simple perspective view on the surrounding area with no clues on the steepness of roads or nearby hills. Digital globes, in contrast, just map aerial photos on the terrain and overlay the road network as line vectors. This is fine for giving an overview over larger areas, and they don't have high accuracy terrain data that is needed for close-up views, for instance from a pedestrian's or driver's point of view. Also, the integration of 3D city models or at least selected landmarks require an accurate representation of the terrain surface, otherwise they become floating in or over the ground.

For our city model of Heidelberg, the local land surveying office kindly provided us with a 5 meter elevation raster that was captured using airborne laser scanning. A 1 meter raster would be also available. Google is currently updating its virtual globe with 5 meter terrain data for the US. In the future much better resolutions can be expected which will be well suited for very detailed city and landscape models. In the Netherlands for instance the AHN-2 (Actueel Hoogtebestand Nederland) with a point density of 10 points per square meter is currently in work.

In order to visualize the road network, we chose to perform a geometrical integration of the road surface into the triangulation of the DTM which is represented by a set of Triangulated Irregular Networks (TINs). This means that the road surface becomes a part of the TIN. The street network is treated as a layer consisting of a collection of polygons representing all the individual network segments. The borders of the polygons are integrated into the TIN as fully topological edges so that we can distinguish between triangles that are part of the street surface and the remaining triangles. Different colors are applied to the triangles so that they become visible. See [16] for a detailed description of the geometrical operations. This approach has the following advantages:

- No image data needs to be produced which may be costly in terms of network bandwidth and graphics memory.
- Simple colors or generic textures can be applied which gives the scene a more map like appearance. In our case the material and texture properties are defined as visualization rules using OGC Styled Layer Descriptor (SLD) documents, a specification for adjusting the map appearance to the use case or personal preferences. SLD has been extended by us in order to support also 3D objects and terrain (3D-SLD, see [11]. In particular, this allows for the definition of the appearance of the DTM dynamically per request from the client side.
- Parts of the surface can be corrected, if the underlying terrain does not have the required accuracy. This will be described in the next section.

8.4 Correcting Street Surfaces within the DTM

After integrating the street network as surface layer, we realized that the quality of the underlying terrain is partly not sufficient. Although a 5 meter raster is in general sufficient for capturing all kinds of surface structures and for the integration with a 3D city model, linear features like ditches, smaller dikes, walls, the rims of terraces, and especially the hard border edges of roads can be only represented insufficiently (see Fig. 8.2). Sometimes the road sidelines seem to be frayed. At steep hillsides the road surface is inclined sidewards. The situation is of course even worse with lower resolution DTM data sources. Similar observations have been made by [5].

Fig. 8.2. Comparison between the original terrain surface (left) and with the flattened road segment (right).

A common way to support also linear features is to include break lines during the terrain triangulation, e.g. using the Constrained Delaunay Triangulation (CDT). However, break lines are seldom available. Therefore we correct the parts of our surface representing areas that should be actually more or less flat. A comparison between the situation before the correction and afterwards is shown in Fig 8.2. It is much more likely that the middle line takes a smooth course between the river and the hillside with approximately the same height, and that the profile is nearly horizontal, as can be seen on the right side, instead of being very bumpy and uneven.

For the flattening process we assume that all TIN edges that are part of the area that needs to be corrected or are connected to it can be represented as elastic springs. Springs store mechanical energy when they are compressed or extended. The force that a simple coil spring is exerting depends on the spring constant (equivalent to the stiffness) and is proportional to the distance that the spring has been stretched or compressed away from the equilibrium position. This relation is described in Hooke's Law [19]. A network of springs will find to equilibrium where all spring forces will cancel out each other. Modeling elastic material as spring networks is a common technique in material science for simulating physical deformations, tension, and crack propagation in solids. Solids are represented as Finite Element data structures.

The Finite Element Method (FEM) is also used in software for performing complex geotechnical analysis such as load carrying capacity, plastic modeling of soils, deformations within the regolith material, etc. [15].

Although we don't deal with solids, we can as well represent our TIN mesh as interconnected springs. For all edges that need to be considered during the calculation we attach additional information on the physical spring properties. As spring constant we choose two different static values. The first is applied to all edges inside or at the border of the area that needs to be flattened and the second is applied to all other edges. The relation between the two values is determining how stiff the area is and how much it will be flattened. As equilibrium length (where no energy is stored in the spring) we take the projected edge length in R^2 on the x-y-plane (as seen from above) for all interior and border edges and the usual length in R^3 (Fig. 8.3). The different edge properties will generate a tension in the mesh and because of the shorter spring lengths of the interior edges these will move to a more horizontal position.

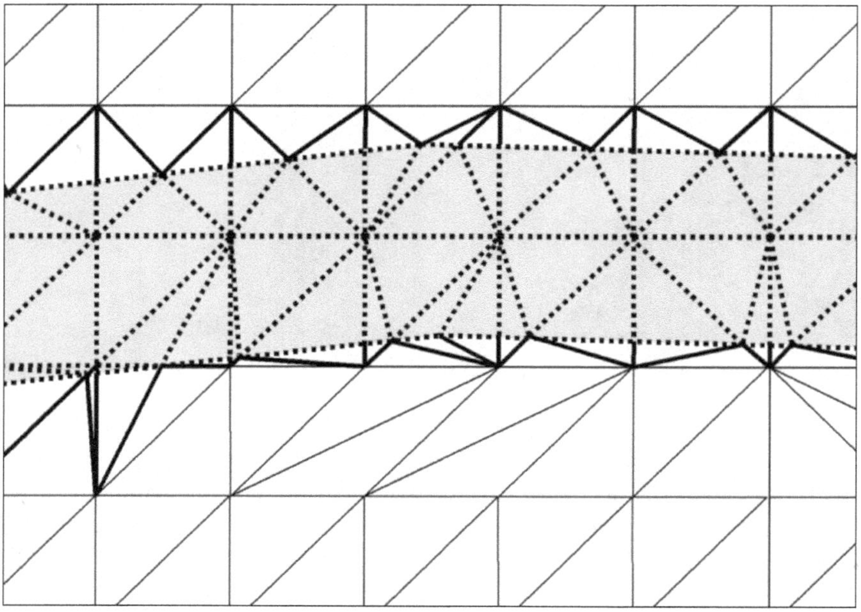

Fig. 8.3. Road segment (in grey) integrated into the DTM partly modeled as spring network. During the flattening process the z value of all border and interior nodes is corrected. Length in R^2 is attached to border and interior edges (dashed) as spring constant, length in R^3 to exterior edges (bold).

The next step is to find the overall equilibrium of the spring network. In order to find this we allow each interior and border node to move up and down only along the z axis so that the shape of the integrated area is not distorted. The force that acts on each node is computed as follows:

$$F = f(N_z) = \sum_{i=0}^{n} \frac{(\vec{e_i} \cdot \vec{u}) \cdot D \cdot (\|\vec{e_i}\| - L)}{\|\vec{e_i}\|} \; ; \; \vec{e_i} = \overrightarrow{M_i N} \quad (1)$$

with

F: force on the node in up direction. Downward forces have negative values.
M: adjacent node
N: current node
e_1: edge i connected to the node
u: up vector (z axis)
D: spring constant
L: spring equilibrium length

The force F can be seen as function of the node's z value: $f(N_z)$, see(1). (Fig. 8.4) shows the forces exerted by the individual edges connected to N in relation to the vertical offset of N and the summarized force acting on N. The intersection of the curve with the axis of abscissa (root) represents the local equilibrium. Since we are not interested in the actual physical simulation or temporal dynamics, we try to find the intersection point immediately using the Newton-Raphson method. This method computes an approximation of the root, which is improved quadratically at each iteration. Starting at the original node position (offset = 0) we limit the number of iterations to 4, which is already good enough.

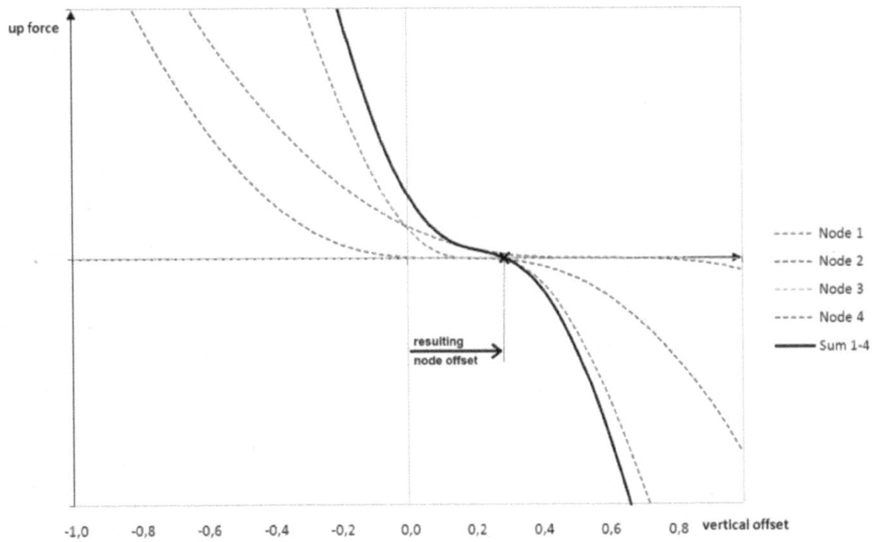

Fig. 8.4. forces exerted by edges connected to a single node measured against the vertical offset. The intersection of the cumulated curve (Sum 1-4) with the abscissa defines the local equilibrium.

The local equilibria are computed for all interior and border nodes of the area to be corrected. The values are interpreted as vertical offsets which are then applied to each node. The maximum offset over all nodes can be seen as error value indicating the difference between the current situation and the desired situation. A single pass of this correction process will not yield the global equilibrium where all forces cancel out each other because major correction values will propagate though the mesh. Usually it takes about 20 iterations in order to reach an error value of 1 cm, which is in our case sufficiently accurate.

8.5 Automatic 3D Network Creation

The main characteristics of 3D Navigation is that the course of the route and the route instructions must be presented in combination with a landscape and city model and that the actual network geometry must exist as 3D line set. The 3D network can be derived from commercial or public domain street data using elevation data. The height of each node in the network is taken from the terrain model which must be available as grid or TIN. Also each vertex of the network geometry must be adjusted and additional vertices must be inserted, because the resulting network geometry must be exactly lying on the terrain surface. Intersections with the surface sometimes occurring at very long network segments should be avoided. The topology of the resulting network remains the same.

The actual route calculation is carried out within the OGC OpenLS Route Service [9]. Because the distance between adjacent nodes must be present as weight attribute attached to the network segment, we can calculate this weight either as absolute travel distance or as travel time or another measure defined by attributes of the network edges. This allows to calculate either the shortest or the fastest route, or according to another criterion. Because we take these values from the 3D network we can also consider the steepness. Although this approach is not new in general, it is still not widely used in popular route planners. After the route calculation is complete, each segment is replaced by the according 3D line string representing the actual geometry which is stored in a lookup table.

For creating the 3D network geometry, we take the network data from the open source project Open Street Map (OSM) and the surface data from the Shuttle Radar Topography Mission (SRTM), which is also freely available. OSM is still incomplete but generally has very good coverage for Germany. We observed a growth of nearly 100.000 streets per month alone for Germany recently, so a nearly complete coverage can be expected in the foreseeable future. SRTM data is available as 3 arc-seconds (approx. 90 m) raster for Germany. For our project area in Heidelberg, we use mostly the more accurate 5 m grid.

As a first step, all elevation data is triangulated as TIN. This step takes the most time and memory. After that we iterate through all the network segments and convert them into 3D by collecting the height information at the vertices and by adding intermediate vertices in order to reflect the surface structures. This can be done very quickly. Travel distance and time is generated for the new segments. The result is stored as ESRI shape file so that it can be imported into a PostGIS database.

Although this approach is correct in most of the cases, it is not sufficient in situations where several street levels exist. This is the case at bridges, skyways, underpasses, and tunnels. Fig. 8.5 shows an example of a route going over a bridge and then making a turn and crossing underneath.

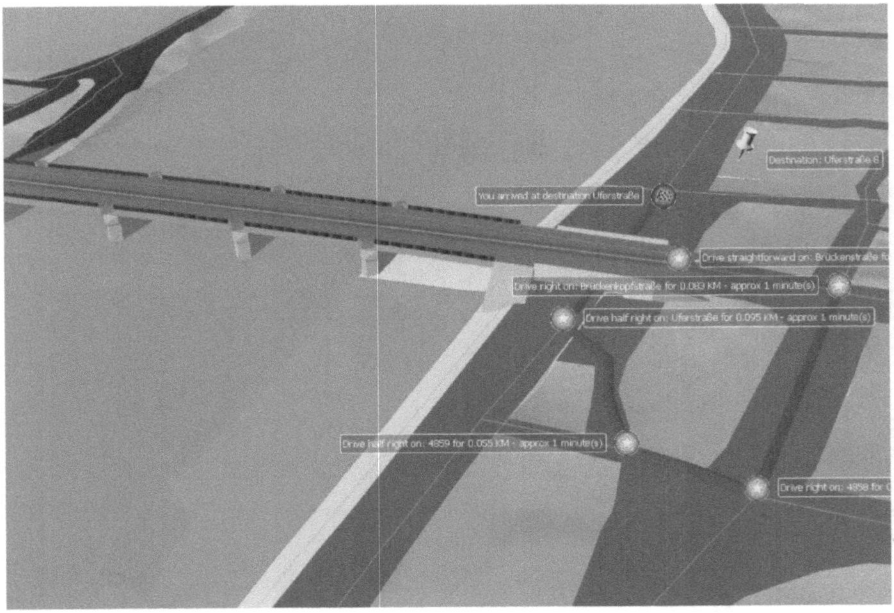

Fig. 8.5. Example for routing with our 3D routing service at several streets levels: route crossing underneath a bridge. The network is shown as thin lines.

Tunnels also require a modification of the network since the height information cannot be taken from the surface. The question how routing through tunnels might be presented to the user is discussed later in the outlook. Therefore, we mark such segments that cannot be converted using the terrain model with an additional attribute. In OSM various properties are attached by key value pairs, e.g. "bridge=yes", "tunnel=yes". There is also a tag called "layer" with values from -5 to +5 which can be used for identifying under ground and above ground layers.

Such parts could also be manually corrected using 3D modelling software and later on copied back into the route network. However, this makes updates more complicated. Therefore we define a second surface representing all bridges and tunnels. This second surface is stored as a collection of sample points in the vicinity of these structures and must be somehow measured or digitized. The sample points are triangulated as an additional TIN. Fig. 8.6 shows the normal terrain and the Karl Theodor Brücke ("Old Bridge") in Heidelberg together with the additional TIN rendered as wireframe model. The second TIN represents the longitudinal profile of the bridge. In cases with multiple layers on top of each other, this needs to be extended to multiple TINs or similar approaches, but we stick here for the sake of clarity to the case with one additional level.

All segments that are marked as bridges or tunnels collect their height information from the second TIN instead of the usual TIN for the terrain. Where no second TIN is available, a straight connection line will be drawn.

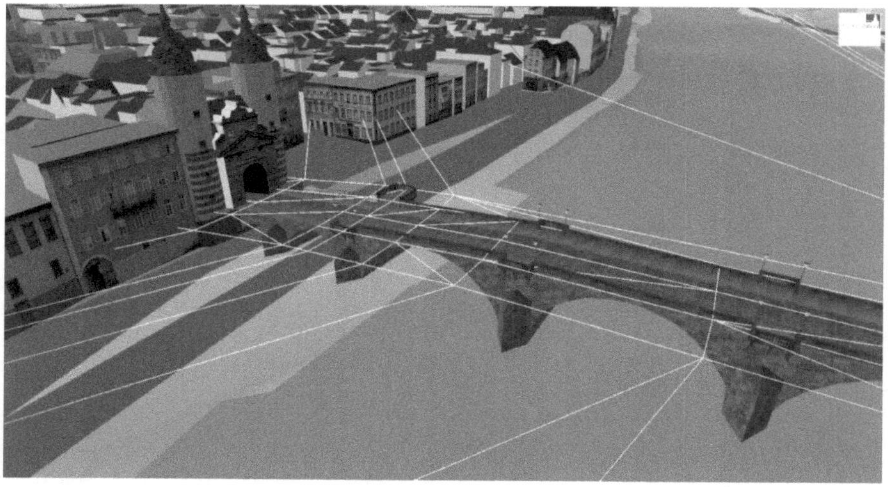

Fig. 8.6. For bridges and tunnels a second surface model is used in addition to the usual terrain (here rendered as wireframe model).

8.6 SDI Integration with Web Processing Services

The method for generating the 3D street network as described above implies that we have a triangulated terrain model for the whole area which we can use for deriving the 3D line strings. This can be extremely memory intensive. In our case, the 5 m DTM grid for Heidelberg contains about 13 Mio Measuring points. The 3-arc seconds SRTM grid for Germany sums up to more than 44 Mio. Points. In the latter case a fully topological TIN would require at least 8 GB RAM plus spatial index data. A TIN data structure is preferred because it enables a better reconstruction of the earth surface and the integration and correction (flattening) of other 2D data becomes possible. Modern server hardware is barely capable of processing such an amount of data. However, it is better to divide the whole processing task into many smaller tasks that can be delegated to other nodes within our spatial data infrastructure (SDI).

For this purpose, we are currently implementing several services for DTM processing tasks which are all compliant to the OGC Web Processing Service (WPS) standard. The WPS specification has been adopted in February 2008 as official standard. It was developed for offering any kind of GIS processing functionality by a standardized interface. The processes can "include any algorithm, calculation or model that operates on spatially referenced raster or vector data" [12]. This can be very complex computations like simulations on a climate model or very simple ones such as computing a buffer. Examples of WPS processes for different domains can be found in [17], [18].

According to the specification there are three mandatory operations performed by a WPS, namely GetCapabilities, DescribeProcess and Execute:

- *GetCapabilities* returns a brief service metadata document describing the resources of the specific server implementation, especially the identifiers and a short description of each process offered by the WPS instance.
- *DescribeProcess* returns a detailed description of a process including its required input parameter (together with the allowed formats) and outputs that are produced.
- *Execute* finally carries out the process execution.

Within the project SDI-GRID (Spatial Data Infrastructure-Grid, www.gdi-grid.de) we have already implemented some basic WPS for DTM processing that greatly improve the performance of often used 3D GIS operations within our 3D SDI, such as DTM triangulation, mesh reduction, polygon-in-mesh integration, and surface tiling [7]. Other operations, like relevance sorting and generalization of 3D geo data will follow, but they are of minor importance within this context.

There is an ongoing discussion on the possible categorization of WPS Processes [3]. Due to the broad range of possible processes, the specification does not contain any directives in this matter. We have identified several services that are necessary or useful within the workflow of 3D network generation, terrain generation, and model preparation for visualization.

8.6.1 Geotessellation Service

This service is responsible for the CPU- and memory-intensive task of creating triangle meshes from input sample points, which may come from SRTM, raw laser scanning data or any other data source. The service implements two different triangulation modes. Delaunay triangulation is the fastest algorithm for the triangulation. The disadvantage is that it works actually only 2 dimensional. Therefore, the service provides also another triangulation algorithm that takes also the surface curvature into account and produces much more effective reconstructions of the earth surface which is important for close-up views (e.g. route animations).

The service contains also optional subsequent operations such as 2D layer integration, surface flattening, and mesh reduction/generalization, and tiling. All these operations are used for integrating the street network into the terrain.

8.6.2 DEMService

This service is used for retrieving height information from the terrain at a specific location and for converting lines into 3D using a TIN. It is connected to a PostGIS database containing all sample points. The very simple *GetHeight* process returns the height value at a specific coordinate. It collects sample points near the coordinate and lets the Geotessellation Service compute a small TIN. The height is picked from the TIN over the coordinate. The *ConvertLineString* process derives a 3D LineString from the input 2D LineString and the underlying terrain. The sample points are collected using a buffer around the 2D LineString as spatial constraint. Subsequently the collected sample points are also forwarded to the Geotessellation Service. The resulting TIN is then used for collecting height values along the LineString. The process also adds additional vertices so that the 3D LineString will follow the terrain surface.

8.6.3 3DNetworkGenerator Service

This service is used as an access point for converting complete shapefiles and creating 3D networks. The process is comprised of two steps. First the z value of each node (where the individual LineStrings meet) is computed by the *GetHeight* process of the DEMService. Then all the LineString geometries are converted into 3D. The nodes are connected again using the new 3D vertices of the LineString. The separation into two steps is necessary in order to avoid discontinuities between the LineString that cause problems at the network topology computation and the 3D visualization. The response of the service contains a shapefile with z values.

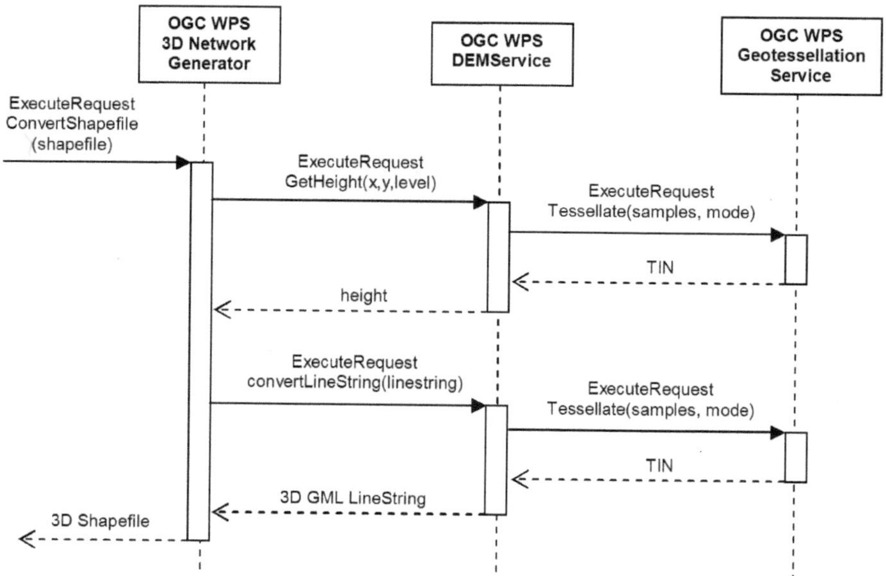

Fig. 8.7. Flow chart of the 3D network generation process using OGC Web Processing Services.

8.7 Results

An OpenLS Route Service is already a part of our SDI. It was extended so that it can also handle 3D shapefiles as input [8]. The response to a DetermineRoute request returns an overview map, a route summary, route instructions and the actual route geometry. Since the geometry is still provided as GML linestring, but with additional z values, the service does still fully comply with the OGC OpenLS specification.

The easiest way to benefit from the additional information is to display an elevation profile along the route. This might be interesting for cyclists and hikers when planning a tour. Fig. 8.8 shows an example of a combined display of route map, summary, and elevation profile. The latter can easily be computed from the 3D points defining the route that are returned by the RouteService.

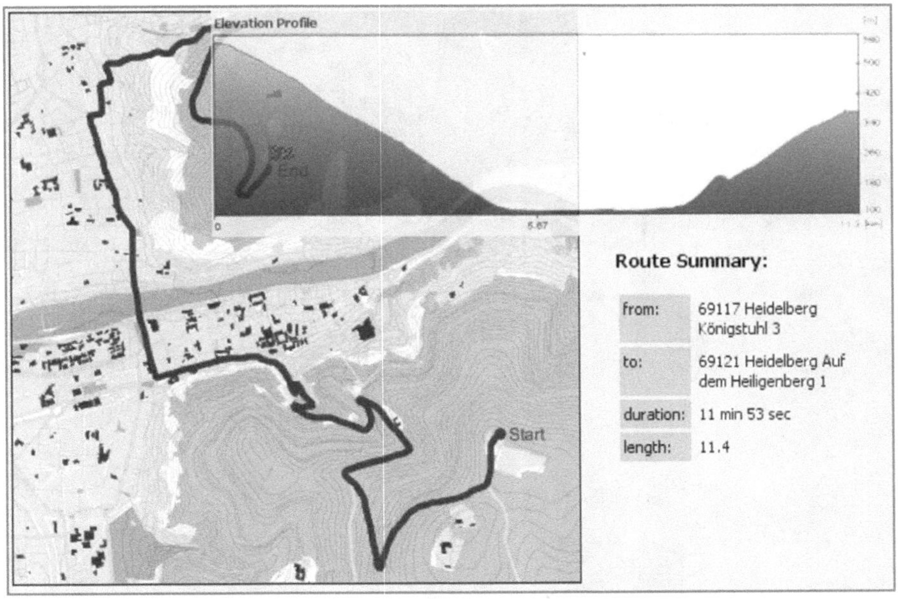

Fig. 8.8. Route overview with map, summary, and elevation profile.

The Route Service is also used by our 3D visualization software XNavigator (see http://www.gdi-3d.de). The latter is based on Java3D and Java Webstart and enables web based access to OGC Web3D Services [14] providing DTM and 3D city models. The possibilities for displaying 3D routes and navigation support are very versatile. Depending on the actual use case this could involve texture overlay on terrain, generating additional 3D route geometries, display of relevant landmarks along the way, 3D display of route instructions, animations, avatars, display of slopes, and much more. Fig. 8.9 shows an example including a detailed 3D city model of Heidelberg. The route is displayed as a pipe and instructions are overlaid as 2D labels. Street names are displayed as additional objects floating directly over the ground. The perspective is almost a street level one, which makes for a very accurate display of the surface and route.

The previously described methods for correcting the road surface and including additional levels for 3D network generation clearly improve the quality of the presentation. The original DTM did not capture the road very well, resulting in frayed sidelines and bumpy surfaces. Although the correction is based on assumptions that might not always be true, it seems to be a valid cartographic method for improving the readability of 3D maps.

Fig. 8.9. Display of a route integrated into a 3D city model with corrected road surface. The route goes right onto the bridge.

Fig. 8.10 – Fig. 8.12 show examples of OSM road and land use data that has been combined with a SRTM 3 arc-seconds (ca. 90 m) terrain model. The scenes show the city of Freiburg im Breisgau and the surrounding mountains. In order to apply the vector-based method for integrating areas into the TIN, buffers have been computed around the road network. The buffer size varies according to importance. Also the values for flattening the street surface depend on the importance and width. Primary and secondary roads are flattened more than cycle ways and minor forest tracks. Thus a more natural appearance is achieved. The street names have been overlaid as geometries on top of the terrain. They have been directly derived from the 3D network by taking the edges as middle lines for stringing together the character geometries.

8.8 Outlook

The previously described methodologies address only a small part of the broad research field of routing and navigation. However, the quality of the route network is very important for many applications. Since non-experts are increasingly involved in creating map data (e.g. OSM), quality standards cannot always be guaranteed for such projects. Further (applied) research topics should feature better or more appropriate presentation techniques that may include text, images, speech, 3D graphics, animations, Virtual Reality, etc. In our case, a solution for navigating through tunnels still needs to be developed. Many prefer an elevated viewpoint above the route course and not the actual driver's viewpoint [6]. In order to adapt the viewpoint so that the user is actually guided through a tunnel and not over it, additional meta information is necessary for controlling the animation. Also, speed limits could be used for the animation. This would require an OpenLS extension. OSM also contains many other tags (if captured) like limits, permissions, track types, elevators, and access ramps for disabled people that could be used for route calculations.

Another important aspect will be the identification, selection, and integration of landmarks depending on the current route. This is currently being investigated in the project MoNa3D (Mobile Navigation 3D, http://www.mona3d.de).

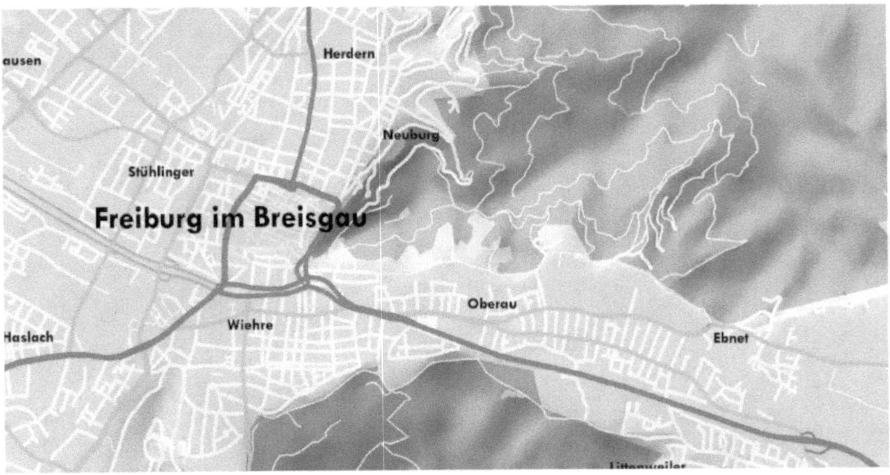

Fig. 8.10. Typical OSM data (city of Freiburg im Breisgau and surrounding area, Germany) combined with the SRTM terrain model. The width of the streets is larger than in subsequent figures to provide a generalized overview.

Fig. 8.11. Close-up view showing all available OSM roads and land use areas. The street names are additionally overlaid as geometries on top of the terrain.

Fig. 8.12. Certain mountain roads have been integrated into the terrain and flattened. Major roads have a higher priority and were flattened more than smaller forest tracks.

References

1. Basanow J, Neis P, Neubauer S, Schilling A, Zipf A (2007) Towards 3D Spatial Data Infrastructures (3D-SDI) based on Open Standards - experiences, results and future issues. In: Oosterom P et al. (eds.)(2008) Advances in 3D Geoinformation Systems. Springer, Berlin Heidelberg. ISBN 978-3-540-72134-5, pp. 65-86.

2. Buchholz H, Döllner J, Ross L, Kleinschmit B (2006) Automated Construction of Urban Terrain Models. In Proceedings of the 12th International Symposium on Spatial Data Handling (SDH 2006), pp 547-562.
3. Goebel R, Zipf A (2008) How to define 3D Geoprocessing Operations for the OGC Web Processing Service (WPS)? Towards a Classification of 3D Operations. International Workshop on Computational Geoinformatics (CompGeo 2008). The 2008 International Conference on Computational Science and Its Applications (ICCSA 2008). Perugia. Italy.
4. Haase M, Zipf A, Neis P , de Camargo V (2008) Interoperable Routing Services in the Context of Evacuation Schemes due to Urban Flooding. EnviroInfo 2008 Conference. Environmental Informatics and Industrial Ecology. Lueneburg, Germany.
5. Hatger C (2002) Assigning the 3rd Dimension to Roads by Use of Laser Scanning Data. In: IAPRS Vol. 34, Part 4 "Geospatial Theory, Processing and Applications", Ottawa, Canada, 2002.
6. Kray C, Elting C, Laakso K, Coors V (2003) Presenting route instructions on mobile devices. In: Proceedings of the 8th international conference on Intelligent user interfaces, Miami, Florida, USA, pp 117-124.
7. Lanig S, Schilling A, Stollberg B, Zipf A (2008) Towards Standards-based Processing of Digital Elevation Models for Grid Computing through Web Processing Service (WPS). The 2008 International Conference on Computational Science and Its Applications (ICCSA 2008). Perugia. Italy.
8. Neis P, Schilling A, Zipf A (2007) 3D Emergency Route Service (3D-ERS) based on OpenLS Specifications. GI4DM 2007. 3rd International Symposium on Geoinformation for Disaster Management. Toronto, Canada.
9. Neis P, Zipf A (2007) A Web Accessibility Analysis Service based on the OpenLS Route Service. AGILE 2007. International Conference on Geographic Information Science of the Association of Geographic Information Laboratories for Europe (AGILE). Aalborg, Denmark.
10. Neis P, Zipf A (2008) OpenRouteService.org is three times "Open": Combining OpenSource, OpenLS and OpenStreetMaps. GIS Research UK (GISRUK 08). Manchester.
11. Neubauer S, Zipf A (2007) Suggestions for Extending the OGC Styled Layer Descriptor (SLD) Specification into 3D - Towards Visualization Rules for 3D City Models. Urban Data Management Symposium. UDMS 2007. Stuttgart. Germany.
12. OGC (2008) OGC(R) Approves Web Processing Service Standard. Press Release on Fri, 2008-02-22.
13. OSM (2008) Main web site http://www.openstreetmap.org.
14. Quadt U, Kolbe TH (2005) Web 3D Service Implementation Specification, Discussion Paper. Open Geospatial Consortium, Doc. No. 05-019 2005.
15. Pruska J, (2003) Comparison of geotechnic softwares - Geo FEM, Plaxis, Z-Soil. Proceedings of the XIII ECSMGE, Vanieek et al. (eds). EGtS, Prague, ISBN 80-86769-01- 1, (Vol. 2).
16. Schilling A, Basanow J, Zipf A (2007) Vector based Mapping of Polygons on Irregular Terrain Meshes for Web 3D Map Services. 3rd International Conference on Web Information Systems and Technologies (WEBIST). Barcelona, Spain. March 2007.
17. Stollberg B, Zipf A (2008) Geoprocessing Services for Spatial Decision Support in the Domain of Housing Market Analyses - Experiences from Applying the

OGC Web Processing Service Interface in Practice. 11th AGILE 2008 Conference on GI Science (AGILE 2008). Association of Geographic Information Laboratories for Europe. Girona. Spain.
18. Stollberg B, Zipf A (2007) OGC Web Processing Service Interface for Web Service Orchestration - Aggregating Geo-processing Services in a Bomb Threat Scenario. W2GIS 2007: Web&Wireless GIS Conference 2007. Cardiff, UK.
19. Symon KR (1971) Mechanics. Third Edition. Addison-Wesley Publishing Company, Reading, Massachusetts.

Chapter 9
Using a B-Rep Structure to Query 9-Intersection Topological Relationships in 3D GIS – Reviewing the Approach and Improving Performance

Claire Ellul and Mordechai Muki Haklay

Abstract. A key component of a three-dimensional Geographical Information System (3D GIS) toolkit is the ability to identify binary (between two objects) topological relationships. These include adjacency (are objects next to each other), containment (is one within another) and intersection (do they interact in any way). Determining such relationships allows the GIS to answer questions such as "what is directly underneath this building?" and "if a tunnel is dug here, what rock types will be cut?" Frameworks are used to fully list the possible relationships between objects and the prevalent approach to the determination of such relationships in 3D GIS takes advantage of structures based on Boundary Representation (B-Rep). The first part of this paper thus presents a review of the 9-Intersection Framework and the use of B-Rep in 3D GIS. The second part of the paper outlines modifications to a B-Rep structure to improve binary relationship query performance, particularly relevant when the increasing size of 3D datasets is considered. An alternative structure is proposed to overcome limitations when using standard B-Rep. Tests using both structures show that the modified structure is between 11 and 15 times faster, and the paper concludes by identifying additional features of the modified structure and suggesting further research into its use.

9.1 Introduction

Mainstream three-dimensional (3D) Geographic Information Systems (GIS) do not yet offer full capabilities corresponding to those available in two dimensions (2D). This is despite the fact that databases such as Oracle [33], IBM DB2 [22] and PostGIS [40] permit, at minimum, the storage of 3D coordinate information [54] and

Dept. of Civil, Environmental and Geomatic Engineering
University College London,
clairee@ge.ucl.ac.uk, m.haklay@ucl.ac.uk

authors including [1, 4, 6, 11, 18, 29, 30, 35, 36, 37, 49, 50, 53] have carried out research into 3D GIS.

One component of functionality critical to the uptake of 3D GIS is topology, which has been "a central and important concept in GIS since the mid 1970s" [46]. Binary topological relationships (between two 3D objects) are the subject of this paper. They include adjacency (are objects next to each other), containment (is one within another) and intersection (do they interact in any way).

In a 3D context, binary topological relationships can be used to answer questions including "which artifacts can be found within a particular layer of stratigraphy?", "which geological regions are cut by a particular fault?", "find any water pipes that intersect with the trench to be dug in this road" and "find the shortest escape route through this 3D building". These queries involve the determination of relationships between large numbers of pairs of objects. Further details of such applications can be found in [16].

Frameworks are used to fully list the possible relationships between objects, with a total of 512 relationships identified by the 9-Intersection framework [14].

Prevalently in 3D GIS a data structure approach is used to model the relationships identified by the Framework - "inherent spatial connectivity and adjacency relationships are explicitly stored" in the structure [46]. Research is ongoing [5, 10, 12, 29, 38, 39, 43, 53] and structures offer the advantage of a 'calculate once, query many times' approach.

The majority of such structures are based on Boundary Representation (B-Rep) of 3D objects, which allows objects to be stored unambiguously (i.e. their interior can be determined). The first part of this paper presents a systematic review of the 9-Intersection framework and the use of B-Rep in 3D GIS topological relationship queries. This review addresses the question:

How are 9-Intersection relationships determined from a B-Rep structure, and what modifications are required to allow the structure to model all the relationships idnetified by the 9-Intersection framework?

B-Rep was originally developed for Computer Aided Design and Computer Vision [21]. Given this, performance may not be optimized for relationship determination, a topic of particular importance when increasing sizes of 3D GIS datasets are considered alongside the generally larger number of objects involved of binary topological queries such as those listed above. The second part of this paper thus seeks to answer the question:

Is the B-Rep structure optimised for binary relationship identification? If not, what modifications are required to improve performance?

A description is then given of tests undertaken to validate that both structures can model all required relationships and to compare their query performance. The results of these tests are then presented. We conclude by describing further work that could be carried out using the modified B-Rep structure.

As object-relational databases now form the backbone of data storage in GIS, the discussion will take place in this context. It is assumed that the reader has some familiarity with the representation of spatial data within such databases, and other the general concepts – such as Structured Query Language (SQL) querying and primary/foreign key relationships between tables. Further information can be found in [42].

9.2 Background

Topological relationships between objects are those that do not change when the space the objects are in is subject to continuous transformation (such as stretching and distortion, without any folding or tearing [52, page 120])[1]. This section presents a review of the 9-Intersection Framework, used to enumerate topological relationships in 3D GIS, and of the Boundary-Representation approach used to model these relationships in a database.

9.2.1 The 9-Intersection Framework

Binary topological relationships between objects are described in GIS by frameworks, which enumerate all possible topological relationships between two objects. The 9-Intersection framework, developed by Egenhofer and Herring [14], is "perhaps the most well-known topological formalism in geographic information science" [26]. This framework also forms part of the ISO 19107 standard issued by the Open Geospatial Consortium (OGC) [32] in collaboration with the International Standards Organization (ISO)[2].

The 9-Intersection framework distinguishes three components of each object - the interior and boundary and exterior - and examines the intersections of each of the components of object A with each of the components of object B, resulting in the 3 by 3 matrix shown in Fig. 9.1.

$$R(A,B)=\begin{pmatrix} Int(A)\cap Int(B) & Int(A)\cap Bnd(B) & Int(A)\cap Ext(B) \\ Bnd(A)\cap Int(B) & Bnd(A)\cap Bnd(B) & Bnd(A)\cap Ext(B) \\ Ext(A)\cap Int(B) & Ext(A)\cap Bnd(B) & Ext(A)\cap Ext(B) \end{pmatrix}$$

Fig. 9.1. 9-Intersection Relationship Matrix

In Fig. 9.1 Int(A) represents the interior of object A, Bnd(A) represents the boundary of A and Ext(A) represents the exterior of A. The symbol $\cap$ represents the intersection operator. Each intersection can be TRUE or FALSE giving a total of 512 possible relationships, although many of these are not achievable in practice.

This framework has been extended to support 3D objects [51] and a complete enumeration (diagrammed by [53], Chapter 6) is available to model the comprehensive set of binary topological relationships between simple objects in 3D. Work undertaken by [25] (cited in [53]) has led to the development of a simple, three-digit, numeric representation for each 9-Intersection relationship. An example of

[1] Note that *topological relationships* between MULTIPLE objects should be differentiated from the *topological validity* of a SINGLE object (as described in [21]). They can also be contrasted with *metric relationships* such as distance, area and volume, which do change as transformation occurs.

[2] The Open Geospatial Consortium is a non-profit, international, voluntary consensus standards organization involved in the development of standards for spatial and location based services

the relationships and corresponding Decimal Encoding (R-Code) values taken from [53] is shown in Fig. 9.2 below.

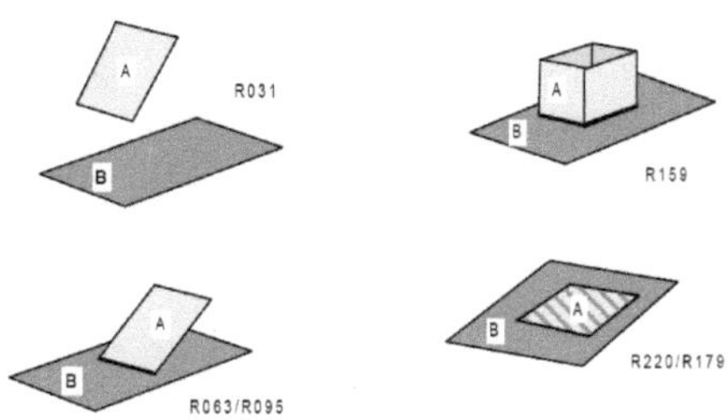

Fig. 9.2. Examples of Surface/Surface Relationships and corresponding R-Codes (taken from [53])

A summary list of the 9-Intersection relationships in 3D is given in Table 9.1.

Table 9.1. Summary of 9-Intersection Binary Topological Relationships in 3D, identified by [53]

Dimension of Embedding Space	Dimension of Objects	Number of Relationships
1	Line and Line	8
2	Line and Line	33
3	Surface and Line	31
2	Surface and Surface	8
3	Body and Body	8
3	Surface and Surface	38
3	Body and Line	19
3	Body and Surface	19

The 9-Intersection framework has been further developed by the original authors [13, 15, 28], with a number of other authors basing their frameworks on 9-Intersection [8, 20]. A number of applications have also been developed around this framework [17, 27, 34].

9.2.2 Boundary Representation

B-Rep is a method for representing a 3D solid using its constituent Surface components – Nodes (0-dimensional), Edges (1-dimensional) and Faces (2-dimensional), collectively known as primitives. The interior of the solid is represented by the space enclosed by the constituent Faces. Nodes are used to define the Edges, which in turn are ordered to define each Face [21, page 37]. One example of the implementation of a B-Rep structure is shown in Fig. 9.3.

Given the GIS and relational database context, the Feature object shown in Fig. 9.3 can represent Point, Line, Surface or Body object types, joining directly to one or more Node, Edge or Face primitives. This extends traditional B-Rep structures such as [3, 19]. In this case, 1 Feature has many Primitives, but each primitive is associated with only 1 Feature.

Node_Edge and Edge_Face tables join the primitives. An ordered list of Edges around a Face can be derived using an ORDER BY query on Edge_Order (intrinsic to SQL), removing the requirement to maintain a list of previous and successive Edges shown in [3]. Similarly, left and right Faces can be derived from the Edge_Direction field (for the same Edge_ID, there will be two different Face_IDs listed in the Edge_Face table). Conventionally, counter-clockwise ordering is used for the Edges around a Face, thus allowing the interior and exterior of a 3D Feature to be determined.

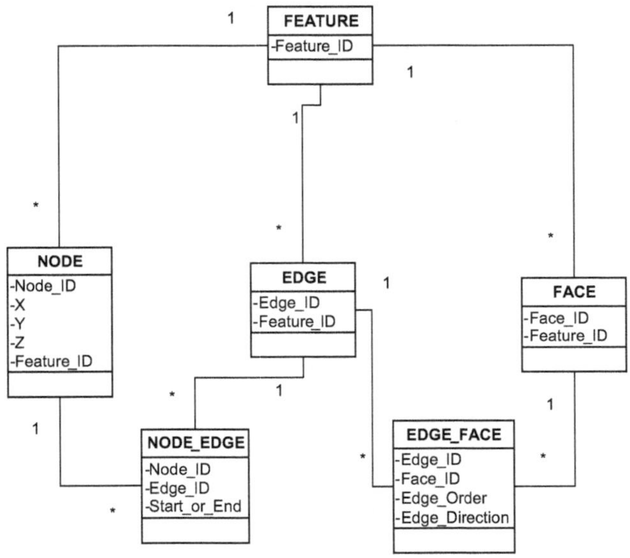

Fig. 9.3. Boundary-Representation Structure in a Relational Database.

9.3 Identifying 9-Intersection Relationships from a B-Rep Structure

The first question posed above regarding the B-Rep structure relates to its ability to model the full set of 3D 9-Intersection relationships. To determine this, it is first necessary to define the general process and queries required to determine binary relationships from B-Rep.

Fig. 9.3 above represents the B-Rep structure required to model a single object, allowing a single Feature_ID to be associated with each primitive. This does not take into account the fact that, in a GIS context, primitives are generally shared between objects (for example, two adjacent buildings share a party wall). Moreover, to determine the 9-Intersection relationship between two objects, the list of primitives constituting the interior and boundary of each object must first be identified. Comparing (intersecting) these lists will identify any shared primitives and hence intersecting interiors, boundaries and exterior (this process is described in more detail below).

To support shared primitives, many:many relationships between the Features and the primitives are required, as shown in Fig. 9.4 below. Note that it is now no longer possible to determine the interior of an object by examining the orientation of the constituent Faces, as both the left and right of each Face could represent the interior of adjacent objects. Thus Left_Body_Feature and Right_Body_Feature fields are added to the Face table.

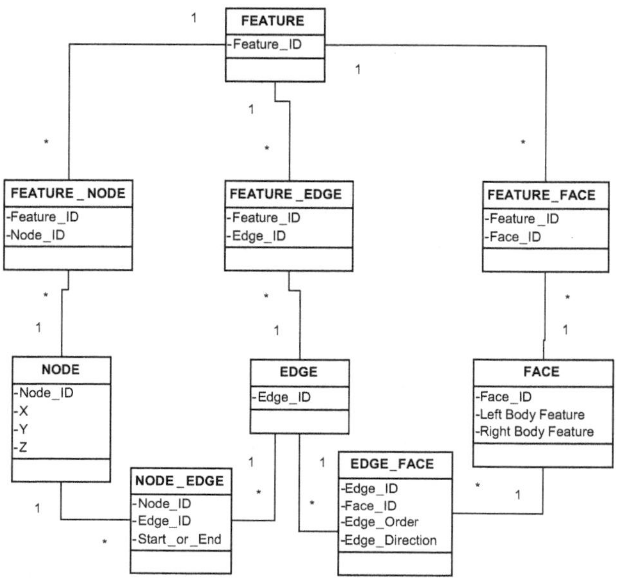

Fig. 9.4. B-Rep Structure modified to store multiple Features sharing the same primitives. This shows the Many:Many relationship between the Features and primitives.

9.3.1 Identifying Interior Primitives

The identification of interior primitives in B-Rep depends on the dimension of each object. The interior of a 2D object (a surface or area) is represented by one or more 2D topological primitives (Faces). The interior of a 1D object (a line) is composed of a number of Edge primitives. The interior of a 3D Body is defined by the space enclosed by the Face primitives forming the Body. For example, to identify the Int(A) • Int(B) relationship between two Surfaces A and B, the interior of each must first be identified by querying the constituent Face primitives, following the joins from Feature to the Feature_Face table. If the Faces reconstructing Surface A have at least one Face primitive in common with Surface B then Int(A) • Int(B) returns the shared primitives. The interior of 3D Bodies will be discussed further below (Requirement for a Volume Primitive).

Note that the Node, Edge and Face tables and joins between them (Node_Edge, Edge_Face) are not queried by this process – the list of required primitives can be obtained from Feature_Node, Feature_Edge or Feature_Face depending on the dimension of the 3D Feature. It is the lists of primitive IDs that are intersected – the geometry (coordinate points for a Node, Start and End points of an Edge, ordered lists of coordinates forming a Face) of the primitives is not required for the topological relationship query.

9.3.2 Identifying Boundary Primitives

The boundary of an object consists of any topological primitives that have dimension less than the maximum dimension of the object. For example, the boundary of a Body can consist of Faces, Edges and Nodes. To identify the Bnd(A) • Bnd(B) relationship between a Line A and a Body B, the Node boundary primitives of A must first be identified (by following joins from the Feature table through Feature_Edge to the Node_Edge table). Similarly, the Node boundary primitives of Body B are identified by following the relationship from Feature to Feature_Face, then to Face_Edge and Edge_Node. If the Node primitives returned for the boundary of the Line A are shared with those returned for Body B then Bnd(A) • Bnd(B). A flow diagram of the process required to identify the intersection between the Interior of Feature A and the Boundary of Feature B is given in Fig. 9.5 below. In this case, the dimension of each Feature is used to identify the corresponding boundary and contained primitives, which are then intersected to find shared primitives. If none exist, then Interior A and Boundary B do not intersect.

With reference to Fig. 9.4 above, it can again be noted that the Node, Edge and Face tables are not queried by the relationship identification algorithms, although in this case Node_Edge and Edge_Face are. For example, it is possible to directly join from the Feature_Edge table to the Node_Edge table to identify a list of Node_IDs forming part of the interior or boundary of a Line Feature. Similarly, it is possible to generate a list of Edge_IDs forming the boundary of a Surface from the Feature_Face and Edge_Face table.

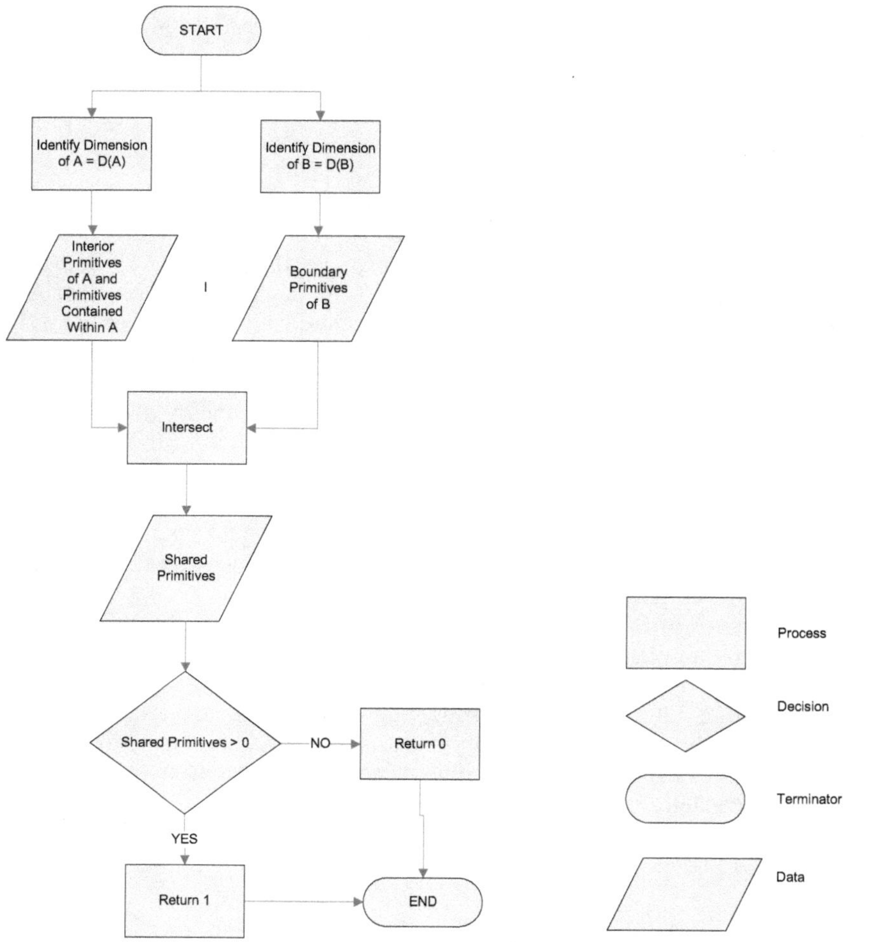

Fig. 9.5. Flow Diagram for Interior of A intersecting Boundary of B.

9.3.3 Identifying the Exterior of an Object

In a B-Rep structure the exterior an object is not explicitly described – instead, it is taken to be all the space not occupied by the interior or boundary of the object. In practice, if a primitive of A is not shared with a primitive of B then it is taken to be outside B (i.e. intersecting with the exterior of B). This is valid in most cases.

However, for the topological relationship of containment, as illustrated in Fig. 9.6 above and Table 9.2 below, the approach to identifying the shared interior and boundary primitives described above would return an incorrect result for the intersection of the Exterior of the outer Body A and the Boundary of inner Body B.

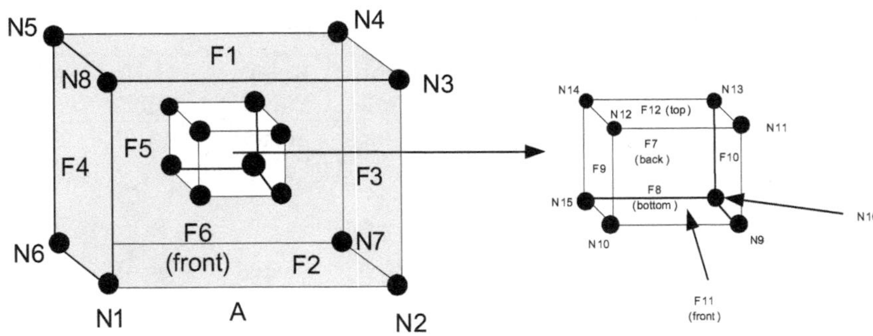

Fig. 9.6. Body/Body Containment –R220. A is the outer Body and B is the inner Body. A does not have an interior cavity.

Table 9.2. 9-Intersection Matrix for R220.

	INT (A)	BND (A)	EXT (A)
INT (B)	TRUE	FALSE	FALSE
BND (B)	TRUE	FALSE	FALSE
EXT (B)	TRUE	TRUE	TRUE

The boundary primitives of inner Body B (Faces, Edges and Nodes) are not shared with Body A, as the latter has no cavity. Using the above test, they would thus be identified as intersecting with the exterior of A. Therefore Ext(A) • Bnd(B) would be incorrectly TRUE. Additionally, the information detailing the left and right Body objects for each Face is now incorrect – Faces F7 to F12 have Body A on the left and right, and also Body B on the right.

Further investigation reveals additional containment exceptions that cannot be handled using the B-Rep structure – Point in Body, Line in Body and Point on Surface and Line on Surface. Conventionally, where they are modeled in GIS, these containment cases (excluding the Body in Body situation) are generally handled by the inclusion of one or more containment exception tables in the B-Rep schema. For example, [30] calls these tables ArcOnFace, NodeOnFace, ArcInBody and NodeInBody, [53] models NodeInBody and NodeOnFace. Modifications required to support Body/Body containment are discussed below (Requirement for a Volume primitive).

9.3.4 Requirement for a Volume Primitive

B-Rep structures do not directly represent the interior of a 3D Body. Although the interior of a Body can be reconstructed from its constituent Faces using the left and right Body information held on the Face, a number of situations arise where this approach is insufficient – i.e. where more than one Body is on the same side of a Face. Table 9.3 shows left and right Body values for shared Faces for each 9-Intersection Body/Body relationship, with the *Relevant Faces* column identifying the Faces that are of interest, and the *Left Body* and *Right Body* columns identifying the Bodies on either side of these Faces.

Table 9.3. 9-Intersection Relationships between Simple Bodies

Relationship Diagram	Relationship and R-Code	Relevant Faces	Left Body	Right Body
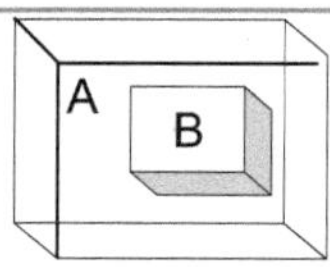	R179 – Contains	Shared Faces	A, B	A
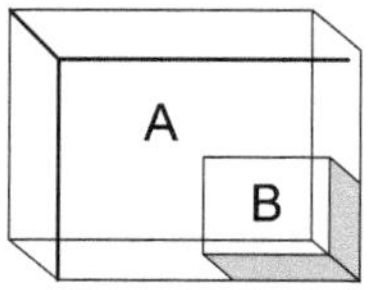	R476 – Covered By	Shared Faces, Bounding A	A, B	0
		Shared Faces, Internal to A	A, B	A,B
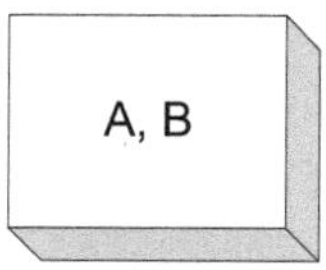	R400 – Equal	All Faces	A, B	0
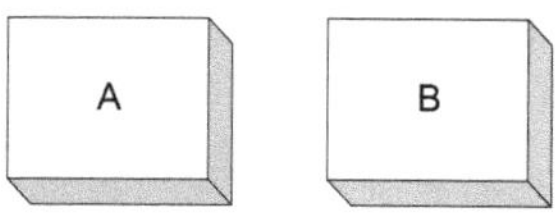	R031 – Disjoint	All Faces	N/A	N/A
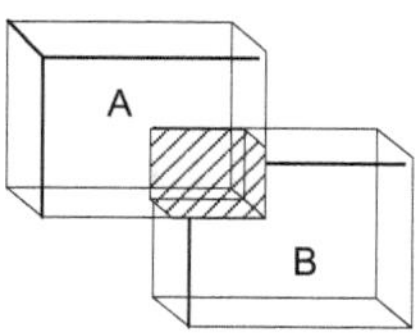	R511 – Overlap	Shared Faces	A, B	A, B
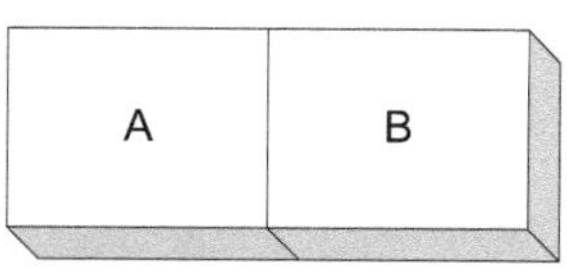	R287 - Meet	Shared Face	A	B

Using the many:many relationship between the Feature and the Face primitive shown in Fig. 9.4, a list of shared Faces can be identified between Body A and Body B. However, there is no means of determining that the shared Faces in fact form an enclosed 3D space. The identification of this enclosed space is required to determine intersections of the interior of these Bodies (the intersection of two 3D bodies is also a 3D space).

The inclusion of a primitive to explicitly represent the interior of a 3D Body resolves this issue. This primitive, known as Volume primitive, allows a set of Faces to be marked as forming an enclosed space. One Feature may be decomposed into to many Volume primitives (for example, Body A for the Overlap relationship in Table 9.3 is comprised of two Volumes) and one Volume may form part of many Features (the shared Volume in this Overlap relationship forms part of both Body A and Body B). Each Face is now related to at most two Volume primitives[3].

9.3.5 Binary B-Rep

Fig. 9.7 shows the resulting Binary B-Rep (BB-Rep) structure, with the four containment exception tables and the additional Volume table and Many:Many relationship between Feature and Volume. Tables shown in red are those required to identify binary topological relationships.

It can be noted that BB-Rep is, in fact, very similar to [29, 37], although the latter does not include the containment exception tables. However, these structures do not appear to have been systematically validated for their ability to model a complete set of 3D 9-Intersection relationships listed by [53]. This validation is described below (Validating the Structures).

Additionally, Fig. 9.7 (and also Fig. 9.8 below) includes spatial information on both the primitives and the Feature itself. This contrasts with traditional B-Rep structures which hold spatial information on the Node primitive alone and also results in duplicate storage of coordinate information. However, it takes advantage of new spatial data types not available in a database when B-Rep structures were initially designed, and facilitates the visualization of the 3D object or higher level primitives as these do not need to be reconstructed from the coordinate information held on the Node.

As these tables are not required for binary relationship determination (only those tables shown in red are required) a trade-off must be made on implementation between the additional time required to visualize the data from coordinates on the Node primitive and the additional time required to maintain duplicate coordinate information.

[3] Note that a similar argument can apply to the requirement for an Edge primitive (excluded by [10] and [53]). This is required to identify the interior of a 2D line as it may not be possible to assume that if two Lines share the same pair of Nodes, they also share the Edge that joins these Nodes.

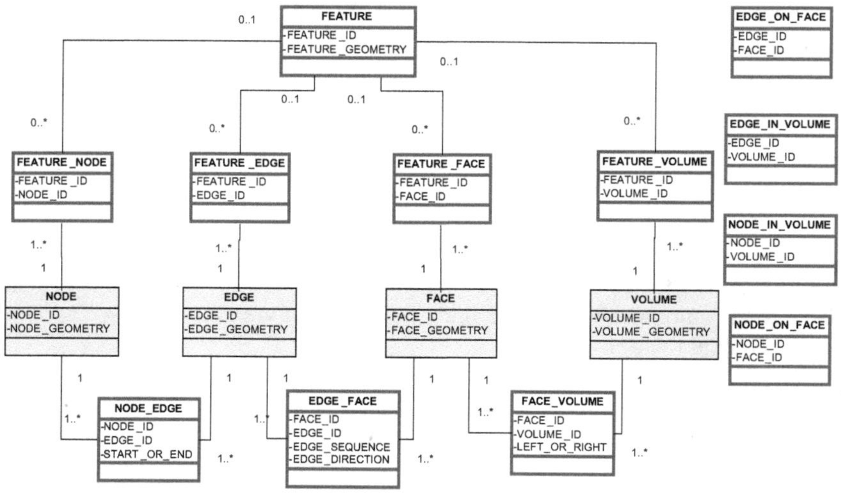

Fig. 9.7. Binary B-Rep (joins between the containment exception tables and the primitives are not shown). Tables shown in with a white background and a thicker border are those required for the determination of a Binary 9-Intersection Relationship.

9.4 Potential 9-Intersection Query Performance Issues

The second question identified in relation to the B-Rep structure relates to its performance in terms of query execution time. As described above, the process to determine the 9-Intersection relationship between two objects initially involves determining the dimension of each object. Once this is done, primitives are flagged as interior or boundary primitives and contained primitives identified. Finally, the process determines the intersection between each set of primitives. Applying this algorithm to the structure shown in Fig. 9.7 above, the queries involved in the second step of this process may require a relatively high number of join relationships to be followed. For example, to determine the Node primitives associated with a Body object, the query must follow joins from Feature to Feature_Volume, to Face_Volume, to Edge_Face and then to Node_Edge. Similar (although more compact) queries are required to determine constituent Nodes for Surface and line objects.

The authors in [2, page 337] note that relational joins are one of the more costly operations in terms of query performance. Joins take two candidate sets of objects and attempt to find matches according to the criteria given in the query – their high cost lies in the possibility of large numbers of potential candidates on either side of the join, greatly increasing the number of comparisons that are required[4].

---

[4] The use of Object-References, which behave like foreign keys [45], may mitigate join performance issues. This approach is applicable both for the BB-Rep structure described in Fig. 7 and for the MBB-Rep structure in Fig. 8 and may provide performance improvements for both.

In addition to the number of join operations, four containment exception tables must also be queried before all shared primitives are identified for each pair of objects. This complicates the queries to identify shared primitives, as well as adding to the number of joins to be traversed.

To address these issues, we propose a Modified Binary B-Rep (MBB-Rep) structure as shown in Fig. 9.8 below. This structure differs from B-Rep in two ways.

Firstly, all levels of primitives associated with a Feature are directly linked to that Feature. By querying the Feature_Node table, for example, all the Nodes associated with a 3D body can be directly identified, where as in BB-Rep, these would have to be identified via the Feature_Volume, then Volume_Face and Face_Edge and Edge_Node tables[5].

Secondly the use of the Is_Boundary flag allows the removal of the containment exception tables, incorporating the information into the main structure. This results in a far more compact structure with the requirement to query only 5 tables to indentify binary topological relationships as opposed to 12 in BB-Rep.

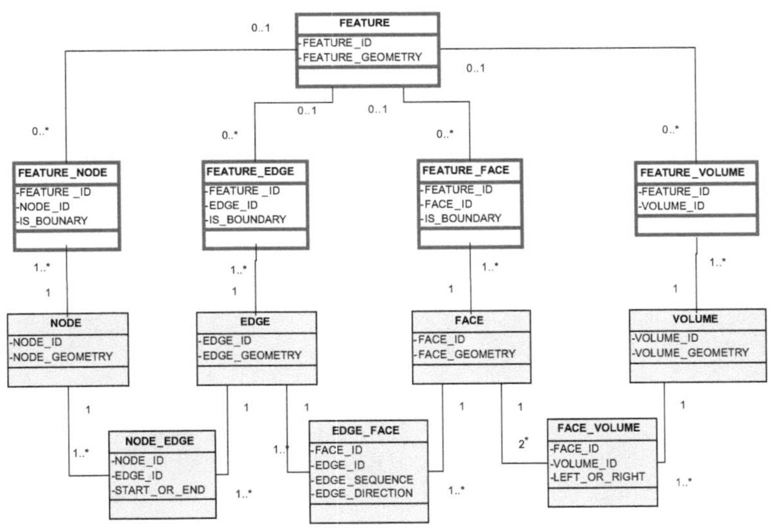

Fig. 9.8. Modified Binary B-Rep. Tables shown in with a white background a thicker border are those required for the determination of binary 9-Intersection Relationships

The algorithms to identify shared primitives using the MBB-Rep structure are identical to those used for the BB-Rep structure in that they query the structure to identify the primitives forming the interior and boundary of each Feature and then determine whether any of these are shared. However, rather than identifying the Nodes

[5] In a standard B-Rep approach (such as BB-Rep), an entry is only created in the Feature_Node table where the 3D Feature is a point, an entry in Feature_Edge table is only created when the 3D feature is a line, in Feature_Face when the 3D Feature is an area or surface and in Feature_Volume when the feature is a 3D body. All lower level primitives are identified through the Volume_Face, Face_Edge and Edge_Node primitives.

associated with a 3D Body through a series of joins (Feature, Feature_Volume, Volume_Face, Face_Edge, Edge_Node tables are queried for BB-Rep), lists of Node_IDs forming a Feature are directly derived from the Feature_Node table, minimising the number of joins to be followed. Thus only the tables shown in with a white background (and a thicker border) in Fig. 9.8 are queried. Two examples (R220 and R287) are given here.

For R220 (Body/Body containment, no cavity) the Is_Boundary flag is used to model containment exceptions, as shown in Table 9.4 (only Feature_Volume and Feature_Face tables are shown - a similar approach applies for Feature_Edge and Feature_Node). All shared primitives are flagged as Is_Boundary = FALSE for Body A and Is_Boundary = TRUE for Body B).

Table 9.5. (a) and (b) Feature_Volume and Feature_Face tables for R220 (Body/Body containment, where A is outside B and A has no cavity)

(a)

Feature_ID	Volume_ID
A	V1
B	V2
A	V2

(b)

Feature_ID	Face_ID	Is_Boundary
A	F1	TRUE
A	F2	TRUE
A	...	TRUE
A	F6	TRUE
A	F7	FALSE
A	F8	FALSE
A	..	FALSE
A	F12	FALSE
B	F7	TRUE
B	...	TRUE
B	F12	TRUE

Ext(A) • Bnd(B) will return FALSE as the Faces making up the boundary of B are shared with A. However Faces F7 – F12 are marked as non-boundary for A but boundary for B, giving Int(A) • Bnd(B) as TRUE. Additionally, due to the shared Volume V2, Int(A) • Int(B) is also TRUE - A references this shared Volume object, and the interior of A is now formed from the sum of V2 and V1.

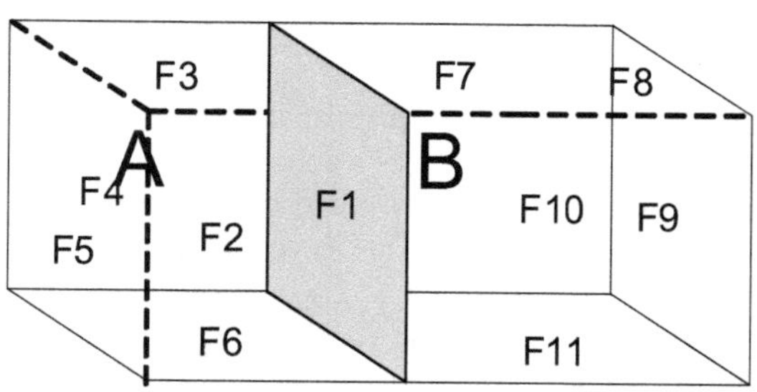

Fig. 9.9. Body/Body Meet (R287). A and B are adjacent and share Face F1.

Table 9.4. 9-Intersection Matrix for R287

	INT (A)	BND (A)	EXT (A)
INT (B)	FALSE	FALSE	TRUE
BND (B)	FALSE	TRUE	TRUE
EXT (B)	TRUE	TRUE	TRUE

Table 9.5. (a) and (b). Feature_Face and Feature_Volume tables for R287

(a)

Feature_ID	Face_ID	Is_Boundary
A	F1	TRUE
A	..	TRUE
A	F6	TRUE
B	F1	TRUE
B	F7	TRUE
	...	
B	F1	TRUE

(b)

Feature_ID	Volume_ID
A	V1
B	V2

Table 9.6 shows the Feature_Face (a) and Feature_Volume (b) tables for the R287 relationship illustrated in Fig. 9.9 Table 9.5 shows the corresponding 9-Intersection matrix for this relationship. In this case, the shared Face F1 is marked Is_Boundary = TRUE for both A and B and they do not share any Volume primitives.

9.5 Validating the Structures

Structure validation can be divided into two areas – can the structure model all the required 9-Intersection relationships, and how do the structures compare for binary relationship query performance?

9.5.1 Modeling the 9-Intersection Relationships

To confirm the theoretical validation of the structures shown in Fig. 9.7 and Fig. 9.8, a test dataset was constructed, based on the comprehensive set of diagrams detailing possible 9-Intersection relationships in 3D between simple objects, documented by [53, Chapter 6]. The dataset was initially created manually using hand-coded SQL scripts (in an Oracle database).

Three query types were identified for validating the structures:

- *Find Intersecting Objects* – given object A, find any objects that have a non-disjoint relationship with A.
- *Find Objects with Relationship* – given Object A and the possible 9-Intersection relationships, find any objects that have a specific relationship with A
- *9-Intersection Pairs*- given two Objects A and B, find the R-Value for the 9-Intersection Relationship between them.

The first two query types listed correspond directly to the "queries on whether two objects satisfy a set of topological relations" proposed by [9], with the generic *Find Intersecting Objects* providing a rapid mechanism to identify any non-disjoint objects without investigating the specific nature of the relationship. The third type of queries directly map to the second type proposed by [9] - "queries on the topological relations between two objects."

Given the multiple steps required to determine each component of a 9-Intersection relationship (described in Fig. 9.5 for one of the 9 components) and the subsequent summation to R-Value (Table 9.1), PL/SQL procedures were written to run the three types of 9-Intersection queries automatically[6].

9.5.2 Performance Comparison

A series of performance tests were carried out to validate the hypothesis that the MBB-Rep Structure provides improved performance for binary relationship identification over BB-Rep. The basic set of Features modelling each relationship pair (used for the validation process described above) was replicated to a total of 1.08 million Features to provide a more robust basis for performance comparisons. Queries were run against the replicated dataset to identify storage requirements for each structure.

A test harness then was created to run repeated iterations of the three query types identified above (*Find Intersecting Objects, Find Objects with Relationship* and *9-Intersection Pairs*).

Using the replicated dataset and the PL/SQL algorithms described above, the performance of each structure was measured. The tests were designed to emulate, as far as possible, a typical database environment, as follows:

- Tests were carried out using increasing numbers of users (2, 4, 6 and 8 concurrent users).
- The tests for each query type were run three times, with 100 iterations of each query per test, to allow the database to pin the execution plan for the query in memory.
- Each query iteration identified the relationship between a random pair of Features to minimize the impact of data caching (pre-loading the data would result in artificially fast query performance).

---

[6] PL/SQL (Procedural Language/SQL) is a procedural language extension within the Oracle Database Management System that allows the development of programs in the SQL environment within the database itself, providing a tighter link to the database engine than other development languages such as Java or Visual Basic. PL/SQL makes use of native oracle data types, including the Object-Relational types such as SDO_GEOMETRY, which removes the requirement to cast data into programming language variable types before operations are performed. This makes PL/SQL an ideal programming language where intensive SQL querying is required, as in this case. PL/SQL was thus selected over other available options (such as C++ or Java), which would be more advantageous where complex coordinate geometry calculations are required

- Query performance results were automatically logged in the database by the test harness.

9.5.3 Tests Results

The tests described above were designed to answer the two questions identified in the Introduction. The first of these asks if B-Rep structures can model the 9-Intersection relationships identified by [53]? The PL/SQL code was run on the test dataset, with results of each query written to the database. The use of this dataset validated the fact that both the BB-Rep (with described modifications) and the MBB-Rep model all the required relationships identified by [53].

Storage requirements for both structures were also compared, and it was determined that a total of approximately 50MB additional storage is required for the MBB-Rep structure using the 1.08 million Feature dataset.

The second question asked if a B-Rep structure is optimised for binary topological relationship identification queries. Results of performance comparisions between BB-Rep structure and MBB-Rep are given in Fig. 9.10, 9.11 and 9.12.

Fig. 9.10 shows the results obtained for an increasing number of concurrent users for the *9-Intersection Pairs* queries (which determine the 9-Intersection relationship between two randomly chosen Features from the dataset). As can be seen, the query execution time obtained for the MBB-Rep structure is indeed less than that for BB-Rep as predicted, showing that the impact of the additional join queries and the exception tables is significant.

Fig. 9.11 shows results obtained for the *Find Objects with Relationship* queries – given Feature A, find any objects having a specific binary topological relationship with A. Again, as predicted, the MBB-Rep structure showed faster query performance than BB-Rep.

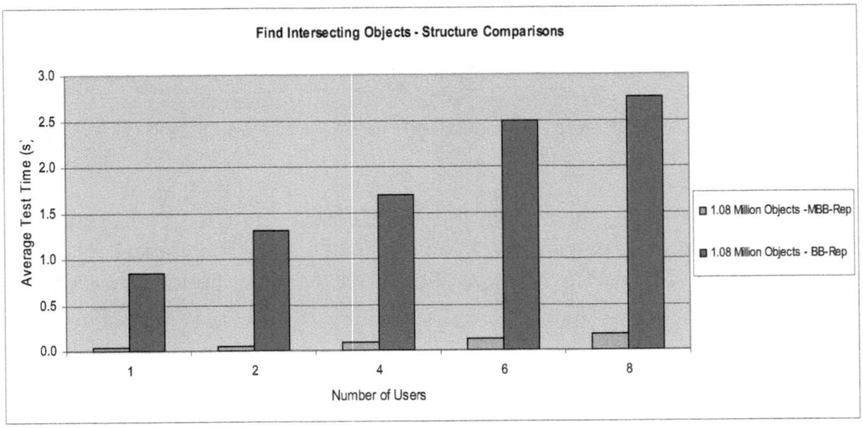

Fig. 9.10. Performance Comparison for MBB-Rep and BB-Rep– *9-Intersection Pairs.* Graph shows Average Test Time for different numbers of concurrent users.

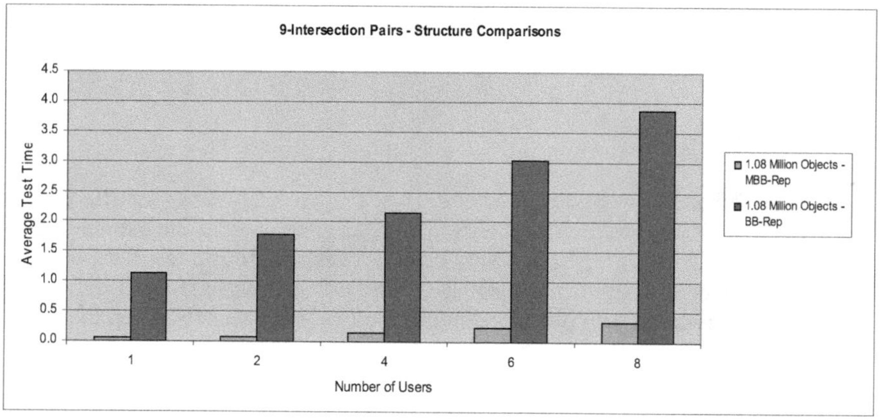

Fig. 9.11. MBB-Rep and BB-Rep – *Find Objects with Relationship.* Graph shows Average Test Time for different numbers of concurrent users.

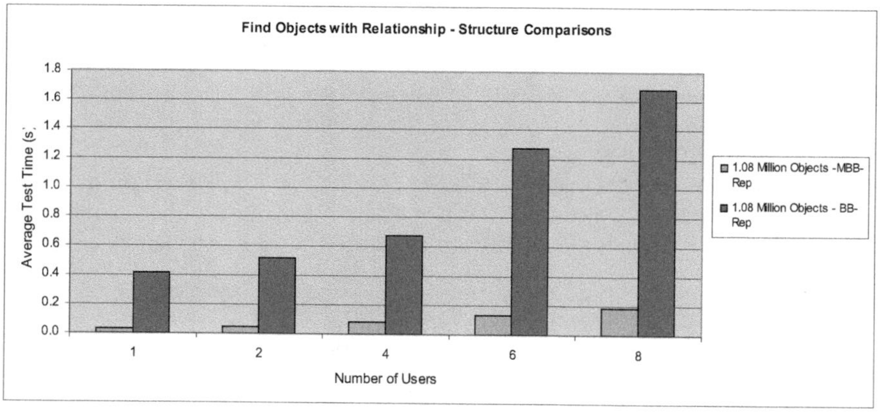

Fig. 9.12. MBB-Rep and BB-Rep – *Find Intersecting Objects.* Graph shows Average Test Time for different numbers of concurrent users

Fig. 9.12 shows results obtained for the *Find Intersecting Objects* queries – given Feature A, find any objects having any non-disjoint binary topological relationship with A. Again, the MBB-Rep structure showed faster query performance than BB-Rep.

9.6 Comparing the Structures

The work described above shows that B-Rep requires considerable extension to model a complete set of 3D binary topological relationships as defined by the 9-Intersection framework. This is to be expected, as the original B-Rep structure was designed to support CAD and computer visualization applications, rather than for use in 3D GIS. Thus for the original structure, it is the integrity of each single object that is important, as well as the ability to reconstruct the object from its primitives, rather than relationships between objects and sharing of primitives.

Once extended, both BB-Rep and MBB-Rep are able to model the complete set of 9-Intersection relationships. However, for MBB-Rep, fewer tables are queried to determine the relationships, and fewer joins followed.

An additional 50MB storage required for MBB-Rep is due to the additional Feature_Node, Feature_Edge, Feature_Face and Feature_Volume records, and also takes into account the reduced storage gained by excluding the data stored in the four containment exception tables.

It is also anticipated that the time to required insert or maintain data will be similar for both structures, as it is the determination of shared primtives from the underlying geometry (using algorithms such as [7, 24, 31, 41, 44, 47]) that is the most time-consuming element of this process, rather than the population of the database tables with the results of these algorithms.

9.7 Advantages of MBB-Rep

MBB-Rep structure benefits from features of BB-Rep including the fact that it can represent a solid unambiguously by describing its Surface and topologically orientating it such that at each Surface point the side on which side a particular solid lies can be identified. Both structures model topologically valid objects (as described in [23] and [21]). MBB-Rep and BB-Rep also make use of the same primitives.

For binary relationship determination, MBB-Rep is a much more compact structure than BB-Rep, with a total of only 5 tables being queried for binary topological relationship determination, as opposed to 12 for BB-Rep. This simplifies data maintenance and the algorithms required for query implementation.

The approach taken for containment exception handling for MBB-Rep is consistent across all dimensions where as in BB-Rep separate containment exception tables are used for specific cases. The MBB-Rep approach can easily be extended to dimensions higher than 3.

Additionally, the MBB-Rep structure allows the system to easily discriminate between specific elements of the components of the 9-Intersection relationships – for example if the user is only interested in relationships that include shared Nodes these can be rapidly and directly identified from the Feature_Node table rather than via joins from higher dimension primitives. It may also be possible to use MMB-Rep to model other frameworks such as those listed in [55].

As predicted, for binary relationship queries, the MBB-Rep structure out performs the BB-Rep structure in all cases. Table 9.7 summarizes the results for all 3 queries for 8 concurrent users and 1.08 million Features. Depending on the nature of the query, the MBB-Rep structure results in a performance improvement of between 11 and 15 times over BB-Rep. This improvement in performance is particularly significant when it is considered that 3D datasets are increasing in size, topological relationship queries are rarely run in isolation of other queries (such as metric measurements, attribute queries) and are also rarely run on a single pair of objects.

Table 9.6. Summary Test Results, 1.08 million objects

Test	Number of Users	MBB-Rep (s)	BB-Rep (s)	BB-Rep/ MBB-Rep
9-Intersection Pairs	8	0.34134	3.86793	11.3316
Find Intersecting Objects	8	0.17271	2.75457	15.9488
Find Objects with Relationship	8	0.18642	2.87999	15.4490

9.8 Factors Impacting the Test Results

A number of factors could reduce the absolute query performance times obtained when similar tests are applied to real-world data using both structures. These include:

- The disconnected nature test dataset (which comprised a series of Feature pairs), leading to a higher number of DISJOINT relationships
- The relative simplicity of the test Features (planar Faces, few primitives per Feature), leading to very short lists of primitivies to be intersected
- The artificial execution of 100 identical queries run in rapid succession against different Features, leading to possible query or data caching

However, it is anticipated that the performance of both structures would be equally reduced and thus that the comparative results would not be impacted.

9.9 Further Work

The availability of 3D spatial object types within an object-relational database allows GIS developers to move away from a traditional B-Rep structure (where the spatial configuration of a Feature is derived from the X, Y and Z coordinates held on a series of Node primitives) towards a more direct approach where the 3D Feature itself holds the required spatial information.

Both BB-Rep and MBB-Rep can take advantage of this approach, eliminating the requirement for multiple join queries to retrieve the coordinate information, and allowing the GIS to take advantage of spatial indexing.

Using this approach, it is also possible to reduce the number of Node, Edge and Face primitives by storing Edges as multi-segment objects and only creating Nodes where Edges shared between two objects actually meet. Fig. 9.13 illustrates this reduction, with Boxes A and B on the left hand side are formed from a total of 12 Nodes, whereas on the right hand side only 4 Nodes are used. Reducing the primitive count may also counter the impact of more complex real-world datasets on the absolute query performance times shown above.

The removal of the requirement to reconstruct a Feature from its primitives could also further the optimisation of structures towards binary relationship query performance. Further work is required to examine the possibility of removing the joins between the Node, Edge, Face and Volume primitives in the MBB-Rep structure. Although this removes the intrinsic topological validation of objects provided by the B-Rep structure, it may be possible to validate the correctness of the data as part of the compuational geometry process of identifying shared primitives, prior to structure population.

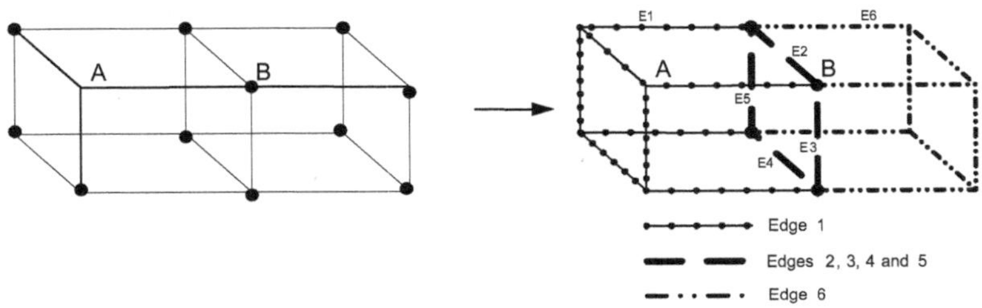

Fig. 9.13. Reducing Primitive Count by taking Advantage of Spatial Object Types.

Removing these joins also opens the possibility of modelling binary topological relationships between invalid objects (such as those listed in [23]) – providing that some process can be identified (perhaps using manual intervention) to allow the system to set the Is_Boundary flags correctly for each primitive. Additionally, in contrast to trditional BB-Rep, where all coordinate information is held on the Node primaitive, it may also be possible to model curved lines and surfaces.

Additional work is also required to identify the potential of integrating the MBB-Rep structure with the time primitives identified by [48]. Two such primitives are noted – point-in-time and time-interval. These could be represented as two non-spatial primitive types, where point-in-time forms the boundary of time-interval. The resulting structure (with joins removed) could be diagrammed as follows:

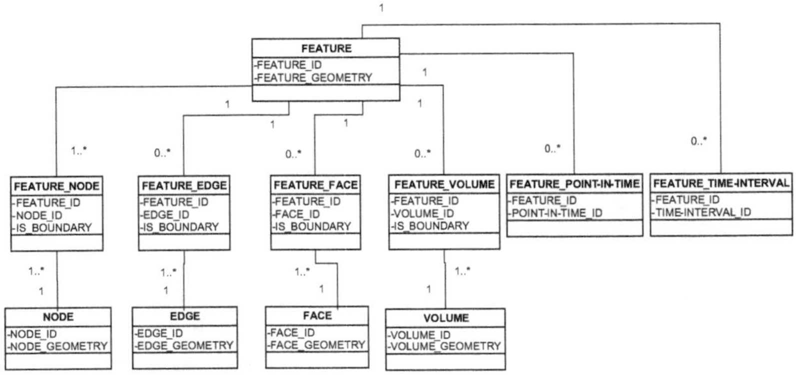

Fig. 9.14. MBB-Rep extended to include with Time Primitives.

The debate between the various approaches to binary topological relationship determination is ongoing in both 2D and 3D GIS, and further work is required to compare approaches, both for the test dataset described here and for real-world 3D data. Such approaches include an as-required determination of relationships direct from feature geometry. These comparisons will be facilitated as database vendors improve on the basic 3D functionality currently offered and move towards a richer range of binary topological queries. The work presented here contributes to this debate, and is of particular relevance to applications where the performance of such queries is important.

9.10 Acknowledgements

The authors would like to thank the Engineering and Physical Sciences Research Council (UK) and 1-Spatial Ltd. (Cambridge, UK) who have both contributed towards the funding of the work described in this paper.

References

1. Arens C, Stoter J, Van Oosterom, P (2005), 2005, Modelling 3D Spatial Objects in a geo-DBMS using 3D Primitive, Computers & Geosciences, 31, 165-177
2. Atzeni, P, Ceri, S, Paraboschi, S, Torlone, R, (1999), Database Systems – Concepts, Languages and Architectures, London, The McGraw Hill Companies
3. Baumgart B (1975) Wing-Edge Polyhedron Representation for Computer Vision, [online] Available from: http://www.baumgart.org/winged-edge/winged-edge.html [Accessed 10th May 2008]
4. Billen R, Zlatanova S (2003), Conceptual Issues in 3D Urban GIS, GIM International, 17(1), 33-35
5. Billen R, Zlatanova S, Mathonet P, Boniver F (2002), The Dimensional Model: a framework to distinguish spatial relationships, in Proceedings of ISPRS Commission IV Symposium on Geospatial Theory, Processing and Applications, Ottawa, Canada
6. Breunig M, Zlatanova S (2005), 3D Geo-DMBS, in Zlatanova S, Prosperi D (eds), Large-scale 3D Data Integration, Taylor and Francis, London
7. Chazelle B, Dobkin D (1987), Intersection of Convex Objects in Two and Three Dimensions, Journal of the Association of Computing Machinery (ACM), 34,1, 1-27
8. Chen J, Li Z, Li C, Gold C, (2001), Describing Topological Relations with Voronoi-Based 9-Intersection Model, in Fritsch, M, Englich, M, Sester, M, (eds), International Archives of Photogrammetry and Remote Sensing, 32(4), 99-104
9. Clementini E, Sharma J, Egenhofer M (1994) Modelling Topological Spatial Relations: Strategies for Query Processing, Computers and Graphics, 18(6), 815-822
10. Coors V (2003) 3D-GIS in Networking Environments, Computers Environment and Urban Systems, 27, 345-357

11. De La Losa A, Cervelle B (1999) 3D Topological Modelling and Visualisation for 3D GIS, Computers & Graphics, 23, 469-478
12. Egenhofer M, Clementini E, Di Felice P (1994) Topological Relations between Regions with Holes, International Journal of Geographical Information Systems, 8(2), 129-142
13. Egenhofer M, Franzosa R (1994) On the Equivalence of Topological Relations, International Journal of Geographical Information Systems, 8(6), 133-152
14. Egenhofer M, Herring J (1990) Categorizing Binary Topological Relations between Regions, Lines and Points in Geographical Databases, NCGIA Technical Report
15. Egenhofer M, Sharma J, Mark D (1993) A Critical Comparison of the 4-Intersection and 9-Intersection Models for Spatial Relations: Formal Analysis, in Mc Master, R, Armstrong, M, eds., Proceedings of Autocarto 11, Minneapolis
16. Ellul C, Haklay M (2006) Requirements for Topology in 3D GIS, Transactions in GIS, 10(2)
17. Grigni M, Papadias D, Papadimitriou C (1995) Topological Inference, in Proceedings of the 14th International Joint Conference on Artificial Intelligence (IJCAI), Morgan Kaufmann, 901-906
18. Grunwald S, Barak P (2003) 3D Geographic Reconstruction and Visualization Techniques Applied to Land Resource Management, Transactions in GIS, 7(2), 231-241
19. Guibas L, Stolfi, J (1985) Primitives for the Manipulation of General Subdivisions and the computation of Voronoi diagrams, ACM Transactions on Graphics 4(2)
20. Haarslev V, Moller R (1997) SBox – A Qualitative Spatial Reasoner – Progress Report, in IRONI, L, ed., Proceedings of the 11th IEEE Symposium on Qualitative Reasoning, Cortona, Italy
21. Hoffman C (1989) Geometric and Solid Modelling: an introduction, Morgan Kaufman Publications
22. IBM (2008) IBM DB2 Spatial Extender – User's Guide and Reference, [online], Available from: http://publibfp.boulder.ibm.com/epubs/pdf/c2712260.pdf, page 14, [Accessed 6th June 2008]
23. Kazar B, Kothuri R, Van Oosterom P, Ravada S (2008) On Valid and Invalid Three-Dimensional Geometries, In: P. van Oosterom, S. Zlatanova, F. Penninga and E. Fendel (Eds.); Advances in 3D Geoinformation Systems, Springer, 2008, Chapter 2, pp. 19-46
24. Konidaris G, Mehlhorn K, Shell D (2004) An Optimal Algorithm for Finding Edge Segment Adjacencies in Configurations of Convex Polygons, [online], Available from: http://domino.mpi-inf.mpg.de/intranet/ag1/ag1publ.nsf/ListPublications?OpenAgent&author=Mehlhorn,+Kurt, [Accessed 2nd April 2007]
25. Kufoniyi O (1995) Spatial coincidence modelling, automated database updating and data consistency in vector GIS, PhD thesis, ITC, The Netherlands, cited in Zlatanova, S, (2000), 3D GIS for Urban Development, ITC Dissertation Series No. 69
26. Li S (2006) A Complete Classification of Topological Relations using 9-Intersection Method, International Journal of Geographical Information Science, 20, 6

27. Lin T, Ward M, Power L, Landy D (1995) From Databases to Visualisation – Providing a User Friendly Visual Environment for Creating 3D Solid Geology Models, in Proceedings of Application of Computers and Operations Research in the Minerals Industries, (APCOM), 11-20, [online], Available from http://web.cs.wpi.edu/~matt/research/pubs/sectionstar3_1.html, [Accessed 27th September 2006]
28. Mark D, Egenhofer M (1995) Topology of Prototypical Spatial Relations between Lines and Regions in English and Spanish, in PEUQUET, D, ed., Proceedings of AutoCarto12, North Carolina, 245-254
29. Molenaar M (1990) A Formal Data Structure For Three-Dimensional Vector Maps. In Brascel, R, Kisomoto, H, Eds., Proceedings Of The Fourth International Symposium on Spatial Data Handling, Zurich, Switzerland, 830-843.
30. Molenaar M (1992) A Topology for 3D Vector Maps, in STEWART A, ed., ITC Journal, 1, 25-33
31. Nguyen V, Parent C, Spaccapietra S (1997) Complex Regions in Topological Queries, in Proceedings of the Conference on Spatial Information Theory (COSIT), Lecture Notes in Computer Science, 1329, 175-192
32. OGC, 2006, Open Geospatial Consortium – Open GIS Specifications (Standards), [online], Available from http://www.opengeospatial.org/docs/01-101.pdf, [Accessed 12 April 2006]
33. Oracle (2008) Oracle Spatial 11g, Advanced Spatial Data Management for the Enterprise, [online], Available from: http://www.oracle.com/technology/products/spatial/pdf/11g_collateral/spatial11g_datasheet.pdf, [Accessed 6th June 2008]
34. Papadias D, Theodoridis Y (1997) Spatial Relations, Minimum Bounding Rectangles, and Spatial Data Structures, International Journal of Geographical Information Science, 11(2), 111-138
33. Penninga F, Van Oosterom P (2007) A Compact Topological DBMS Data Structure For 3D Topography In: S.I. Fabrikant and M. Wachowicz (Eds.) The European Information Society - Leading the Way with Geo-Information, Lecture Notes in Geoinformation and Cartography, Springer, 455-471
34. Penninga F, Van Oosterom P, Kazar B (2006) A TEN-based DBMS approach for 3D Topographic Data Modelling, in: Riedl A. Kainz W, Elmes G, (eds.). Progress in Spatial Data Handling, 12th International Symposium on Spatial Data Handling, 581-598
35. Pfund M (2001) Topologic data structure for a 3D GIS, Proceedings of the 3rd ISPRS Workshop on Dynamic and Multidimensional GIS, Bangkok, Thailand, 34 (2W2), 233-237
36. Pigot S (1995) A Topological Model for a 3-Dimensional Spatial Information System, PhD Thesis, University of Tasmania, Australia
37. Pilouk M (1996) Integrated modelling for 3D GIS, PhD Thesis, ITC, The Netherlands.
38. PostGIS (2008) PostGIS Manual, [online], Available from: http://postgis.refractions.net/docs/postgis.pdf, page 13, [Accessed 6th June 2008]
39. Preparata P, Shamos M (1985) Computational Geometry: An Introduction, Springer-Verlag, New York
40. Rigaux P, Scholl M, Voisard A (2002) Spatial Databases: With Application to GIS, Morgan Kauffman, London

41. Rikkers R, Molenaar M, Stuiver J (1994) A query oriented implementation of a topologic data structure for 3-dimensional vector maps, International Journal of Geographical Information Systems, 8(3), 243 – 260
42. Schneider P, Eberly D (2003) Geometric Tools for Computer Graphics (The Morgan Kaufmann Series in Computer Graphics), Morgan Kaufmann, San Francisco
43. Silberschatz A, Korth K, Sudarshan S, (2002) Database System Concepts – 4th Edition, McGraw-Hill Higher Education, New York
44. Theobald D (2001) Topology revisited: representing spatial relations, International Journal of Geographical Information Science, 15, 689-705
45. Van Der Most (2004) An Algorithm for Overlaying 3D Features Using a Tetrahedral Network
Master's Thesis TU Delft, 2004, 96 p
46. Van Oosterom P, Ploeger H, Stoter J, Thomson R, Lemmen C (2006) Aspects of a 4D Cadastre: A First Exploration, Proceedings of the XXXII International FIG Congress, Munich
47. Van Oosterom P, Vertegaal W, Van Hekken M (1994) Integrated 3D Modelling within a GIS, Proceedings of Advanced Geographic Data Modelling (AGDM), Delft, The Netherlands
48. Wang X, Gruen A (1999) The Configuration and the Implementation of a hybrid 3-D GIS for Urban Data Management, Geographic Information Science, 4(1-2)
49. Wei G, Ping Z, Jun C (1998) Topological Data Modelling for 3D GIS, The International Archives of Photogrammetry and Remote Sensing, XXXII (4)
50. Worboys M (1995) GIS – A Computing Perspective. London, Taylor & Francis
51. Zlatanova S (2000) 3D GIS for Urban Development, ITC Dissertation Series No. 69
52. Zlatanova S (2006) 3D Geometries in Spatial DBMS, In: A. Abdul-Rahman, S. Zlatanova and V. Coors (eds); Innovations in 3D Geo Information Systems, pp. 1-14
53. Zlatanova S, Rahman A, Shi W (2004) Topological Models and Frameworks for 3D Spatial Objects, Computers and Geosciences, 30(4), 419-428

Chapter 10
Query Processing in 3D Spatial Databases: Experiences with Oracle Spatial 11*g*

Siva Ravada, Baris M. Kazar and Ravi Kothuri

Abstract. Falling costs in laser scanning equipment have made it easier to acquire 3D representations of real world entities as high-density point clouds, which may then be generalized to compact forms such as TINs, meshes, surfaces and solids. Scaling to the sizes of such large point clouds is becoming increasingly impossible for desktop applications. Sensing the need for a scalable database management of such 3D data, Oracle developed the 3D Spatial Engine in Oracle 11*g*. In this paper, we first present the overall architecture and functionality of the 3D Spatial Engine and introduce new geo-referenced data types for scalable storage and management of point clouds, TINs, and 3D vector geometries. We then focus on efficient processing of queries on these data types using a 2-stage filtering. For the second-stage filtering on 3D data, we propose and evaluate various techniques for efficiently answering different types of 3D queries such as *anyinteract* and *distance*. We note that to compare the points of point-cloud block or TIN block efficiently with the query, a spatial index may be useful but the creation cost is high. Our experiences are as follows: (1) For an anyinteract query, R-tree based processing is only required when both data and query geometries have large numbers of sub-elements. (2) For a *distance* query, an R-tree based evaluation is always helpful. We also propose a novel *vertex-avoiding* ray shooting algorithm for the point-in-solid function.

10.1 BACKGROUND

In recent years falling costs for laser scanning and terrestrial imaging have enabled the use of advanced 3Dimensional scanning equipment for a variety of GIS tasks, thus shifting the focus from 2-D to 3D. Let us look at two popular techniques for acquiring 3D data for city/terrain modeling applications.

Oracle USA, Inc.
{Siva.Ravada, Baris.Kazar}@Oracle.com
Currently at Innerscope Research
RaviKothuri@gmail.com

10.1.1 3D Data from Point Clouds

Fig. 10.1 shows the process of using scanning equipment to acquire 3D data as point cloud data (z, and other values for each pair of x, y). Such point cloud data is first generalized to triangulated meshes and then to higher-order geometries such as 3D surfaces, solids or NURBS (non-uniform rational B-Splines for smooth modeling).

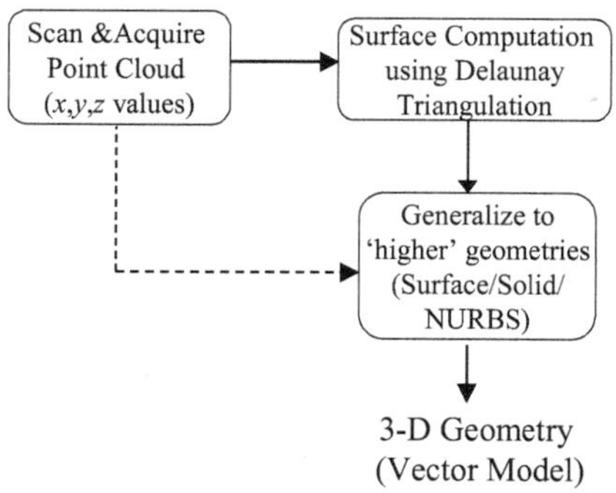

Fig. 10.1. Acquisition Process Workflow for 3D Vector data.

10.1.2 3D Data from Extrusion

Another approach for acquiring 3D building data is by "extruding" from the 2-D footprints. Fig. 10.2 shows an example of extruding a 2-D polygon to a solid (usually by specifying heights for each vertex of the polygon).

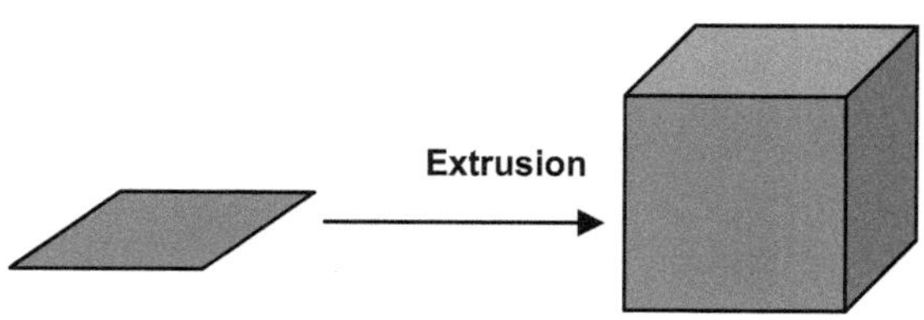

Fig. 10.2. 2-D Rectangle Extruded to a 3D Solid.

Using either of the above-mentioned techniques, the cadastral (property boundary) information in various cities is being modified to incorporate 3Dimensional building models on an experimental basis. Examples of such cities include Berlin, Bonn, and Hamburg [14].

In going from 2-D to 3D, there is a specific distinction between the CAD/CAM and the GIS industry requirements. Most existing CAD products operate mainly in local coordinate spaces and typically do not include a geo-referencing aspect. Besides, the scale and number of objects handled by such CAD/CAM applications is sometimes limited. On the other hand, the 3D objects in the city modeling applications in the GIS industry need to support many coordinate systems suitable to different regions of the Earth and for performing different functions.

Oracle introduced the 3D Spatial Engine in Oracle 11*g* to address the growing need for 3D GIS. In this paper, we first describe the functionality and architecture of this 3D Spatial Engine. We believe this presentation will serve the following purposes: (1) present our experience in building a 3D query engine, (2) and point out new avenues for further research.

The rest of the paper is organized as follows: Section 10.2 describes the architecture and various components of Oracle's Spatial 3D engine. Section 10.3 describes three different types of queries are on the various data types. Section 10.4 examines various alternatives for implementing these queries. The final section summarizes these results and points out open issues for research.

10.2 Functionality of the 3D Spatial Engine in Oracle 11*g*

The Spatial 3D Engine in Oracle 11*g* has the following components.

- 3D Data types include SDO_POINT_CLOUD, SDO_TIN, and the SDO_GEOMETRY to support storage of point cloud, TIN, and vector geometries, respectively.
- Query engine for efficiently processing different types of queries on the 3D data types.
- Adapters and loaders: These include converters between external formats such as GML and KML and the SDO_GEOMETRY data type. Loaders are also provided for storing a set of points into the SDO_POINT_CLOUD or the SDO_TIN data types.
- Utility functions such as validation of individual geometries [20], and extrusion of 2-D footprints to 3D solid geometries [19].

We will look at the data types in more detail and in the following section we will look at query processing on such data.

10.2.1 3D Data types in Oracle

To effectively manage the data at each stage in the workflow of Fig. 10.1, Oracle introduces two different data types, shown in Fig. 10.3—SDO_POINT_CLOUD, SDO_TIN and extends SDO_GEOMETRY to 3D.

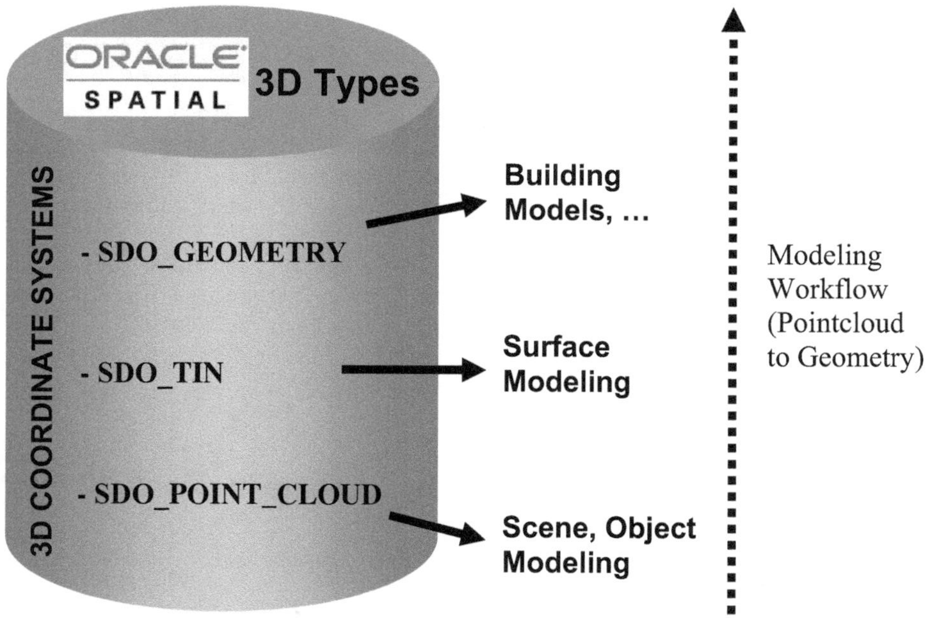

Fig. 10.3. 3D Data types in Oracle 11*g*.

10.2.1.1 SDO_POINT_CLOUD

This data type allows for the scalable storage and management of point clouds obtained from scanner equipment. Users can store up to $4 \cdot 10^{18}$ points in a single point cloud object in Oracle. This is achieved by partitioning the points into fixed-size blocks and storing the blocks as multiple rows in a separate block table. Fig. 10.4 shows the storage schematic for the point cloud.

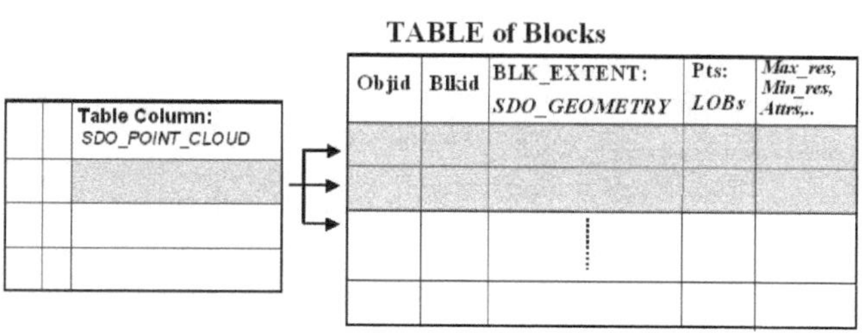

Fig. 10.4. Storing Point Cloud as Multiple Blocks in a "block table". Metadata is stored in the SDO_POINT_CLOUD column

As shown in Fig. 10.4, only the metadata, such as the *block capacity*, the *total number of dimensions* (at least 3: *x*, *y*, *z* and usually more, e.g., color), and the *block table* information is stored in the SDO_POINT_CLOUD column. The data points themselves are partitioned into subsets (using spatial partitioning techniques), each subset being stored in the *points* column of the associated block table. Note that the *blk_extent* column stores the extent of each subset and is spatially indexed. This *blk_extent* column allows for efficient query processing in a 2-stage filter processing as discussed in Section 10.3. The *Max_res*, *Min_res* columns allow point clouds to be stored in multiple resolutions. Such multi-resolution point clouds are useful in some applications that need to access them based on proximity to a viewpoint (closer for higher resolution and farther for less detail). This partitioned or blocked storage of a point cloud ensures scalability.

10.2.1.2 SDO_TIN

This data type, also referred to as the TIN type, allows for the scalable storage and management of triangulated irregular networks (TINs). The storage scheme is very similar to that of point clouds. Each row of the block table contains not only the points for that block but also an additional column to store the triangles that make up the TIN.

A single TIN in Oracle can handle up to $4 \cdot 10^{18}$ points. This is in sharp contrast to most triangulation software packages, which assume that all the points can fit in main memory and cannot scale beyond a few million points.

10.2.1.3 SDO_GEOMETRY

The SDO_GEOMETRY data type has been extended to support arbitrary 3D vector types such as points, lines, surfaces, solids and collections as a single geometry. Fig. 10.5 shows the different type of objects represented by the SDO_GEOMETRY data type.

The SDO_GEOMETRY is represented by an array of one or more elements. The element array can be used to represent a point, line-string, surface, solid, or collection (heterogeneous collection or homogenous collection such as multi-point, multi-curve, multi-surface, or multi-solid). Two or more points form a line. A closed line-string is called a ring. One or more rings (1 outer and 0 or more inner rings) form a polygon. One or more polygons form a surface (called a composite if more than 1 polygon and the surface then should be connected). One outer surface and 0 or more inner (represented as 1 or more in Fig. 10.5) surfaces form a simple solid (SSolid in Fig. 10.5). One or more 'attached' solids form a solid (referred to as composite if more than one). The collection types are formed as one or more elements of the appropriate type (e.g., one or more points form a multi-point, and one or more solids form a multi-solid).

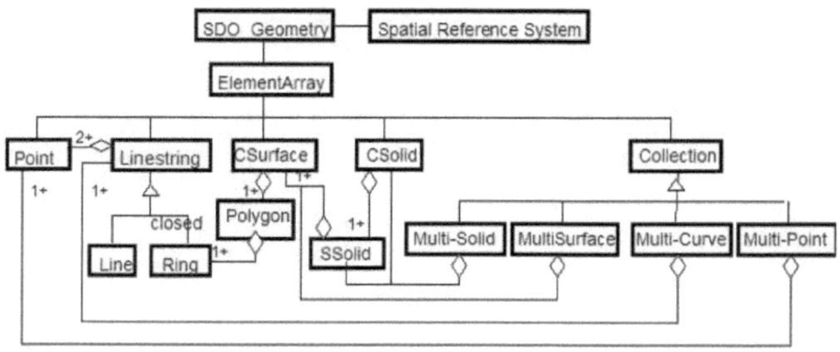

Fig. 10.5. Conceptual Model for 3D SDO_GEOMETRY

10.2.1.3.1 Comparison with GML and KML

Note that the SDO_GEOMETRY is similar to the GML geometry class shown in Fig. 10.5a.

The following similarities can be noted between the two representations from Fig. 10.5 and Fig. 10.5a: The *GML_Aggregate* class is same as the collections class in SDO_GEOMETRY. The Surface type in SDO_GEOMETRY can represent both *GM_Surface* (primitive class) as well as *GM_CompositeSurface*. Likewise, the Solid type in SDO_GEOMETRY can represent both *GM_Solid* (primitive) as well as *GM_CompositeSolid*. The *GM_Point* and the *GM_Line* classes map to the Point, Line classes in SDO_GEOMETRY. Given these similarities, there are some restrictions with the SDO_GEOMETRY type compared to GML geometry class:

- No equivalent for GM_CompositePoint in SDO_GEOMETRY.
- Polygons and solids need to have at least one exterior element in SDO_GEOMETRY
- In GML, aggregates can be nested inside aggregates, but such nesting of collections (aggregates) is not allowed in SDO_GEOMETRY.
- No surface holes are allowed on inner/outer solid geometries in the SDO_GEOMETRY model (whereas they are allowed in GML). However, these can be modeled as collections in SDO_GEOMETRY.

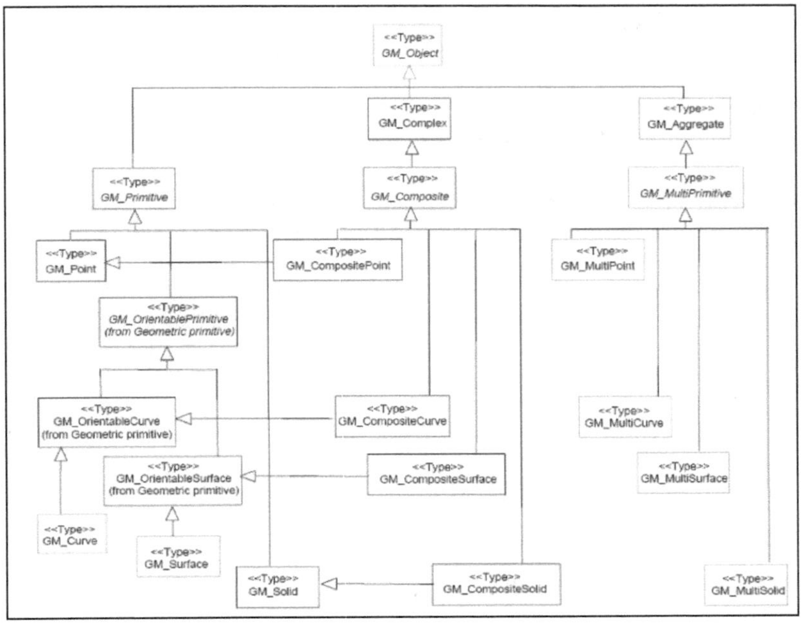

Fig. 10.5a. GML geometry basic classes (Fig. 10.5 of ISO_FDIS_19107).

Even though the SDO_GEOMETRY model is slightly different than GML3.1.1, it can still model real world 3D objects by removing unnecessary elements. Note that while the GML Geometry is a superset of the KML counterpart, the SDO_GEOMETRY is also a superset of the KML geometry.

10.2.1.3.2 Comparison with CAD and other models

There are different methods of 3D object representations that can mainly be classified into boundary representations (Breps) [1] and constructive solid geometry (CSG) [1]. Boundary representations model just the boundary of the 3D object. For example, for a solid shape, only the outer surface of the shape is represented. In CSG, the solid is defined using primitive solid elements as appropriate. Due to the ease of representation, Oracle mainly uses the boundary representation format for modeling 3D data. Given any type of solid (without any parametric curves and parametric surfaces), Oracle can store it by explicitly by specifying the composing surfaces.

10.2.1.4 Spatial Reference System

For 3D data, Oracle Spatial supports the following types of coordinate systems that can be associated with any of the 3D types:

- *Local*: These refer to coordinate systems that are specific to an application. Used in CAD/CAM and most non-GIS based applications.
- *Geocentric 3D*: These refer to 3D Cartesian coordinate systems.

- *Geographic-3D*: These refer to coordinate systems that perform calculations on the 3D ellipsoidal data that represent the terrestrial surface.
- *Compound*: These refer to coordinate systems that combine Geographic-2-D systems such as projected and geodetic systems with vertical coordinate systems that measure height (from a reference point such as sea level)

Whereas the *Local* coordinate systems can be used in any application, the rest are typically used to model data on the surface of the Earth (e.g., buildings in a city model). Since the Earth's surface is not a regular ellipsoid geometric shape, different data and the associated coordinate systems use different 'ellipsoids' to maximize the accuracy for specific regions and specific tasks (such as area and distance computations etc.). Oracle supports a rich set of 3D coordinate systems that are based on the European Petroleum Standard Group's (EPSG) model.

10.3 Indexing and Querying

In this section, first we present how queries are processed on the SDO_GEOMETRY type. Then, we describe how they are processed on SDO_POINT_CLOUD and SDO_TIN types.

10.3.1 Queries on SDO_GEOMETRY

After storing different types of geometries using the SDO_GEOMETRY type, users can create spatial indexes on the 3D data. Oracle uses the Oracle R-tree [5] for such spatial indexing. The 3D spatial R-tree index can be created by setting the *sdo_indx_dims* parameter to 3 as follows.

```
SQL>CREATE INDEX bldgs_idx ON bldgs(bldg)
    INDEXTYPE is MDSYS.SPATIAL_INDEX
    PARAMETERS ('sdo_indx_dims=3');
```

Users can then perform the following types of queries efficiently by specifying the query in the WHERE-clause of an SQL statement.

- SDO_ANYINTERACT: Given a 3D geometry as the query object, this type of query identifies all 3D geometries in a table that have any spatial interaction (such as intersection) with the query object. In the following example, the query object is a polygon.

```
SQL>SELECT c.mkt_id, c.name
    FROM cola_markets c
    WHERE SDO_ANYINTERACT(c.shape,
    SDO_GEOMETRY(3003, NULL, NULL, -- Query object
    SDO_ELEM_INFO_ARRAY(1,1003,3),
```

```
SDO_ORDINATE_ARRAY(
6.0,6.0,6.0,
5.0,6.0,10.0,
3.0,4.0,8.0,
4.0,4.0,4.0,
6.0,6.0,6.0))) = 'TRUE';
```

- SDO_WITHIN_DISTANCE: Given a 3D geometry as the query object and a distance "*d*", this type of query identifies all 3D geometries in a table that are "*within a distance d*" from the query object.
- SDO_NEAREST_NEIGHBORS: Given a 3D geometry as the query object and the number "*k*", this type of query identifies all 3D geometries in a table that are "*the k neighbors*" for the query object. The query also has provision for specifying non-spatial predicates such as 'building_type='OFFICE'' to identify the "*k*" OFFICE-type buildings from a *bldgs* table.

Each of these queries requires the table being queried to have a spatial index. Using such an index, the queries are processed in a two-stage filter-and-refine approach [6, 7] as described in Fig. 10.6. Using the 3D minimum bounding volumes (*MBVs*) in the spatial index, the candidate geometries that may satisfy a query window are first identified. These candidates are then sent to the Geometry Engine to refine and identify the final result set. This two-step filtering mechanism ensures that the 'expensive' geometry engine operations are performed only on a few candidate geometries and uses the fast-filtering power of spatial index to weed out irrelevant candidates.

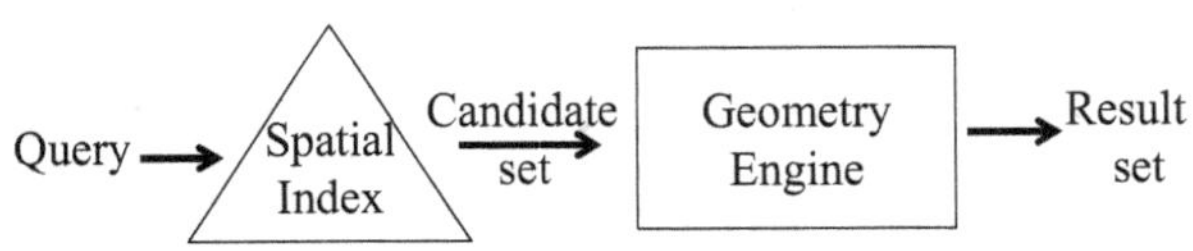

Fig. 10.6. Two-stage Processing of Queries in 3D Spatial Server.

Note that depending on what the query type is, different functions from the Geometry Engine are used. For SDO_ANYINTERACT query, the *anyinteract* function of the Geometry Engine is used. For distance-based queries such as the SDO_NN and the SDO_WITHIN_DISTANCE, the *distance* function in the Geometry Engine is used. Since the first stage was already available in Oracle before, the contribution of this study in Oracle 11*g* in terms of querying 3D data lies in the implementation of different functions in the Geometry Engine, which is the second stage in Fig. 10.6. These functions will be discussed in more detail in Section 10.4.

10.3.2 Queries on SDO_POINT_CLOUD/SDO_TIN

Oracle provides a clip PL/SQL query on the SDO_POINT_CLOUD and the SDO_TIN types. This query takes a *spatial window* (specified as an SDO_GEOMETRY) and identifies all points and/or triangles that have any interaction with the spatial window. In addition to the spatial window, the query also takes a resolution range to identify only those blocks that satisfy the resolution range. This type of query is typically called a resolution-based visibility query and we refer to this as the clip operation. To restrict the scope of the discussion, we refer to the spatial window as the "query" in the rest of the section, ignoring any resolution range portions of the query.

As shown earlier in Fig. 10.4, a point cloud is stored as multiple rows in an associated block table, each row containing a *blk_extent* column of type SDO_GEOMETRY to represent the extent of the points in that row. For TINs, a similar *blk_extent* is maintained in the associated block table. This *blk_extent* is implicitly indexed with an Oracle R-tree index.

Using such an index, the CLIP query on an SDO_POINT_CLOUD or an SDO_TIN is processed in 2-stages as shown in Fig. 10.6a.

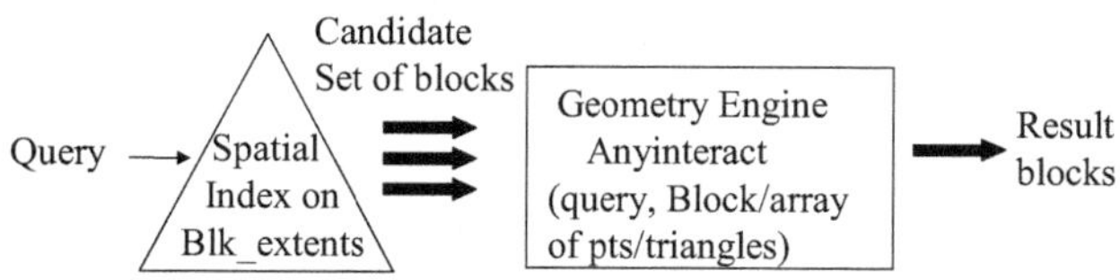

Fig. 10.6a. 2-Stage Query Processing on a Point Cloud/TIN.

1. *First-stage*: Using the spatial index on the *blk_extent*, the query identifies all the relevant blocks of a point cloud or a TIN.
2. *Second-stage*: For each relevant block for the query (identified in first-stage), the associated points (and triangles if the type is an SDO_TIN) are retrieved. Typically the number of such points in a block is of the order of thousands or tens of thousands (typical block capacity). These points are composed as a multi-point SDO_GEOMETRY and compared with the query geometry using the *anyinteract* function. For an SDO_TIN block, the triangles are composed as a multi-surface and compared with the query geometry using the *anyinteract* function. The intersecting points (and or triangles) are returned as part of the result of the CLIP routine. The CLIP routine, implemented as a pipelined table function [19], returns one result block row corresponding to each candidate block identified in the first stage in an iterative manner.

Oracle Spatial creates a spatial R-tree index on the *blk_extent* column of the block table. Using such an index, Oracle quickly identifies the blocks that are relevant for a spatial query window. The actual points and triangles that are "definitely" relevant for the query window are identified using an "*anyinteract*" query between the query-geometry and the points/triangles in the point cloud/TIN in the Geometry Engine (secondary filter).

10.4 The 3D Geometry Engine

The 3D Geometry Engine provides a variety of functions for geometry-geometry relationships, geometry area/length/volume determination and geometry transformation (rotation, scaling, translation, etc.). In this paper, we focus primarily on the geometry-geometry relationship functions such as *anyinteract* and *distance*.

In providing this functionality, the Geometry Engine:

1. Processes inner geometries, i.e., holes.
2. Handles not only convex but also concave geometries that have planar surfaces, and without performing any triangulation process for conversion from concave geometry to convex equivalent(s). While most software applications in the literature do not consider concave geometries, the ones that consider concave geometries convert them into their convex equivalents using a triangulation process, which adds extra costs into the computation.

To the best of our knowledge, there is no software in the literature which supports BREP (boundary representation) with or without inner geometries (holes), and handles concave geometries without triangulation. In what follows, we describe the functionality of the 3D Geometry Engine in more detail.

10.4.1 Geometric Relationship Functions

First we will describe the algorithm for *anyinteract* function. We present various alternatives for the query processing and evaluate using experimental data. We then repeat the same exercise with the *distance* function. Finally, we discuss how holes are processed in this framework.

10.4.2 Anyinteract

The *anyinteract* function takes two arguments: one *query* SDO_GEOMETRY and a second *data* SDO_GEOMETRY. The *anyinteract* function determines whether or not these two input 3D geometries have any spatial interaction (such as intersection) with each other. Each of the input geometries can be any valid SDO_GEOMETRY: i.e., it can be a point, line, polygon, composite surface, solid, composite solid, or a homogenous collection (multi-point, multi-line, multi-surface, or multi-solid) or heterogeneous collection. This type of generic interface is useful in processing the queries for SDO_GEOMETRY, SDO_POINT_CLOUD and SDO_TIN data types. For instance, for a TIN type, the *anyinteract* function will have two arguments, the first one being the query window and the second being a multi-polygon (multi-surface) consisting of the set of triangles in a row of the TIN block table. Most importantly, the minimum bounding volumes (*MBV*s) of the two input geometries in an anyinteract function may already intersect (as the anyinteract function is typically called as a second-stage processing in the 2-stage filtering of Fig. 10.6 and Fig. 10.6a). The *anyinteract* function is called either to eliminate false positives (*MBV*s intersect but not the geometries themselves), or to identify the sub-elements of the geometries that intersect.

The author of [3] deals mainly with 3D topological relationships but assumes that the geometries do not have any inner cores (or *holes)*. In this paper, we consider inner geometries as well (in section 10.4.4), which makes the determination of *anyinteract* relationship more complex. The inner geometries could be surfaces or solid geometries.

Fig. 10.7 shows the overall flowchart for the *anyinteract* determination algorithm using appropriate building blocks [12, 13, 17].

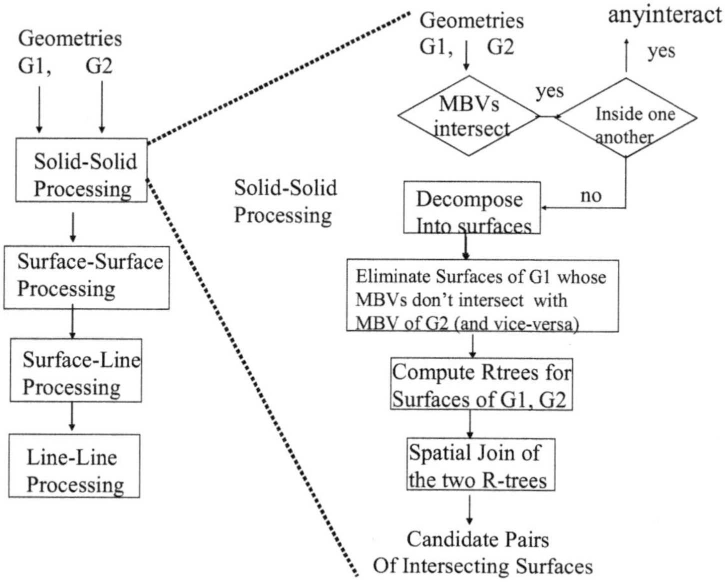

Fig. 10.7. Algorithm for Anyinteract: Determining if Two 3D Solid Geometries Intersect.

Since the input geometries can be collections of primitive elements such as polygons (e.g., triangles for TIN block), lines, or points (as in a point cloud block), in-memory spatial indexes such as R-trees can be constructed on different sub-elements of the geometries and they can be used to eliminate the non-interacting parts of the input geometries, thus ensuring scalability of the algorithm with the complexity (number of surfaces, lines, etc.) of the geometries. Fig. 10.7 shows the typical processing algorithm for the *anyinteract* function assuming two solids, G_1 (query geometry) and G_2 (data geometry) as input. The minimum bounding volumes (*MBVs*) of the two solids are first compared. If they do not intersect, the solids do not intersect. Otherwise (i.e., if the *MBVs* do intersect), the solids are checked to see if one is inside the other (using a *point-in-solid* function, which will be discussed in Section 10.4.2). If they are inside one another, the solids have an anyinteract relationship. Otherwise, the processing is continued by decomposing the solids into lower-level elements such as the bounding surfaces. Observe that the processing is carried out in a recursive manner on lower-level primitives, i.e., solid-solid anyinteract is evaluated using 'surface-surface' anyinteract which is in turn evaluated using surface-line, and line-line processing and so on.

Note that the above algorithm suggests the use of R-trees (or other spatial indexes). Constructing such spatial indexes during query processing may be costly (typically of the order of $O(N/f * \log_f (N/f))$ for N elements and an R-tree of fanout f). If G_1 (query) and G_2 (data geometry) are the same, then all of the sub-elements of these geometries match. This causes the R-tree construction costs to be comparable to the direct comparison cost, which is $O(N^2)$. On the other hand, if both G_1 (query) and G_2

(data geometry) have large number of sub-elements *few of which match*, the R-tree construction cost turns out to be smaller than the $O(N^2)$ comparison cost. Since few sub-elements from query and data 3D solid geometries are expected to match in real-life data, we preferred to include R-tree construction (on the surfaces of G_1 and G_2) for solid-solid processing part in Fig. 10.7. Furthermore, the surface-surface processing part in Fig. 10.7 will activate R-tree construction (on the edges of a surface of G_2) only if the number of edges of surface of G_2 is greater than 10. Even though we are unable to discuss other cases like composite-surfaces due to space limitations, we should note that we applied similar strategies to arbitrary pairs i.e., G_1 and G_2 being different type of geometries.

10.4.2.1 Query has few sub-elements (e.g., solid box)

In most cases (for example, queries from a visualizer such as Google Earth), the query geometry is a simple window (may be a solid box or a simple surface). The data geometry can be either an SDO_GEOMETRY representing a building, or a block of an SDO_TIN representing a triangulated surface, or a block of a SDO_POINT_CLOUD. In these situations, it is not clear whether or not the R-tree construction cost is an acceptable overhead cost. To better understand these issues when one of the geometries (query) has fewer elements compared to the second (data) geometry, we conducted several experiments and evaluated various alternatives.

Our experimental setup is as follows. We compared the *anyinteract* processing cost for a TIN block and a solid box query window. The number of points forms a 2-D grid in the TIN and is varied as follows: 20-by-20, 50-by-50, 100-by-100, 150-by-150 (we chose only these sizes of 400 to 22500 points as these numbers represent a typical block capacity of a TIN or a point cloud). The z value is randomly chosen using a random number within the query range. The query window itself is varied randomly across the extent of the TIN block and chosen to be either 1% or 0.1% of the underlying TIN block extent. Using these data, we compared the following alternatives:

- *Brute-force*: This method compares each triangle in the TIN block to the query window directly.
- *ElemMBV check*: This method performs an *MBV* check of each element (i.e., each triangle) with the query and if successful, performs a direct comparison.
- *R-tree*: This method constructs an R-tree on all the elements (triangles) first, and then identifies relevant candidates using the R-tree and performs lower-level comparison on the candidate results.

Fig. 10.8 plots the query response time for different database sizes (query times on y-axis on logarithmic scale). Note that the *ElemMBV check* approach is slightly faster than the R-tree approach. This is expected, as the R-tree construction cost becomes an overhead.

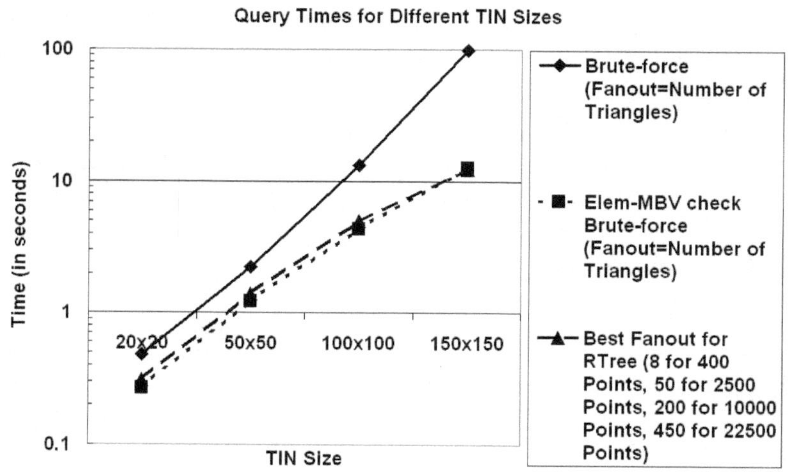

Fig. 10.8. Query times for different TIN sizes where 20-by-20 size TIN with z-values has 722 triangles, 50-by-50 size TIN with z-values has 4802 triangles, 100-by-100 size TIN with z-values has 19602 triangles and 150-by-150 size TIN with z-values has 44402 triangles.

Fig. 10.9 plots the query response time for different R-tree fanouts where the TIN size is 100-by-100 points. You can observe that as the fanout increases the time improves. This is expected because the effect of the R-tree construction cost, $O(N/f * \log_f (N/f))$, decreases as fanout increases and becomes a constant as f becomes comparable to N.

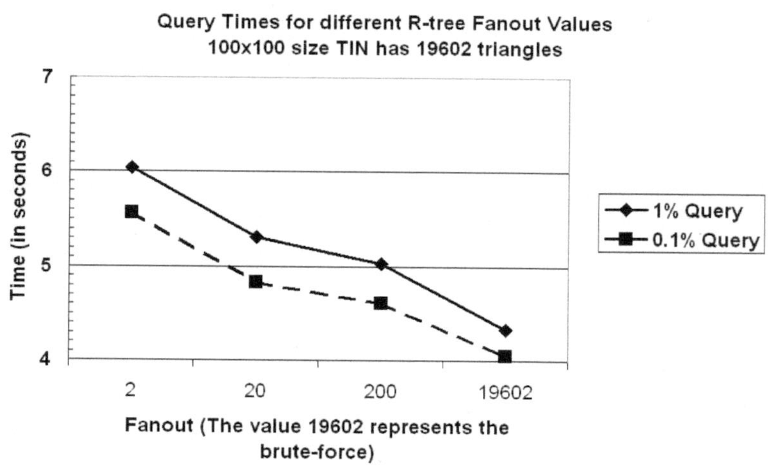

Fig. 10.9. Query times for different R-tree fanout values.

Table 10.1 presents similar results with real data for the city of Berlin from Tele Atlas.

Table 10.1. Query times for a solid geometry (39 faces) vs. a point query.

	Query time (seconds)
R-tree	0.026
ElemMBV check	0.014

10.4.2.2 Query has several sub-elements

Next, we repeat the *anyinteract* query time experiments when the query itself has many sub-elements, i.e., both the query and the data geometries have many sub-elements. This situation is common for radiation plum detection and determining whether a ship can dock into a harbor properly. For this, we compare a building with 39 faces with another building of 24 faces. We observe that the R-tree helps in this case, because of the following: the R-trees add an overhead of only $O(2*N/f * \log_f (N/f))$ but the *ElemMBV check* has to perform $O(N^2)$ checks. Table 10.2 (real data for the city of Berlin) indicates that the *R-tree* strategy is indeed faster than *ElemMBV check* strategy.

Table 10.2. Query time for *anyinteract* of a Building of 39 faces with another Building of 24 faces

	Query time (seconds)
R-tree	0.036
ElemMBV check	0.093

10.4.3 Point-in-Solid Function

This function determines whether a point is inside a solid or not[1]. This is used internally in the *anyinteract* function to determine whether two solids are inside one another (by checking a vertex of one solid with the other solid and vice-versa). The 2-D counterpart of the point-in-solid, which is called point-in-polygon, can be solved using the ray-shooting algorithms [18] in the literature. Given a point *p* and a polygon *S*, the algorithm is as follows. Fig. 10.10 illustrates the algorithm with an example.

- Shoot a ray from point *p* to a point *q* outside *S*.
- If the ray from *p* to *q* intersects the polygon an even number of times, then the point *p* is outside *S*, otherwise it is inside.

[1] If the point is on the boundary of a solid then it is considered not "inside".

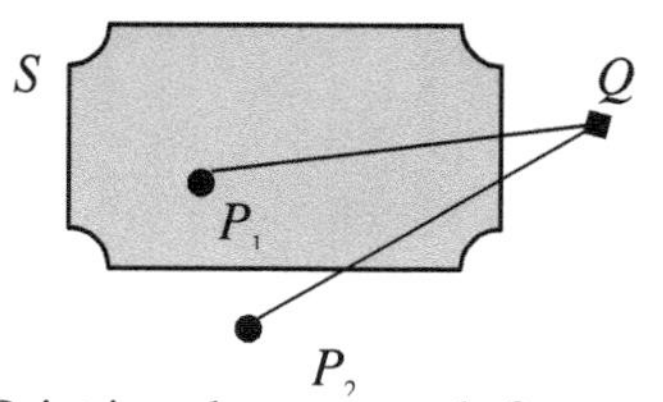

P_1 -> Q: 1 intersection
P_1 is inside S

P_2 -> Q: 2 intersections
$\Rightarrow P_2$ is outside S

Fig. 10.10. Point-in-polygon example for two points P_1 and P_2 and polygon S.

As easy as it may seem, there are several boundary cases for the ray-polygon intersections that need to be accounted for in the above algorithm. These include intersection at a vertex of S, and the ray grazing along an edge. The boundary cases are solved in the literature using a variety of ways such as by counting the number of edges '*above*' the ray, etc. [18]

Generalizing the above algorithm to 3D solids introduces more boundary cases: the ray can graze along a polygonal surface of the solid, too. Besides, the rules for resolving all the boundary cases in 3D solids are not straightforward. Consequently, Oracle takes the series of steps for point-in-solid:

1. Check if p is inside (and coplanar with) any of the boundary polygons of solid S (within tolerance). If so return as 'outside'.
2. Identify a random ray between point p and a point q outside S (q can be any point outside the *MBV* of the solid S).
3. If the ray intersects a vertex, or grazes along an edge, then determine a vertex-avoiding ray calculation as described below.

Algorithm for Vertex-avoiding Ray:

1. From the point p, take any line parallel to any of the three axes, say z-axis. Call this line `Ref_line`.
2. Now connect p to each of the vertices $v_1,\dots,v_n$ of the solid S.
3. For each v_i, compute angle a_i = absolute angle between (`Ref_line`, `Line` $<p, v_i>$).
4. Let a_k be the min of all <u>non-zero</u> a_i. (exclude vertices that are on the `Ref_line`).
5. Now shoot a ray R midway between `Ref_line` and the `line` $<p,v_k>$ (on the plane of `Ref_line` and `line` $<p,v_k>$).

One can prove (by contradiction) that the ray will not intersect any vertex of the solid S (nor will it graze along an edge of S). However, it may graze along (i.e., be coplanar with) a polygon of solid S. Once you find a vertex-avoiding ray from p for a solid S as described above, you can determine whether p is inside or outside of S using the following rules:

1. If the ray is coplanar with a polygon of S, that polygon is removed.
2. Otherwise, count the total number of intersections of the ray with the edges of the solid S.

3. If the total number of intersections is even, then point *p* is outside *S*. Otherwise, it is inside *S*.

The above algorithms provide a clear-cut solution to the point-in-solid problem. Since the number of vertices of a solid is limited, and there is infinite number of rays from point *p*, it is very likely that a random ray does not intersect a vertex of the solid. Even when it does, the vertex-avoiding ray algorithm ensures the next ray chosen does not. We compared the cost of computing the vertex-avoiding ray with that of finding a random ray and checking if it intersects with the vertices. We used the largest building with 783 vertices in the Berlin city model data provided by Tele Atlas. The results for average times of 100 point in solid computations (using 50 random points inside and 50 outside the solid) are shown in Table 10.3. You can observe that the vertex-avoiding technique that we proposed is within 2 times slower than the random ray technique. This makes it quite effective to be combined with random ray shooting as described above in the series of steps taken for point-in-solid in Oracle 11*g*. This novel approach ensures fast point-in-solid determination in Oracle.

Table 10.3. Time (seconds) for computing point-in-solid using a vertex-avoiding ray vs. random ray.

Number of vertices in Solid	*Random Ray*	*Vertex-avoiding Ray*
783	0.013	0.024

10.4.4 Distance

Next, we will describe how distances are computed between a pair of geometries. The *distance* function takes two arguments: one query SDO_GEOMETRY and a second data SDO_GEOMETRY. This function computes the minimum distance between the two input 3D geometries. If they intersect (*anyinteract* with) each other then the distance is 0, otherwise it is non-zero. Each of the input geometries can be any valid SDO_GEOMETRY: i.e., it can be a point, line, polygon, composite surface, solid, composite solid, or a homogenous collection (multi-point, multi-line, multi-surface, or multi-solid) or heterogeneous collection. Unlike the *anyinteract* function, where the *MBVs* of the two input geometries intersect, for a *distance* function that assumption seldom holds.

Fig. 10.11 shows the algorithm for computation of distance between two solid geometries (can be generalized to arbitrary pairs). Observe that the algorithm computes the distance in a recursive manner by identifying the nearest pairs of sub-elements using the nearest-neighbor processing of spatial indexes such as R-trees.

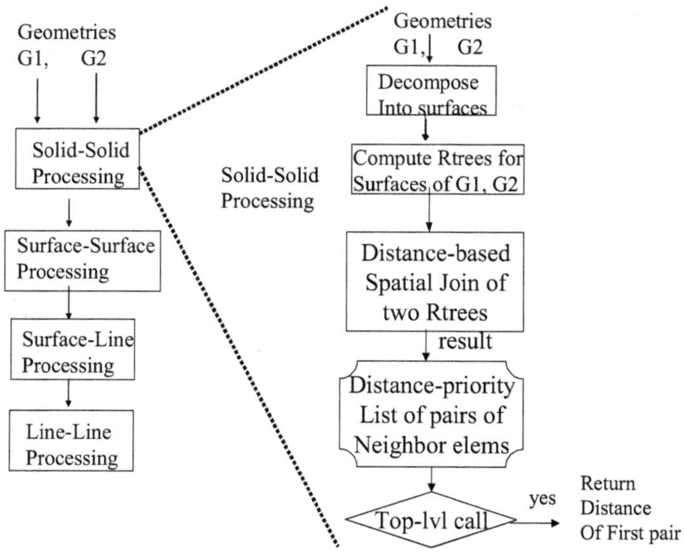

Fig. 10.11. Algorithm for Distance Computation for Two 3D Solid Geometries

To evaluate the performance of the above algorithm, we conducted several experiments using the TIN block data used for *anyinteract*. The queries, however, are chosen to be random points outside the TIN block extent. We compared the above algorithm with R-tree indexes and without R-tree indexes (i.e., brute force: no spatial index based join but a pair-wise join). Fig. 10.12 plots the results for different sizes of the data. As expected, the above algorithm using R-tree is substantially faster than brute force distance computation. Observe that the brute force times increase quadratically ($O(N^2)$ distance computations) as the size of the data increases.

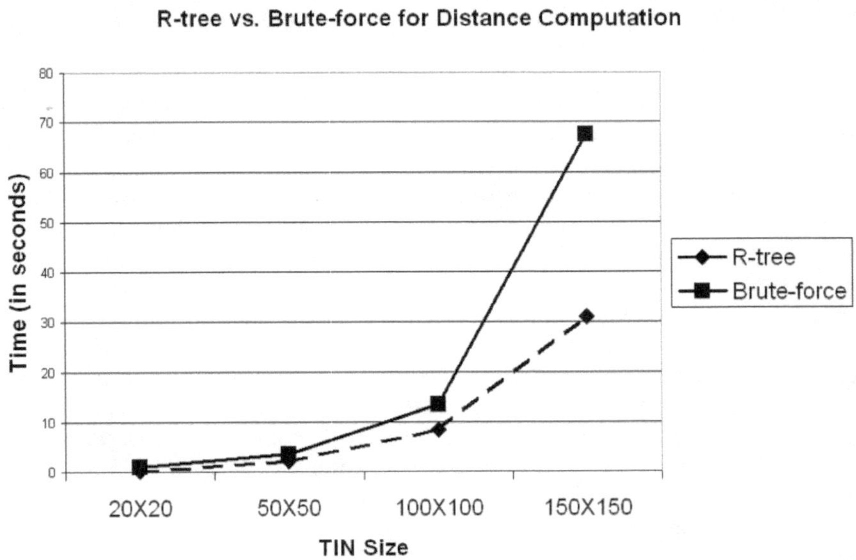

Fig. 10.12. R-tree vs. brute force for distance computation

10.4.5 Relationship Computation When Geometries Have Holes

As described earlier, solids and surfaces represented in Oracle can have holes. The above algorithm for relationship determination can be extended to geometries with holes as follows.

Let the input geometries G_1 and G_2 be composed of outer geometries O_1 and O_2 and hole geometries H_1 and H_2 respectively. G_1 and G_2 intersect if and only if the following holds:

1. O_1 is not inside H_2 and O_2 is not inside H_1, and
2. O_1 intersects O_2.

These criteria can then be generalized to geometries with not one but many holes.

10.5 Conclusions and Future Work

In this paper, we presented the architecture and the functionality of the 3D Spatial Server. We described different data types introduced in Oracle 11*g* for catering to 'data at different stages' of a city modeling or 3D GIS application. The 3D Spatial Server has one of the few TIN creation techniques that can scale to arbitrarily large dataset sizes (> terabytes). The other ones involve alternate and orthogonal strategies such as streaming techniques in [21]. We then examined query processing on these new data types. Again for scalability, the queries are processed in a 2-stage filtering approach. The first stage performs filtering based on spatial indexes. The second stage uses an exact relationship determination between a pair of geometries- one query and one data geometry. Two types of relationship functions are possible in the second stage - *anyinteract* or *distance*. In addition, a *point-in-solid* function is necessary to process the *anyinteract* function.

To efficiently process the *point-in-solid* function, we proposed a novel technique to compute *vertex-avoiding* rays from a point *p* for a given solid *S*. Such rays can be utilized if random ray shooting does not provide clear-cut answers (i.e., intersects a vertex or an edge).

Next, we described algorithms for efficiently processing *anyinteract* and *distance* functions for a pair of geometries. Since the geometries can have many or few sub-elements, we compared two main variants for query processing: (1) constructing and utilizing R-trees on sub-elements of geometries, and (2) just comparing *MBVs* of elements of first geometry with the *MBV* of the second geometry and vice-versa (referred to as *ElemMBV check*). We can summarize our results as follows.

- For *distance* computation, processing using R-trees on sub-elements of geometries is always faster by a factor of 2 or more based on the number of elements.
- For *anyinteract* computation, use of R-trees on the geometries is faster if both geometries have a large number of (such as 20 or more) elements.
- However, for *anyinteract* computation, if one geometry has very fewer number of sub-elements compared to the second, then R-trees are not the right choice. Instead, processing using *ElemMBV check* is faster.

REFERENCES

1. Stoter J. and Zlatanova, S., 3D GIS, where are we standing, In: Joint Workshop on Spatial, Temporal and Multi-Dimensional Data Modelling and Analysis, Quebec city, 2003, Canada, 6p.
2. Penninga, F., van Oosterom, P. & Kazar, B. M., A TEN-based DBMS approach for 3D Topographic Data Modelling, International Symposium on Spatial Data Handling, Vienna, 2006, pages 581-598.
3. Zlatanova, S., On 3D Topological Relationships, 11th International Workshop on Database and Expert Systems Applications (DEXA), 2000, pages 913-919.
4. Mirtich, B., Fast and accurate computation of polyhedral mass properties, Journal of Graphics Tools, 1996, pages 31-50.
5. Kothuri, R. and Ravada, S., Sharma, J., Banerjee, J., Indexing Medium Dimensionality Data in Oracle, ACM SIGMOD, 1999.
6. Beckmann, N., Kriegel, H., Schneider, R. and Seeger, B., The R* tree: An efficient and robust access method for points and rectangles. In Proc. ACM SIGMOD Int. Conf. on Management of Data, pages 322-331, 1990.
7. Brinkhof, T., Kriegel, H., and Seeger, B., Efficient processing of spatial joins using R-trees. In Proc. ACM SIGMOD Int. Conf. on Management of Data, pages 237-246, 1994.

8. Gustavo Alonso, Fabio Casati, Web Services and Service-Oriented Architectures, ICDE 2005, p. 1147.
9. H. Samet, The Design and Analysis of Spatial Data Structures, Addison-Wesley, 1989.
10. The X3D Specification. www.web3d.org/x3d/specifications
11. Kothuri, R., Ravada, S., Efficient processing of large spatial queries using interior approximations, Symposium on Spatio-Temporal Databases, SSTD, 2001.
12. Schneider, P. and Eberly, D., Geometric tools for computer graphics, MorganKaufmann Publishers, 2003.
13. Van den Bergen, G., Collision detection in interactive 3D environments, Morgan-Kaufmann, 2004.
14. Hans Plum. "3D City Models for Berlin, Bonn and Hamburg based on Open Source Software and Open Standards", Free and Open Source Software for Geo-Spatial, Victoria, Canada, 2007.
15. Remondino Fabio. "From Point Cloud to Surface: The Modeling and Visualization problem", International Archives of Photogrammetry, Remote Sensing and Spatial Information Sciences, Vol. XXXIV-5/W10. International Workshop on Visualization and Animation of Reality-based 3D Models, 24-28 February 2003, Tarasp-Vulpera, Switzerland.
16. Ravi Kothuri, Albert Godfrind, Euro Beinat. "Pro Oracle Spatial 11*g*", Apress, 2nd Edition, Nov 2007.
17. David Mount. "On Geometric Intersection". *The Handbook of Discrete and Computational Geometry*, 2nd Edition, eds. J. E. Goodman and J. O'Rourke, Chapman & Hall/CRC, Boca Raton, pages 857-876, 2004.
18. M. Pellegrini, Ray shooting and lines in space, Handbook of discrete and computational geometry, ed. Goodman, O'Rourke, 2nd ed. 2004, pages 239-256.
19. Oracle Database 11*g* Release 1 Documentation. http://www.oracle.com/technology/documentation/database.html
20. Baris Kazar, Ravi Kothuri, Siva Ravada, On Valid and Invalid Three-Dimensional Geometries, 2nd International Workshop on 3D Geo-Information, Netherlands, pages 19-46, 2007.
21. Martin Isenburg, Yuanxin Liu, Jonathan Shewchuk, Jack Snoeyink, Streaming Computation of Delaunay Triangulations, Proceedings of SIGGRAPH'06, pages 1049-1056, July 2006.

Chapter 11
Making Interoperability Persistent: A 3D Geo Database Based on CityGML

Alexandra Stadler, Claus Nagel, Gerhard König and Thomas H. Kolbe

Abstract. Virtual 3D city models are becoming increasingly complex with respect to their spatial and thematic structures. CityGML is an OGC standard to represent and exchange city models in an interoperable way. As CityGML datasets may become very large and may contain deeply structured objects, the efficient storage and input/output of CityGML data requires both carefully optimized database schemas and data access tools. In this paper a 3D geo database for CityGML is presented. It is shown how the CityGML application schema is mapped to a relational schema in an optimized way. Then, a concept for the parallelized handling of (City)GML files using multithreading and the implementation of an import and export tool is explained in detail. Finally, the results from a first performance evaluation are given.

11.1 Introduction

Like many cities in Germany, Berlin is currently establishing a virtual 3D city model. More and more applications require additional height information and object structuring in vertical direction – just think of urban and landscape modelling, architectural design, tourism, 3D cadastre, environmental simulations, radio network planning, disaster management, or navigation. In order to assess the comprehensive 3D geoinformation for a city like Berlin, an appropriate data management component has to be built. Here, data may be collected, compared, adapted, updated, and exchanged. The data is used for urban studies, planning variants, calculation of intervisibility, impacts of vegetation alterations, and semantic data explorations. A necessary precondition is the existence of a standardised data model, ensuring consistent and interoperable data structuring.

CityGML [9] is an international standard for the representation and exchange of 3D city and landscape models issued by the Open Geospatial Consortium (OGC). The

Technische Universität Berlin, Institute for Geodesy and Geoinformation Science,
{stadler, nagel, kolbe}@igg.tu-berlin.de, gerhard.koenig@tu-berlin.de

common information model behind CityGML defines classes and relations for the most relevant topographic objects in cities and regional models with respect to their geometrical, topological, semantic and appearance properties. By covering thematic information and structures, CityGML complements 3D graphics formats like KML and X3D/VRML. CityGML is implemented as an application schema for the Geography Markup Language (GML) 3.1.1 [3] of the OGC and the ISO TC211.

Based on CityGML, a 3D geo database for the official Berlin 3D city model was established. The main development objective was to achieve both the efficient storage and fast processing of CityGML. For this reason, the CityGML data model was mapped to a compact relational database schema. Moreover, an import/export tool was realised to facilitate the high-performance processing of CityGML and GML structures. Both aspects are considered integral parts of the 3D geo data management in Berlin. In this paper, we therefore address the modelling and database design of the Berlin 3D geo database as well as the software engineering aspects of the import/export tool.

11.2 A 3D geo database for Berlin

The 3D geo database has been realised as an Oracle 10G R2 Spatial relational database schema. In the first project phase, the Institute of Geodesy and Geoinformation, University of Bonn, built a first version that was confined to a subset of CityGML [1]. In the second phase, we now have redesigned the existing database schema to fully comply with CityGML 1.0.0. For this upgrade, additional support of interior building structures, the new appearance model and the full set of CityGML's thematic modules is provided.

In detail, the database implements the following key features of CityGML:

- **Complex thematic modelling**
  The description of thematic features includes attributes, relations, and nested aggregation hierarchies (part-whole-relations) between features. Since on the spatial level geometry objects are assigned to features, both a semantic and a geometrical aggregation hierarchy can be employed. The rich semantic information can be used for thematic queries, analyses, or simulations.
- **Five different Levels of Detail (LODs)**
  Following the idea of multi-representation, every geo object (including DTMs and aerial photographs) can be stored in five different LODs. With increasing LOD, objects not only obtain a more precise and finer geometry, but do also gain in thematic refinement.
- **Appearance data**
  In addition to semantic information and geometry, features can have appearances, i.e., information about the observable properties of a feature's surface. Appearances can represent textures and materials, but are not restricted to visual properties. In fact, appearances can transport any surface based theme, such as infrared radiation, noise pollution, etc.
- **Complex digital terrain models (DTMs)**
  DTMs may be represented in four different ways in the 3D geo database: by regular grids, triangulated irregular networks (TINs), 3D mass points and 3D break

lines. For each LOD a complex relief can be aggregated from any number of DTM components of different types. For example, 3D mass points and break lines can be used together to form complex terrain models.

- **Representation of generic and prototypical 3D objects**
  Prototypical objects are used for memory-efficient management of objects that frequently occur in the city model at different locations, e.g., pieces of street furniture like lanterns, road signs or benches. Each instance of a prototypical object may refer to a particular prototype object for each LOD.
- **Free, also recursive aggregation of geo objects**
  Geo objects can be aggregated to groups according to user-defined criteria, e.g., to model a building complex consisting of individual building objects. The groups themselves represent geo objects which can be assigned names and additional classifying attributes. Groups again may contain other groups as members, resulting in aggregation hierarchies of arbitrary depth.
- **Flexible 3D geometries**
  The geometry of 3D objects can be represented through the combination of surfaces and solids as well as any, also recursive, aggregation of these elements.

The previous version of the database added two aspects to the underlying information model which exceed the capabilities of CityGML [1]. These aspects have been retained in the upgraded database:

- **Management of large aerial photographs**
  The database can handle aerial photographs of arbitrary size. Using Oracle 10G R2 Spatial GeoRaster functionality, tiled, homogeneous photographs can be aggregated to one single image.
- **Version and history management**
  The version and history management employs Oracle's Workspace Manager. It is largely transparent to applications that work with the database. For administration of planning areas and embodied planning alternatives, the tool "PlanningManager" was implemented and added to the 3D geo database. Furthermore, procedures saved within the database (Stored Procedures) are provided, which allow for comfortable management of planning alternatives or versions [10].

The following sections explain the different steps of the database development. Important design decisions are pointed out. The three main steps are marked as (a), (b), and (c) in Fig. 11.1:

(a) **Simplification of CityGML's data model** (section 11.3)
In order to achieve a more compact database schema and improve query performance, the complex data model was simplified at some critical points.

(b) **Derivation of the relational database schema** (section 11.4)
The simplified object-oriented data model has been mapped to relational tables. The number of tables was optimized in order to minimize the number of joins for typical queries.

(c) **Creation of an import and export tool** (section 11.5)
The database administration tool allows processing of very large CityGML instance documents (>> 4 GB). Multiprocessor systems or multi-core CPUs are leveraged through a multithreaded architecture.

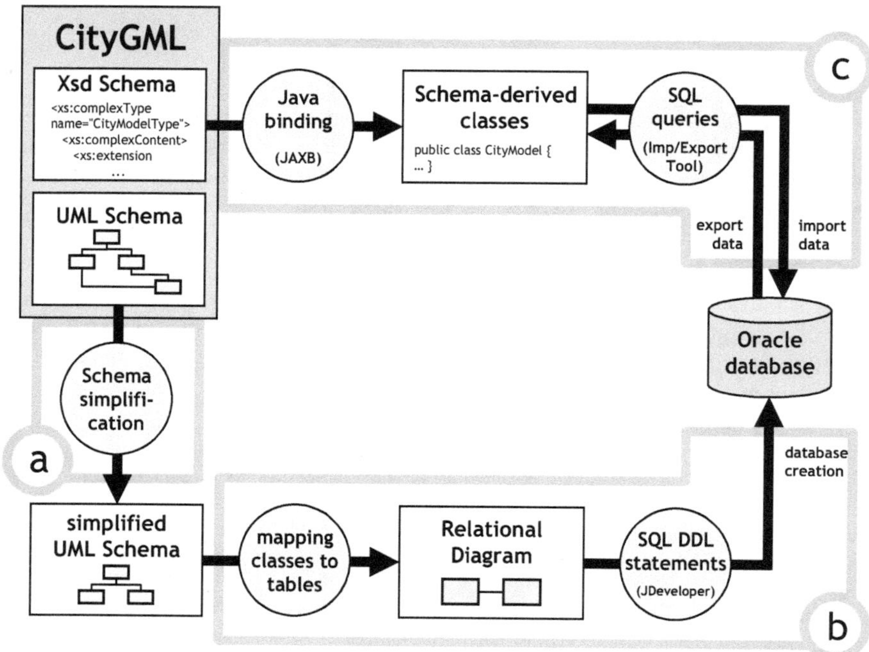

Fig. 11.1. Tasks in the development of Berlin's city geo database:
(a) Simplification of CityGML's data model,
(b) Derivation of the relational database schema,
(c) Creation of an import and export tool for CityGML instance documents.

11.3 Simplification of CityGML's data model

In order to cope with arbitrary CityGML datasets, the 3D geo database has to cover all modelling aspects of CityGML 1.0.0 [9]. However, CityGML uses concepts which are seen as quite complex by some users and software implementors. Complexity is found on different levels:

- CityGML comprises **different thematic models** to represent the topographic objects in cities and regional models. These models are different in structure, e.g., regarding relations and aggregation hierarchies between features, and spatial properties of features. Consequently, there is no simple 'mechanic' way to create a standardized relational model adhering to each thematic component.
- On the **level of single thematic model components** we have to face different relations between objects, allowing for modelling in various levels of complexity. First, the thematic model of CityGML is based on hierarchical decomposition of geo objects. E.g., a building may be decomposed into building parts, rooms, walls, windows, and doors, etc. Second, some cyclic relations exist, resulting in nested

object structures of arbitrary depth. E.g., building parts may be recursively decomposed into nested building parts. Another prominent example is the "generalizesTo" relation for CityObjects which maintains links between corresponding objects in different LOD. Third, multifaceted aggregations are included between thematic classes and their spatial properties. For one object, several geometric representations may be provided simultaneously.

- On the **geometry level**, hierarchies comparable to those on the thematic side can be found. CityGML supports a subset of GML's geometry model comprising mainly polygonal geometries: Surfaces, CompositeSurfaces, MultiSurfaces, Solids etc., all of which may be assigned appearance information. This subset facilitates various ways of composing spatial entities, including nesting of arbitrary depth.

The ability to represent structured urban information is one of CityGML's key features. However, when it comes to data collection at a large scale (e.g., for a big city), some simplifications should be considered to enhance performance of database access. For Berlin, a simplified data model has been created and then used as basis for the derivation of the actual database structure. In the following, modifications are listed and discussed in detail:

- **Simplified treatment of recursions**
  Recursive database queries are very time consuming, especially if the recursion depth is unknown. In order to guarantee high performance, each database object stores both its direct parent element and its root element. The root element is essential for typical high level queries. E.g., all parts of a building can be obtained by simply accessing those elements storing the building's ID as root ID. This operation can be executed on thematic as well as geometry side. The explicit storage of parent elements allows for reconstruction of the whole recursion tree without information loss.
- **Alternative design of GML geometry classes**
  In CityGML, spatial properties of features are represented by objects of GML3's geometry model, consisting of primitives, which may be combined to form complexes, composite geometries, and aggregates. The usage for CityGML is restricted to a subset of the GML3 geometry package, dealing with the representation of polygonal geometries. This may be Points, LineStrings, Polygons, Solids and all valid geometry collections such as CompositeSurfaces or MultiSolids. For topology and appearance information, CityGML requires identification of geometry parts, even if contained in geometry collections. Spatial databases usually provide data types for the aforementioned geometry types. In fact, only those data types enable spatial queries within these databases. Unfortunately, database implementations of geometry collections such as MultiSurface do not allow for naming and referencing of the internal geometry parts. This hinders their use for CityGML.
  The solution is a simplified geometry model for 2D and 3D geometries as shown in Fig. 11.2.

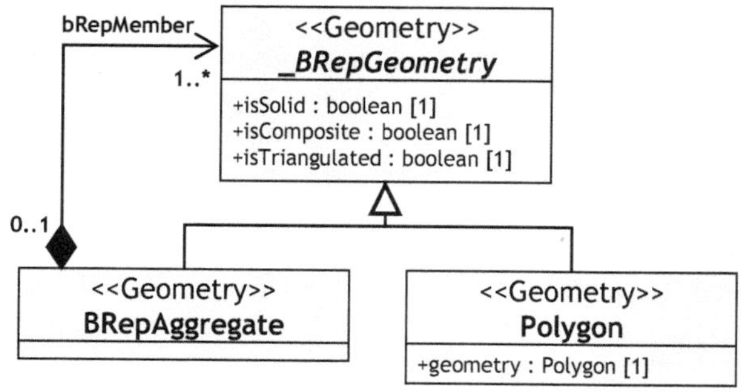

Fig. 11.2. Simplified modelling of polygon-based GML geometry classes.

The abstract root class is called *_BRepGeometry* and acts as basis for all surface-based geometry objects within the city database. It specialises to two concrete classes Polygon and BRepAggregate which hold explicit coordinates or information on aggregation hierarchy respectively. All surfaces are split into individual polygons such that each is identifiable. Polygon uses a native spatial data type (Oracle SDO_GEOMETRY of type Polygon) to enable spatial queries. Polygons are then aggregated using BRepAggregate, again with each being identifiable. Various flags denote the type of aggregation: is Triangulated denotes a TriangulatedSurface, is Solid distinguishes between surface and solid, and is Composite defines whether this is an aggregate (e.g., MultiSolid) or a composition (e.g., CompositeSolid). Geometric aggregates are arbitrary aggregations of geometry elements which are not assumed to fulfill specific topological constraints. In contrast, composite geometries represent a set of topologically connected primitive geometric objects of same dimension, whose interiors are disjoint. The geometric composite itself must be isomorphic to a primitive geometric object [11]. The recursive relation bRepMember enables nesting of geometry collections.

- **Project specific classes and class attributes**
  The city database of Berlin must also store orthophotos, metadata, and version controls. Since in CityGML this information is not represented, appropriate classes and class attributes have been added.
- **Data type customisation**
  Some of the data types specified in CityGML were substituted by simpler ones to enable a more efficient representation within the database, e.g. code lists and colour vectors were replaced by strings and non-polygonal GML geometry types were mapped to their Oracle-specific data type SDO_GEOMETRY.
- **Reducing multiplicities of class attributes**
  Simple attributes with an unbounded number of occurrences (*) are represented in the database by either an array containing a predefined maximum number of elements or by a data type permitting storage of arbitrary values in one object (e.g., several values may be represented by using the data type String with a predefined separator to detach elements from each other). Only then such attributes can be included into a single database column.
- **Cardinalities and types of relations**
  To represent relations of cardinality n:m in a database, an additional table is needed, which contains pairs of linked object IDs. With the simplification to 1:n or

n:1 relations this additional table can be avoided. Therefore, all relations defined in CityGML were tested for more restrictive interpretation. One result was changing the aggregation between rooms and furniture to a composition, since furniture is always bound to a specific room.

11.4 Derivation of the relational database schema

GML is inherently object oriented. The GML application schema of CityGML contains specialization and aggregation hierarchies, nested objects, and complex attributes. In order to map these structures to a relational schema different rules have been proposed and discussed in the past. For example, the mapping of specialisation hierarchies onto database tables may follow one of three different mapping methods [13]. Fig. 11.3 illustrates these methods for the mapping of our simplified geometry representation, introduced in Fig. 11.2 (the bRepMember relation is omitted here).

In order to create space-efficient representations, systems that are capable of automatically deriving relational schemas from GML application schemas typically employ variants (c) or (b). For example, the two commercial systems GO Loader [8, 12] from Snowflake Software and SupportGIS [4] from CPA Geo-Information follow this line.

For the Berlin 3D geodatabase the mapping was carried out manually allowing careful optimisations. For each CityGML class the mapping alternative was chosen individually. The criteria were the expected number of tuples within the tables, the number of joins to reconstruct the CityGML objects, and the overall complexity of the most important queries (with respect to the CityGML object structure). Emgård and Zlatanova [6] provide a more detailed discussion about the mapping of a 3D information model to a relational database. They outline and evaluate two alternative mappings that are optimised for spatial or semantic queries respectively.

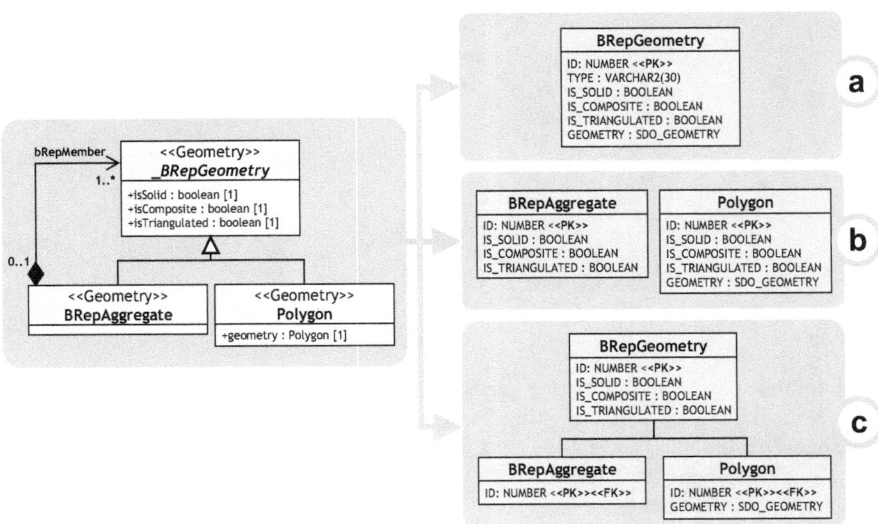

Fig. 11.3. Mapping a class hierarchy to database tables: (a) mapping all classes to a single table; (b) mapping non-abstract classes to their own tables; (c) mapping each class to a single table.

Central CityGML classes were identified that require individual tables. They include CityObject, CityModel, SurfaceData, and major thematic classes such as PlantCover or AbstractBuilding. Remaining classes have been merged into these tables either due to similar structure (e.g., BuildingInstallation and IntBuildingInstallation) or due to inheritance (e.g., Road, Track, Railway, and Square into TransportationComplex). See [9] for CityGML class diagrams.

The overall root class Feature has no associated table. Instead, its attributes GMLID, GML_NAME, and NAME_CODESPACE have been moved to direct subclasses of CityObject. In addition, GMLID has been equipped with an attribute GMLID_CODESPACE. It contains the full path of the originating CityGML instance document or a user-defined value. The codespace is required since GMLIDs are to be unique in single instance documents only. Only the combination of ID and codespace ensures uniqueness within the database. For the class CityObject, the attributes GML_NAME and NAME_CODESPACE have been pushed down an additional level along the inheritance hierarchy. This is due to typical database queries, such as searching for a specific building named "Brandenburger Tor". Storing the attributes directly in table CityObject would require unnecessary table joins to identify affected buildings.

Finally, the attribute CLASS_ID has been added to CityObject. It helps implementing queries starting in table CityObject, such as by GMLID, by spatial extent (e.g., a bounding box), or through meta information. To facilitate further database scanning, the attribute CLASS_ID provides information on the class affiliation of the identified entries in CityObject and enables direct access to relevant tables.

The implementation of geometry in the 3D geodatabase (Fig. 11.4) follows the first approach (Fig. 11.3(a)). By avoiding multiple tables the number of joins is reduced, resulting in higher database performance. As explained in section 11.3 (Simplified treatment of recursions), the bRepMember relation (Fig. 11.2) was reduced to two attributes PARENT_ID and ROOT_ID. Further explanations can be found in the database documentation [2].

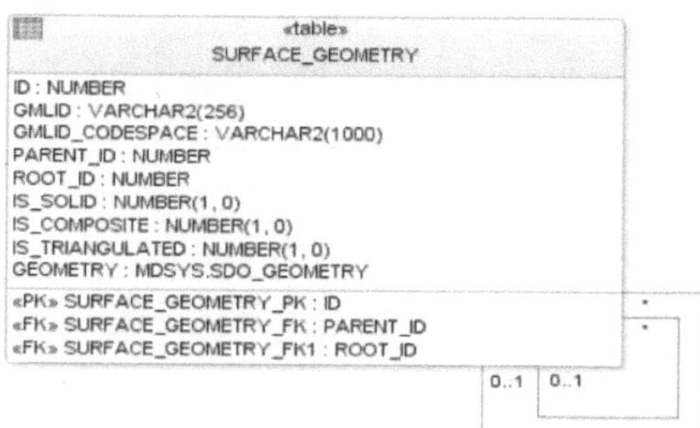

Fig. 11.4. Implementation of 3D BRep geometry in the 3D geo database. This table stores everything from single Polygons over MultiSurfaces, CompositeSurfaces, TriangulatedSurfaces, up to Solids, MultiSolids, and CompositeSolids. By only using Oracle's SDO_GEOMETRY type for single polygons, every polygon is a tuple with an ID value and may be referenced e.g. from appearance information.

11.5 Creation of an import and export tool

In this section, we introduce a software tool for the import and export of CityGML datasets for the 3D geo database. In addition to the relational geo database schema, the tool is considered an integral part of an overall solution for the efficient processing, storage, and retrieval of 3D city models.

In this context, the processing of CityGML and GML structures has to face three main challenges:

1. GML provides a powerful and expressive data model and XML is verbose. This results in voluminous encodings of spatial data. Thus, instance documents quickly grow in file size and may exceed main memory limits (e.g., one million buildings in LOD1 require about 7GB of disk space).
2. Objects may be arbitrarily nested leading to complex data structures which have to be efficiently parsed and mapped to according database tables.
3. Property elements may reference their value using XLinks pointing to remote objects within the same or an external document. These Xlinks have to be resolved in order to correctly represent objects within the geo database.

The import/export tool addresses these challenges by employing strategies for the handling of CityGML instance documents of arbitrary file size and the resolving of XLinks. Furthermore, high-performance data processing is achieved based on a multi-threaded software architecture. Fig. 11.5 gives an overview of the import and export tool's implementation. It is composed of three parts. In the middle, the process of binding the underlying CityGML XML schema definition to a Java object model is illustrated, further explained in section 11.5.1. The top and bottom parts show the process of data import and export, addressed in sections 11.5.2 and 11.5.3.

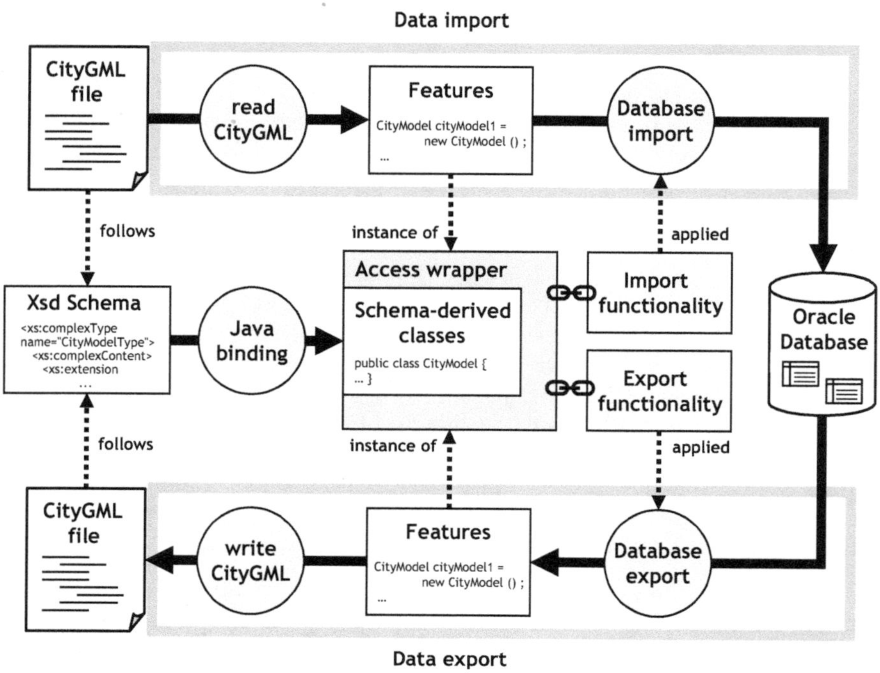

Fig. 11.5. Import and export tool: Overview of the overall data flow.

Special attention is directed to the management of object IDs, because only a systematic and careful usage of a worldwide unique identifier guarantees consistent updating of geo objects stored in the database. As GMLIDs are optional and unique only within one dataset, all imported objects are either assigned a new worldwide unique identifier (UUIDs) or only objects with missing GMLIDs are assigned such UUIDs. In both cases, the attribute GMLID_CODESPACE described in section 11.4 is added to every object (including surface-based geometries).

11.5.1 Software design of the import/export tool

The import/export tool is implemented as a Java application. It employs a chunk processing strategy for XML documents in order to handle instance documents of arbitrary file size. Furthermore, the software design is entirely based on multithreading allowing for high-performance and concurrent data processing.

11.5.1.1 Support for CityGML instance documents of arbitrary file size

The processing of large XML documents is realized through a stream-based XML object binding. This approach utilizes two existing Java frameworks for reading and writing verbose XML files. On the one hand, the Java Architecture for XML Binding (JAXB) is employed to facilitate an object-oriented view of XML data. On the other hand, the Simple API for XML (SAX) is used for the stream-based and event-driven parsing of XML documents. By combining both frameworks, the benefits of each API can be leveraged while limiting their downsides.

The JAXB framework provides a convenient and easy-to-use way to validate and process XML content using Java objects by binding the underlying XML schema definition to a Java object model. It enables storing and retrieving of data in memory in any XML format without the need to implement a specific set of XML loading and saving routines. Moreover, the generated object model captures the structure of XML better than general-purpose APIs like the XML Document Object Model (DOM) or SAX. Like most object-binding APIs, JAXB requires the entire XML stream or file to be present in main memory before data processing can begin. Consequently, memory consumption limits the maximum XML document size.

In contrast, streaming parsers such as SAX, allow for a serial access to XML documents. Each element of the document is passed to the application in sequence of its occurrence via user-defined callback methods. Accordingly, the memory footprint of this event-driven approach is mainly based on the data stored in a single XML element, and thus is always only a small fraction of the size of the entire document. However, stream-based parsing is unidirectional, i.e., previously parsed data cannot be re-read. This hinders a simple and object-oriented handling of complex XML content.

In order to provide both an object-oriented view of XML data and a solution to the data overload problem, the import/export tool implements a two-stage approach: 1) XML documents are parsed using a SAX streaming parser which splits the document into single chunks of manageable size. For each chunk, the occurring SAX events are recorded using a memory buffer. 2) The buffer contents are individually mapped to Java objects for data processing using JAXB. For writing XML documents, the inverse process is applied.

By this means, memory usage is no longer dependent on the document size, but only relates to the amount of data kept in the memory buffer. For CityGML instance documents, reasonable XML chunks are embraced by <cityObjectMember> tags which represent the top-level features within a CityGML model.

In anticipation of future changes, a thin access wrapper is added to the JAXB binding classes. This wrapper abstracts from the underlying binding classes to enable parallel support of CityGML version 0.4.0 and 1.0.0.

11.5.1.2 Concurrency of data processing

The overall architecture of the import/export tool is based on multithreading, i.e., concurrent execution of multiple interacting computational tasks. Generally, each task within the import/export process of CityGML data is carried out by separate threads. The decoupling of compute bound from I/O bound tasks and their parallel non-blocking processing usually leads to an increase of the overall application performance. In a multi-core environment, threads can even be executed simultaneously on different CPUs.

However, a common and simplistic thread-per-task approach faces disadvantages for a large number of active threads such as thread life-cycle overhead and resource thrashing. For this reason, the import/export application reuses threads for multiple but homogenous tasks by applying a pooled threads model. Thread-creation overhead is spread over many tasks, and because a thread already exists within the pool when a task is requested, the delay introduced by thread creation is eliminated.

The thread pool model is implemented as a work queue combined with a fixed group of associated worker threads (Fig. 11.6). The work queue is realized as a blocking queue which is a common pattern in multithreaded programming: one thread produces objects, i.e., single tasks to be performed, and places them on a queue for consumption by other threads, which remove them from the queue. The blocking queue implementation is following the producer-consumer design pattern [7] to avoid deadlocks of the associated worker threads.

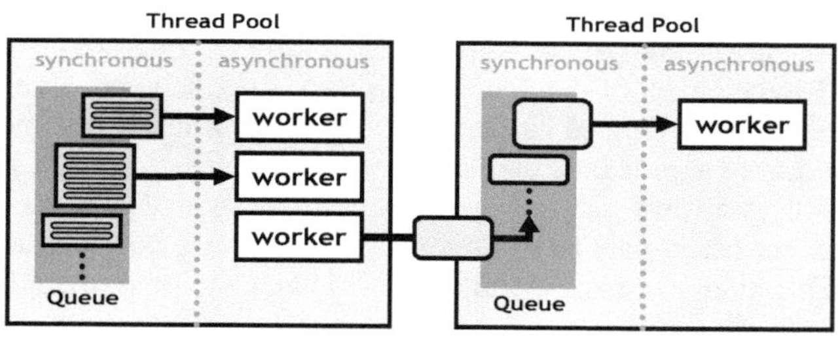

Fig. 11.6. Thread pools are implemented as work queues with fixed groups of associated worker threads.

Single thread pools can be easily combined to entire process chains. The worker threads within a thread pool concurrently process the tasks placed on their associated work queue, and pass their processing results to the work queue of the subsequent thread pool. This allows for the decoupling of process steps in a modular way, allowing for easy extension by further process steps. Inter-process communication between the process steps is realized by an event-dispatcher thread which is implemented as single-worker pool allowing for asynchronous and synchronous message transfer according to the publisher-subscriber design pattern [7].

The optimal number of worker threads within a thread pool mainly depends on the number of processors available, the nature of the task, e.g., I/O bound or compute bound, and the cost of thread context switching. The import/export tool supports user-defined threshold values and a management unit per thread pool for autonomous adaptation based on the overall workload. Furthermore, the queue size is a determining factor for memory consumption, and thus can be adjusted to specific system configurations.

11.5.2 The import process

The process of importing CityGML datasets into the database is illustrated in Fig. 11.7. The workflow is implemented by chaining thread pools which cover single steps of the import process. It comprises thread pools for the chunk-processing of the input XML document (parser thread pool), the conversion of single chunks to JAXB objects representing CityGML top-level features (converter thread pool), and the data processing of these features (importer and XLink thread pools). The associated work queues are also shown in Fig. 11.7.

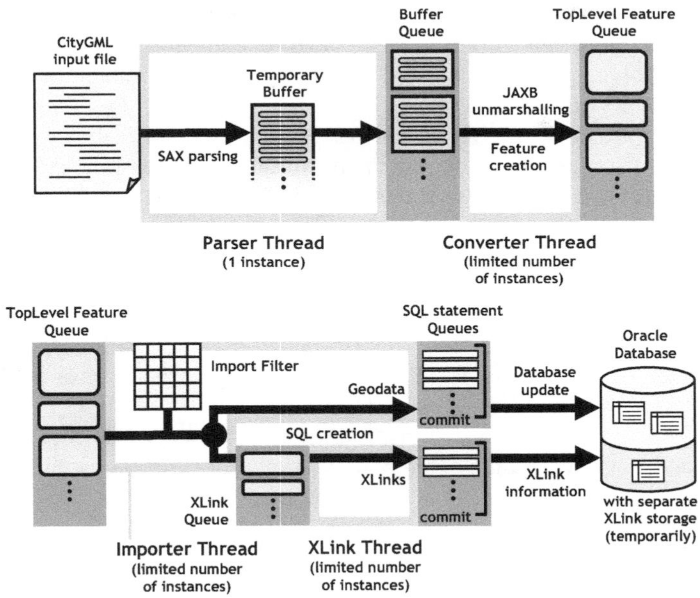

Fig. 11.7. Import Tool (phase 1) – Detailed workflow from input CityGML instance documents to database storage. XLink references are separately stored in temporary tables to enable XLink resolving in phase 2.

The first two steps of the process chain, i.e., chunk processing of the input data and unmarshalling of CityGML top-level features, have already been discussed in section 11.5.1. By analyzing the resulting JAXB objects, the subsequent importer thread shown in Fig. 11.7 derives the final SQL statements using the import functionality, which maps the objects to corresponding tables of the relational database schema (section 11.4). This step can be customized by user-defined import filters. For example, the import may be restricted to a given set of feature types, e.g., only CityGML buildings, or to features located within a geographic bounding box. Filtered objects can be immediately skipped and released from memory.

In order to optimize database response times, SQL statements are precompiled and stored in corresponding Java objects by the importer thread. The precompiled objects can then be used to efficiently execute the same statement multiple times with varying data. Furthermore, SQL statements are collected and only forwarded to the database if a certain amount of statements has been reached (bulk processing). This allows for an efficient batch processing on the database side. The optimal number of statements within a batch depends on the complexity of the imported CityGML dataset, and thus can be customized.

The multithreaded design of the import tool allows for the concurrent execution of all process steps. For example, the parsing of the XML document is never blocked by threads waiting for database response, and multiple JAXB objects can be processed simultaneously. However, the chunk-processing of the input data also introduces a further level of complexity. In CityGML instance documents, properties of features like relations to other features or geometry objects may be denoted using the XLink concept of GML3. Thus, instead of having the XML content inline within the feature element, the content is given by reference to a remote object identified by its GMLID. Since backward references within the same XML document are allowed, a one-stage approach for resolving of XLinks would require the whole document to be present in memory.

To solve this problem, the import tool employs a two-phase strategy. In the first run (Fig. 11.7), features are written to the database neglecting references to remote objects. However, if a feature contains an XLink, an additional entry is written to a temporary table by a separate XLink thread. This entry mainly contains the referencing feature and the referenced GMLID. Furthermore, the import tool keeps track of every previously encountered GMLID and its corresponding object representation in database. In a consecutive run, only these temporary tables are revisited and queried. Since the entire XML document has been processed at this point in time, valid references can be resolved and processed accordingly. The second phase of the import process is depicted in Fig. 11.8.

By means of this two-phase import process, the advantages of chunk-processing can be kept and a correct storage of CityGML instance documents within the database can be ensured.

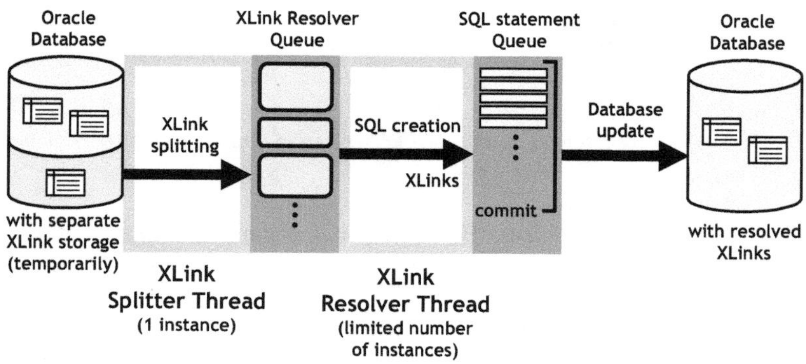

Fig. 11.8. Import Tool (phase 2) – After phase 1 is completed, the information stored in the temporary tables is queried (splitter thread pool) to resolve XLink references to remote objects (resolver thread pool).

11.5.3 The export process

The workflow for exporting CityGML datasets from the database is shown in Fig. 11.9. As for the import, the workflow is realized as a process chain combining several thread pools, each of which covers a separate process step.

The first step of this workflow is to query the database according to user-defined criteria. These export filters are identically defined as the import filters, e.g., allowing for spatial queries of features within a given geographic bounding box. Spatial queries are performed within the database itself using the spatial indexing capabilities of Oracle 10G R2 Spatial. Furthermore, the queries are only executed on a single database table in the first run, which only holds general information for all contained features.

Thus, the queries can be rapidly processed by the database and returned to the export tool.

As soon as the first query results are returned, the splitter thread places them on the work queue of the subsequent exporter thread. By this means, the worker threads of the export pool already start to reconstruct the corresponding CityGML feature objects while the splitter thread still processes the initial database response.

Depending on the type of feature to be reconstructed, the exporter workers have to execute more complex database queries spanning several tables in order to retrieve the necessary feature data. The export functionality provides respective code for each feature type. At this stage, most of the data processing is done by the database server due to efficient use of SQL join statements. Furthermore, if a single worker thread is waiting for database response, it does not block the other threads within the export pool. If all data has been received, it is mapped to corresponding JAXB objects. Also the export tool keeps track of the feature's GMLID in order to avoid duplicate output of objects and to use XLinks instead. The JAXB objects are marshalled to SAX events which are recorded by a temporary memory buffer. The buffered SAX events are placed on the queue of the subsequent thread pool. Afterwards, the export thread continues to work on the next feature.

The last step of the export process, i.e., writing the entire feature data to a CityGML instance document, is carried out by an I/O writer thread. This thread takes the SAX event buffers from its work queue and translates them into XML elements. Again, these file-based processes are decoupled from all other steps of the process chain.

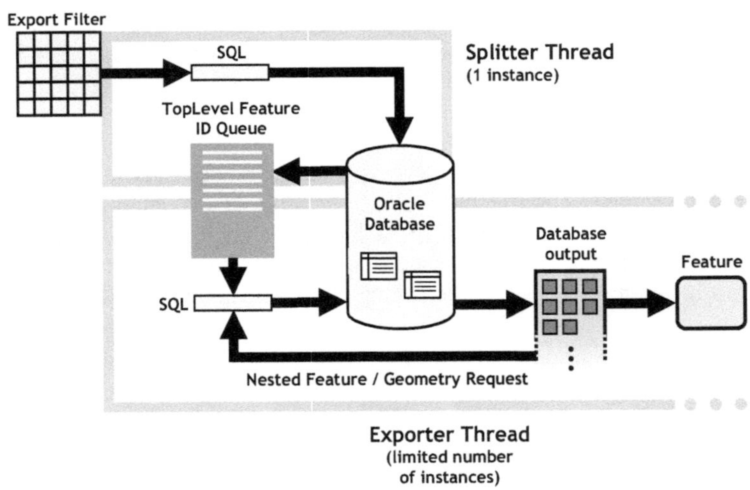

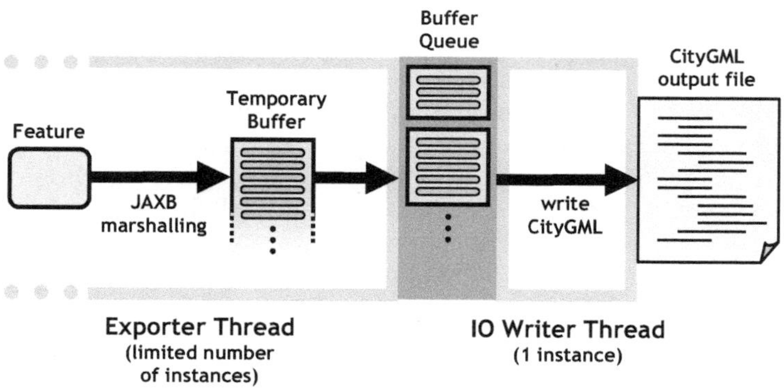

Fig. 11.9. Export Tool – Detailed workflow from database filtering to the export of CityGML instance documents.

11.5.4 Performance

Currently, performance has not been measured thoroughly. However, initial tests exhibit good throughput in the following setting:

- **Database server:** 2 x Intel Xeon Dual Core@3GHz, 6GB RAM, SCSI Raid 5 Disk Array, Windows Server 2003 SP2, 100MBit LAN
- **Client running the import/export tool:** Intel Core2 CPU 6600@2.4GHz, 2GB RAM, WindowsXP SP2, 100MBit LAN

The database server was running a default Oracle 10G R2 Spatial installation without manual tunings or additional settings. All import tests have been performed with spatial indexes disabled. Additionally, indexes on table rows containing VARCHAR2 values (e.g., GMLID, GMLID_CODESPACE) were switched off. Unless stated otherwise, Oracle's table versioning was disabled. The performance measurements for the import of different CityGML instance documents are shown in Table 11.1.

Table 11.1. Performance measurements for the import of CityGML datasets using the developed import tool. Row headers are as follows: [1] File size, [2] Overall import time, [3] Top-level features per second (please note that these may contain nested features which are not counted here), [4] Workers per thread pool, [5] Imported top-level features / geometries / texture images / Xlinks, [6] Versioning using Oracle Workspace Manager

Dataset	**[1]**	**[2]**	**[3]**	**[4]**	**[5]**	**[6]**
1 mio. LOD1 buildings	7.5GB	1:15h	228	4	1055951 / 11511040 / 0 / 0	off
Dataset	**[1]**	**[2]**	**[3]**	**[4]**	**[5]**	**[6]**
1 mio. LOD1 buildings	7.5GB	2:25h	121	1	1055951 / 11511040 / 0 / 0	off
209 complex LOD3 objects	40.1MB	17s	12	4	209 / 73823 / 0 / 0	off

-“-	40.1MB	24s	9	1	209 / 73823 / 0 / 0	off
-“-	40.1MB	3:20h	0.02	4	209 / 73823 / 0 / 0	**on**
10927 fully textured LOD2 buildings	163MB + 57 MB textures	25min	7	4	10927 / 434714 / 43348 / 391006	off
-“-	163MB + 57 MB textures	42min	4	1	10927 / 434714 / 43348 / 391006	off
-“- (w/o textures)	163MB	6:30 min	28	4	10927 / 434714 / 0 / 391006	off

Corresponding export tests are illustrated in Table 11.2.

Table 11.2. Performance measurements for the export of CityGML datasets using the developed export tool. Row headers are as follows: [1] File size, [2] Overall export time, [3] Top-level features per second, [4] Worker per thread pool, [5] Exported top-level features / geometries / texture images / XLinks

	[1]	[2]	[3]	[4]	[5]
1 mio. LOD1 buildings	7.5GB	38min	455	10	1055951 / 11511040 / 0 / 0
-“-	7.5GB	1:57h	150	1	1055951 / 11511040 / 0 / 0
209 various and complex LOD3 objects	40.1MB	7s	30	10	209 / 73823 / 0 / 0
-“-	40.1MB	17s	12	1	209 / 73823 / 0 / 0
10927 fully textured LOD2 buildings	163MB + 57 MB textures	4:20min	42	4	10927 / 434714 / 43348 / 391006
-“- (w/o textures)	163MB + 57 MB textures	2min	91	4	10927 / 434714 / 43348 / 391006

11.6 Conclusions and Future Work

In this paper a relational 3D geo database for the storage of CityGML data was presented. Optimizations on the object model and its mapping to a relational schema were discussed. Furthermore, a concept for multithreaded processing of large (City)GML files was proposed. The first performance evaluations showed that the usage of concurrent worker threads leads to significant speedups.

The 3D geodatabase has been developed in the course of the project “Geo data management for the Berlin government – The official Berlin 3D city model” [5]. By relying on CityGML’s comprehensive data model, members of different communities, such as city planners, architects, surveyors, etc. are provided with a database for semantically rich city models. With the project close-out at the end of this year the entire software will be released as Open Source on [2].

In the future, the database administration tool will be extended by data matching functionality, i.e. buildings in the database that represent the same real world object will be detected and linked automatically. This allows for the exchange of thematic information attached to either of the buildings as well as automated updating procedures. In case of equivalent geometries, one of the objects will be deleted to avoid database inconsistencies.

Finally, there is room for performance optimisations. Important topics not yet covered are logical table and index partitions as well as their optimal distribution on physical volumes.

References

1. 3D city database, version 1, 2006. http://www.3dcitydb.org
2. 3D city database, version 2, 2008. Accessible via http://www.citygml.org
3. Cox, S., Daisey, P., Lake, R., Portele, C., Whiteside, A, 2004. OpenGIS Geography Markup Language (GML) Implementation Specification, Version 3.1.1, OGC Doc. No. OGC 03-105r1, Open Geospatial Consortium
4. CPA Geo-Information, SupportGIS product website, 2008. http://www.supportgis.de
5. Döllner, J., Kolbe, T.H., Liecke, F., Sgouros, T., Teichmann, K., 2006. The Virtual 3D City Model of Berlin - Managing, Integrating, and Communicating Complex Urban Information, In: Proc. of the 25th Urban Data Management Symposium UDMS, Aalborg
6. Emgård, L., Zlatanova, S., 2007. Implementation alternatives for an integrated 3D Information Model, In: Advances in 3D Geoinformation Systems, Springer
7. Gamma, E., Helm, R., Johnson, R.E., 1995. Design Patterns. Elements of Reusable Object-Oriented Software, Addison-Wesley Longman, Amsterdam
8. Snowflake Software, GO Loader product website, 2008. http://www.snowflakesoftware.co.uk/products/goloader/index.htm
9. Gröger, G., Kolbe, T.H., Czerwinski, A., Nagel, C., 2008. OpenGIS City Geography Markup Language (CityGML) Encoding Standard, Version 1.0.0, OGC Doc. No. 08-007r1, Open Geospatial Consortium
10. Gröger, G., Kolbe, T.H., Schmittwilken, J., Stroh, V., Plümer, L., 2005. Integrating versions, history and levels-of-detail within a 3D geodatabase, In: Proc. of Int. Workshop on Next Generation City Models, Bonn, EuroSDR publications
11. Herring, J., 2001. The OpenGIS Abstract Specification, Topic 1: Feature Geometry (ISO 19107 Spatial Schema), Version 5. OGC Document Number 01-101
12. Müller, H., Curtis, E., 2005. Extending 2D interoperability frameworks to 3D, In: Proc. of Int. Workshop on Next Generation City Models, Bonn, EuroSDR publications
13. Rumbaugh, J., Blaha, M., Premerlani, W., Eddy, F., Lorensen, W., 1991: Object-Oriented Modelling and Design, Prentice Hall

Chapter 12
Use of Finite Arithmetic in 3D Spatial Databases

Rodney, James Thompson

Abstract. Most of the spatial theory that underpins geographic information repositories is based on the assumption of Euclidean space, which requires the support of the real number system. The actual implementation of this theory on computer hardware then requires the use of finite precision arithmetic, meaning that some approximations must be made during the processing of certain calculations. This paper considers these approximations, and the effect they have when made at different critical points in the processes.

12.1 Introduction

Most of the spatial theory that underpins geographic information repositories is based on the assumption of Euclidean space, which requires the support of the real number system. The actual implementation of this theory on computer hardware then requires the use of finite precision arithmetic, meaning that some approximations must be made during the processing of certain calculations.

This paper considers these approximations, and the effect they have when made at different critical points in the processes.

12.2 Approaches Taken

There is an inherent and unavoidable loss of accuracy when any measurements of real-world phenomena are taken and stored. This obvious fact is common to all approaches discussed in this paper. The issue being discussed here is the small inaccuracies that accumulate in the database representation of spatial features as those representations are processed and manipulated during the life of the data.

Department of Natural Resources and Water, Queensland, Australia
Rod.Thompson@nrw.qld.gov.au
Delft University of Technology, OTB, GIS Technology, The Netherlands.
rthompson@tudelft.nl

In order to compare the different approaches, a comparison will be made using as examples:

- The validation of a polygon in 3D as a planar object.
- The validation of polyhedra.
- The operation of forming the union of polyhedra.
- A test for equality of volumetric features in space.

In addition, the important topological operations: union, intersection and complement [4], the predicates intersects and overlaps, and the Region Connection Calculus predicates: connected, disconnected, tangential proper part etc. [12] are considered.

12.2.1 Validation of a Polygon as a Planar Object

Any three points may have a plane that passes exactly through them, and if the points are not collinear or coincident, this plane will be uniquely defined. Where four or more points are involved, it may not be possible to find a plane that passes through them. Many 3D objects are represented as having planar faces, where these faces have more than three vertices. It is important in these representations that the faces are truly planar.

This issue can be sidestepped by using a triangulation of the surfaces, where each surface is broken into triangular facets. This does, however lose the definition of the original surfaces, and does not document the intention that they be planar.

12.2.2 Validation of Polyhedra.

In addition to the requirement for planar surfaces, there are other validation requirements being suggested for 3D objects. One of the more important is that the volume be "watertight". That is to say, if the volume is defined in terms of the bounding faces, then the faces must meet exactly along all edges. In practice, given the finite precision of point representations, this means that additional vertices must be inserted in some faces. For example in Fig. 12.1, a point at p has to be inserted into face A, to prevent mismatch of the approximated position with the edge.

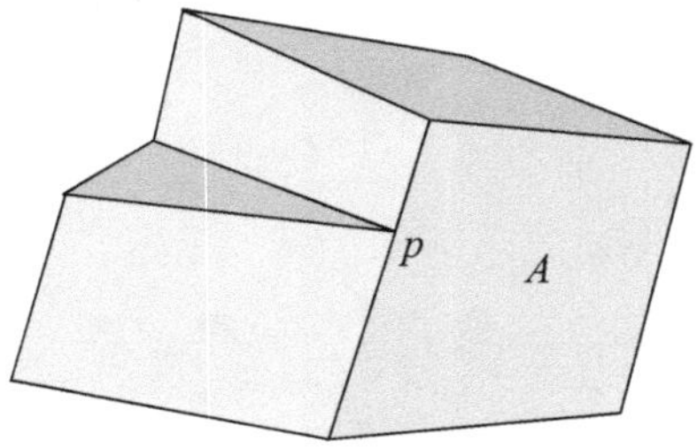

Fig. 12.1. A redundant point in face *A* of polyhedron

Another requirement is that the outer surface should be simple, while the internal volume should be connected. [9] give a number of specific cases of polyhedra which fail this requirement. In the cases depicted in Fig. 12.2, the finite precision calculations may result in their being detected as valid or invalid by different software.

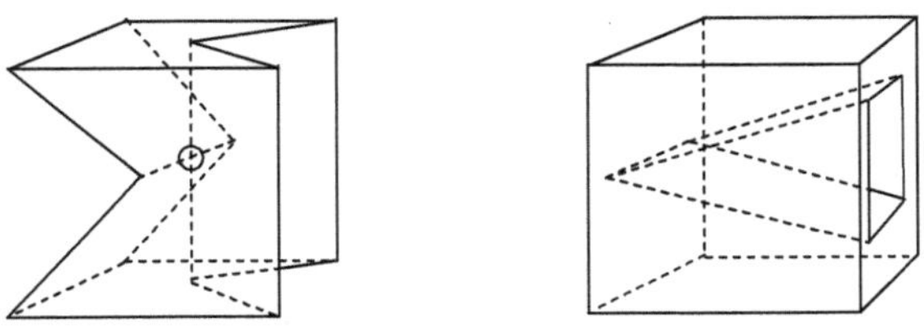

Fig. 12.2. Polyhedra with non-simple boundary surfaces

12.2.3 Union of Polyhedra

The operation of forming the union of two polyhedra can be problematic, when calculated using finite precision arithmetic. For example, if a region is connected to another region, it should be connected to the union of that region with any other. For example, in Fig. 12.3, since *A* is connected to *B*, it should be connected to $B \cup C$, however, small inaccuracies in the calculation of the points of intersection of the boundaries can cause this to break down. For simplicity, this issue is discussed using 2D cases, but applies in any dimensionality.

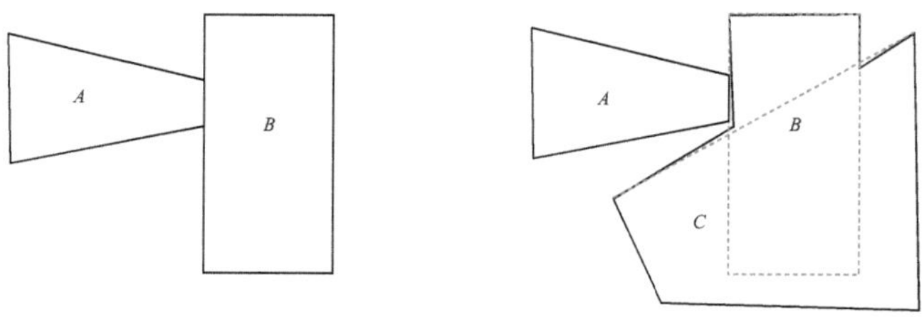

Fig. 12.3. Connectivity is not necessarily conserved in calculations. Here A is strongly connected to B, but due to limited accuracy of calculations (shown exaggerated) is not connected to $B \cup C$.

Further, the calculation of the union or intersection of regions can lead to non-associativity of operations. In the same way that floating point arithmetic is not necessarily associative, it is possible that the result of calculation of the union or intersection of spatial regions may differ depending on the order that calculation is carried out. For example in Fig. 12.3, since A is connected to B, the union $A \cup B$ is a simple, connected polygon. Forming $(A \cup B) \cup C$, the result would be a simple connected polygon. If, however, $B \cup C$ is calculated, small inaccuracies can result in A not being connected to $B \cup C$, so that $A \cup (B \cup C)$ is a multi-polygon.

12.2.4 Equality of Volumetric Objects in Space

The simple test of determining if A is equal to B should be the most basic of operations in any system. It is therefore surprising that it is rarely implemented correctly if at all, and that the ISO 19107 definition of equality of spatial objects would not satisfy the requirements of the Object class in Java.

The ISO 19107 definition of equals() (ISO-TC211 2001) uses the phrase "shall return true if this GM_Object is equal to another GM_Object", but qualifies this definition with: "Since an infinite set of direct positions cannot be tested, the internal implementation of equal must test for equivalence between two, possibly quite different, representations. This test may be limited to the resolution of the coordinate system or the accuracy of the data. Application schemas may define a tolerance that returns true if the two GM_Objects have the same dimension and each direct position in this GM_Object is within a tolerance distance of a direct position in the passed GM_Object and vice versa" (ISO 19107 Section 12.6.2.2.18.3).

There is also a need for a stronger form of equality. As an example, consider the implementation of geometric objects as subclasses of the Object class of Java. This requires the implementation of "equals" and "hashCode" methods [13]. A hash-keyed structure is used where large numbers of objects are to be stored, with the expectation that the hashCode method generates a key value that can be used to provide fast access to any object in the collection. It is acceptable (and unavoidable) for different objects to generate the same key value ("collisions"), with the "equals" method being used to distinguish between them. The requirements of the hashCode method are that for two objects a and b:

```
a.equals(b) ⇒ a.hashCode() = b.hashCode()

a.hashCode() = b.hashCode() "nearly always implies"
a.equals(b)
```

The "nearly always implies" determines how useful the algorithm is. If many unequal objects generate the same hash code, the algorithm is inefficient, leading to multiple collisions.

It is difficult to imagine any hashCode routine that is useable in conjunction with an "equals" test which allows a tolerance.

12.3 Approximate the Results of Every Step

This is the most common approach taken currently. The problem is that at any stage of an operation it is possible for the logic to break down, because the rounding implied in any stage can invalidate the results of earlier stages.

12.3.1 Floating Point Arithmetic

At each step in an operation the arithmetical unit of the computer performs a rounding. This may be an explicit rounding to a selected accuracy, or an implied rounding that is generated by the floating point arithmetic unit. This is the approach taken by many of the systems that operate in floating point.

The IEEE standard for floating point operations IEEE 754 [5, 6] is a partially successful attempt to ensure that a calculation will give consistent results in differing computational contexts. The remaining inconsistencies seem to be in the handling of overflow, underflow and "divide by zero" [8].

On the other hand, the results of certain arithmetic calculations are exactly specified – down to the mandatory rounding to be used, for any computer hardware that advertises IEEE compliance. These include addition, subtraction, division, multiplication, square root and binary to/from decimal conversions. Significantly, they do not include the trigonometric functions (sin, cos etc).

It might be thought that it would be possible to take a similar approach to the calculation of geometric results, but this is out of scope of this paper. The difficulties of this approach are legion, but include the same types of issues that prevent the exact standardisation of trigonometric functions.

One non-technical reason for the reluctance of standardisation bodies to take such an approach is that the details of actual calculations may be commercial in confidence, and particularly well designed algorithms can be valuable distinguishing features of software, giving a competitive commercial advantage.

In summary, it may be assumed that the calculation of a result, such as the point of intersection of three planes, will result in a good approximation to the true point regardless of the platform used, but that the same result cannot be expected on different platforms.

It should be expected, however, that the re-calculation of a function from identical inputs should have exactly the same result, but even this can be jeopardised by "over-enthusiastic" optimisation in some programming environments [8].

Arithmetic operations between floating point numbers are not necessarily associative. That is to say $(a+b)+c$ may not give the same answer as $a+(b+c)$.

For example, $10^{10} + (-10^{10} + 10^{-10})$ gives zero, while $(10^{10} - 10^{10}) + 10^{-10}$ gives 10^{-10}. This can be significant in many calculations, and must be considered where repeatability of results is desired.

12.3.2 Validation of a Polygon as a Planar Object

In general, any polygon in 3D with more than 3 vertices cannot be represented using floating point coordinates such that the vertices are exactly coplanar. If the calculation of the vertex coordinates were carried out using infinite precision arithmetic, and the results rounded according to the IEEE 754 rules, the amount of deviation from planarity could be kept to a predictable (and very small) distance.

If the vertices are the result of calculations which are carried out using floating point arithmetic, the deviation can be more extreme. For example, if a vertex is calculated as the point of intersection of three planes, defined by parametric equations:

$$\begin{aligned} a_1x + b_1y + c_1z + d_1 &= 0 \\ a_2x + b_2y + c_2z + d_2 &= 0 \\ a_3x + b_3y + c_3z + d_3 &= 0 \end{aligned} \quad (1)$$

then the point of intersection $p = (x, y, z)$ can be calculated as $x = p_x/q$, $y = p_y/q$, $z = p_z/q$, where:

$$q = \begin{vmatrix} a_1 & a_2 & a_3 \\ b_1 & b_2 & b_3 \\ c_1 & c_2 & c_3 \end{vmatrix} \quad (2)$$

$$p_x = \begin{vmatrix} -d_1 & -d_2 & -d_3 \\ b_1 & b_2 & b_3 \\ c_1 & c_2 & c_3 \end{vmatrix}, p_y = \begin{vmatrix} a_1 & a_2 & a_3 \\ -d_1 & -d_2 & -d_3 \\ c_1 & c_2 & c_3 \end{vmatrix}, p_z = \begin{vmatrix} a_1 & a_2 & a_3 \\ b_1 & b_2 & b_3 \\ -d_1 & -d_2 & -d_3 \end{vmatrix} \quad (3)$$

The calculations of these intermediate results require the addition of six terms, each of which are the product of three variables. Thus, the final accuracy of x, y, and z will be dependent on the initial values. This can be reduced by normalising the parameters in various ways (e.g. by replacing d_1 by $d_1 / \sqrt{a_1^2 + b_1^2 + c_1^2}$, a_1 by $a_1 / \sqrt{a_1^2 + b_1^2 + c_1^2}$ etc.. thus converting to Hessian normal form) [16], but there will still be a deviation of the vertices from the strictly correct position. The non-associativity of floating point arithmetic means that the results may not be repeatable on different platforms.

Typically, the approach taken to the issue of planarity of a polygon is to state that a polygon is considered planar if all points are within some tolerance of the plane.

This is not, however a useful form of statement for standardisation. It is necessary, in deciding that all points are sufficiently close to a plane to determine the plane first. It is quite possible for such a plane to exist and be known to the sender, but if the parameters of the plane are not passed to the receiving system, there is no guarantee that it can be found. (The sender may have used a more sophisticated form of interpolation than was available or known to the recipient).

Fig. 12.4. Attempting to find a plane close to vertices

For example, in Fig. 12.4, the vertices marked as x's are all within a reasonable tolerance distance of the plane (drawn from the side as a solid line). However, the plane shown as a dashed line, the result of a least squares fitting does not pass sufficiently close to all points. The plane generated, even by the same algorithm, cannot be guaranteed to be the same on all platforms. It is not common practice for the parametric forms of planar surfaces to be included in interchange specifications.

12.3.3 Validation of Polyhedra

In order to ensure that this validation is computable, it is common practice when making the initial data load approximations, to apply a normalisation[1] in the form suggested by [11]. This specifies a tolerance ε, such that the distance between a point and a line can be calculated with accuracy $\frac{\varepsilon}{10}$. Note that the polyhedra in Fig. 12.2 highlight the need in 3D for additional rules to those of Milenkovic, that no two lines may be closer than ε except at their endpoints, no line may be nearer than ε to a plane, etc.

Given the complexity of the calculation of the minimum distance between two lines, and the non-associativity of floating point arithmetic, even the determination of Milenkovic normalisation cannot be used as a validity criterion, unless we are willing to specify the exact calculation to be used, and the order of individual operations

[1] This is usual when building a topologically-encoded data structure.

within that calculation. It is not even clear whether the order of the two lines, or the direction of the two lines affect the calculation (e.g. for points *a,b,c,d* does the distance calculated between line *ab* and line *cd* equal the distance calculated between *cd* and *ab*, and between *ab* and *dc* etc).

That is to say that even if an object is valid, and Milenkovic normal[2] to tolerance ε (including the additional criterion), there is no guarantee that it is Milenkovic normal as determined by other software.

12.3.4 Union of Polyhedra

Since the calculation of the intersections of lines and surfaces is of limited precision, rounding effects will apply as described in section 12.2.3. Thus it cannot be assumed that many obvious results of spatial theory can apply. For example, using the terminology C(*A*, *B*) to mean *A* is connected to *B*:

C(*A*, *B*) does not imply C(*A*, $B \cup C$).

C(*A*, $B \cap C$) does not imply C(*A*, *B*).

$A \cup (B \cup C)$ is not necessarily equal to $(A \cup B) \cup C$.

$A \cap (B \cap C)$ is not necessarily equal to $(A \cap B) \cap C$.

And so on. In such a system, it is difficult to create a toolkit of operations and predicates that can be relied on in all cases.

The situation in topologically-encoded data stores is different. Once a complete polygon coverage of space has been built, the topological and connectivity operations between polygons are clearly rigorous and repeatable. On the other hand, the later addition of polygons to an existing coverage is not so clear, and it is possible that the order of adding these polygons may be significant.

12.3.5 Equality of Volumetric Objects in Space

The implementation of the ISO 19107 definition of equality in a repeatable and standard form is highly problematic. Note that the standard does not specify the tolerance, but further, it does not specify how the tolerance is to be applied.

Determination of an algorithm is also problematic, since the representations of two objects may, indeed be different, while they are equal by this definition. For example, in Fig. 12.5, two polyhedra are depicted. Due to small differences in positioning, the polyhedron on the left is topologically homeomorphic to a toroid, while the polyhedron on the right is simply connected. Under the ISO 19107 definition, if the gaps are sufficiently small, the two polyhedra can be equal.

[2] Note also that the “winding number” defined by Milenkovic needs to be extended to 3D in the obvious way.

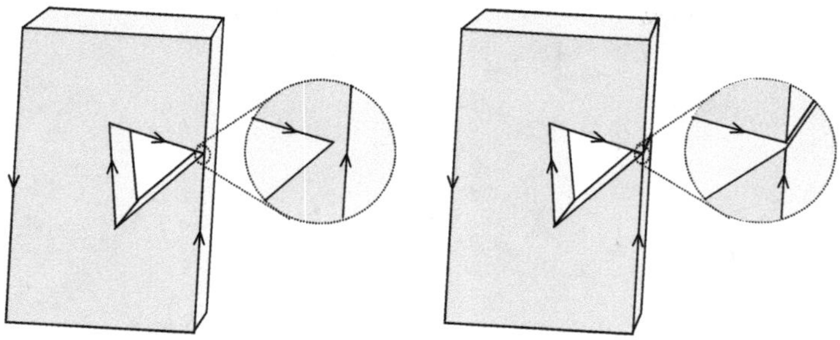

Fig. 12.5. Equal polyhedra of differing forms. The object on the left has a hole right through it. The one on the right has a simple interior.

In a similar way, a single polyhedron can be equal to a disconnected set of polyhedra. Thus, in a system that does not permit multiple geometries, a feature may be valid, but equal to a feature which is invalid.

12.4 Delay Rounding Until End of Operation

Again, when data is loaded, there is an implied or explicit rounding/truncation.

In this approach, the rounding at every step in the process is delayed until the whole operation is completed. For example, in the case depicted in Fig. 12.3, only when the full calculation of $(A \cup B) \cup C$ or $A \cup (B \cup C)$ is completed is any rounding permitted.

This can be achieved in one of two ways:

- Using rational number arithmetic (unlimited precision) during the calculation, and then rounding back to integer or floating point coordinates at the completion of the calculation, or
- Using floating point or integer arithmetic, with "lazy evaluation" of point positions.

The use of infinite precision rational arithmetic will be discussed in Section 12.5, while the lazy evaluation approach will be discussed here. The essence of this approach is that any calculation which involves rounding is delayed as long as possible. Alternatively, a result is calculated in rounded form, but the original data are retained so that subsequent calculations can be carried out with the unrounded values.

Returning to the example of Fig. 12.3 (see also Fig. 12.6), in calculating $B \cup C$, the point t may be calculated in approximate form, but the original line pq is retained, so that when the union of A with $(B \cup C)$ is calculated it can be used to calculate points r and s (in approximate form). Using this approach, it makes no difference to the final result whether the union is formed as $A \cup (B \cup C)$ or $(A \cup B) \cup C$.

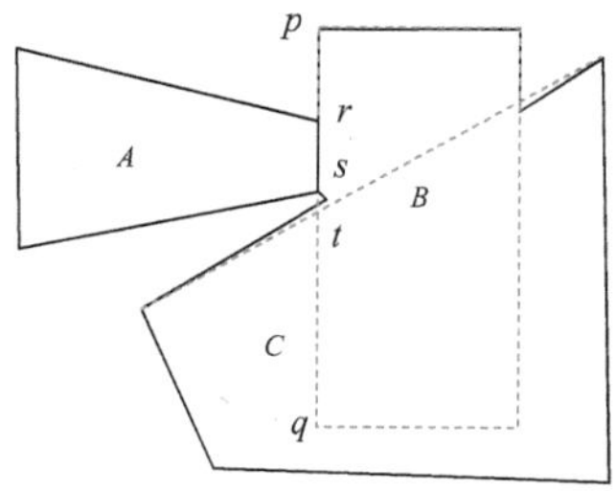

Fig. 12.6. Forming the union of regions using lazy evaluation of point positions

This is only true for as long as the original points are retained, and if the operation is carried out as two steps, the result will vary – e.g. if a new region is calculated as $D = (B \cup C)$, then $A \cup D$ may not equal $(A \cup B) \cup C$.

The essence of this approach is that the imprecision can be seen as applying only at the end of each operation, while the operations themselves are rigorous. Thus it can be modelled as if the output of the calculation is then applied to an imprecision function. Thus in the example above it is clear why APPROX(APPROX($A \cup B$) $\cup$ C) is not necessarily equal to APPROX($A \cup$ APPROX($B \cup C$)).

12.5 Round on Data Load Only

In this approach, rounding is avoided as far as possible, apart from the rounding implied at the time of data load. Typically, a rational number representation is used, or the points are represented by homogenous coordinates. In this discussion, homogenous integer coordinates are used. Rational coordinates have a similar behaviour.

Some rounding may be necessary, for example in datum shifts or point positional adjustments, and the geometry validity may be compromised in this situation.

12.5.1 Homogeneous Coordinates

In section 12.3.1., the formula for the point of intersection of three planes is given in terms of p_x, p_y, p_z and q, with the point being represented as $p = (x, y, z)$, where $x = p_x/q$, $y = p_y/q$, $z = p_z/q$. An alternate storage for rational points is the homogeneous coordinate form, as defined in the discipline of Projective Geometry [3]. Here the q is retained, and the point (in 3D) is represented as a quadruple of numbers $p = (p_x, p_y, p_z, q)$. The advantage is that integers can be used throughout, since in calculations such as determining the plane through three points, or the intersection of three planes, no division need be done. (In the discussion below, capital letters will be used for integers, lowercase letters for rational, floating point or real numbers).

The formula for the plane through three points (X_1,Y_1,Z_1,Q_1), (X_2,Y_2,Z_2,Q_2), (X_3,Y_3,Z_3,Q_3) becomes:

$$\begin{vmatrix} X & Y & Z & Q \\ X_1 & Y_1 & Z_1 & Q_1 \\ X_2 & Y_2 & Z_2 & Q_2 \\ X_3 & Y_3 & Z_3 & Q_3 \end{vmatrix} = 0 \tag{4}$$

This is used, for example in the LEDA library of geometric tools [10]. In this approach, the rational points are stored as a tuple of integers (X, Y, Z, Q) with $Q > 0$. The point is interpreted as having rational coordinates (x, y, z) where $x = X/Q$, $y = Y/Q$, $z = Z/Q$.

The point of intersection of three planes defined as

$$\begin{aligned} A_1X + B_1Y + C_1Z + D_1Q &= 0 \\ A_2X + B_2Y + C_2Z + D_2Q &= 0 \\ A_3X + B_3Y + C_3Z + D_3Q &= 0 \end{aligned} \tag{5}$$

can be calculated as $p = (X, Y, Z, Q)$ where:

$$Q = \begin{vmatrix} A_1 & A_2 & A_3 \\ B_1 & B_2 & B_3 \\ C_1 & C_2 & C_3 \end{vmatrix}, \tag{6}$$

$$X = \begin{vmatrix} -D_1 & -D_2 & -D_3 \\ B_1 & B_2 & B_3 \\ C_1 & C_2 & C_3 \end{vmatrix}, Y = \begin{vmatrix} A_1 & A_2 & A_3 \\ -D_1 & -D_2 & -D_3 \\ C_1 & C_2 & C_3 \end{vmatrix}, Z = \begin{vmatrix} A_1 & A_2 & A_3 \\ B_1 & B_2 & B_3 \\ -D_1 & -D_2 & -D_3 \end{vmatrix} \tag{7}$$

There is a pleasant symmetry to this approach, where a point is stored as a tuple of four integers, $p = (X, Y, Z, Q)$ and a plane is also stored as four integers (A, B, C, D). This parallels the fact that three points define a plane, while three planes define a point.

12.5.2 Validation of a Polygon as a Planar Object

Using homogeneous coordinates, the vertices that are calculated as the points of intersection of planes can be exact, so that there is no need to allow a tolerance in the planarity of surfaces. (except at initial data load).

12.5.3 Validation of polyhedra.

Returning to the example in section 12.2.2, Fig. 12.1, Fig. 12.7, the point at *p* does not need to be inserted into face *A*, since it can be calculated to fall exactly on the boundary of *A*. On the other hand, if it is made a vertex of *A*, it can be guaranteed to fall exactly on the plane of *A*.

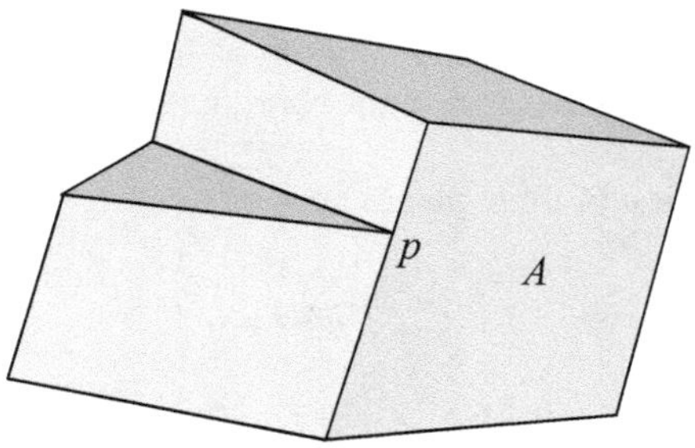

Fig. 12.7. A redundant point in face *A* of polyhedron (reprint of Fig. 12.1)

Returning to the second example in Section 12.2.2, Fig. 12.2, the exact calculations of intersections of planes allow repeatable determination of the validity of complicated surfaces.

12.5.4 Union of Polyhedra

The difficulties in forming the union of two polyhedra are also solved by using homogeneous coordinates, since all calculations are exact.

12.5.5 Equality of Volumetric Objects in Space

The equality of polyhedra can be determined exactly using homogeneous coordinates, but if a tolerance is to be allowed, the complexity of the issues as described for floating point representations still apply. It would be expected that in this type of approach, exact equality would be mandated, or that a clear distinction would be made between equality, and an "approximately equal" condition based on the ISO 19107 definition.

12.5.6 Operations Requiring Imprecise Calculations

Although this approach covers many of the operations that make up the toolkit of geographic information systems, there are some that it does not. Any operation that involves rotation, projection or adjustment to a different datum can best be handled by approximated transformations. This is not necessarily a problem, provided it is realised that these do not necessarily preserve the full integrity of the database. Ideally, following such an operation, the validity of all features should be re-checked. For example, it is possible that some contiguous geometric objects may become disconnected.

Frequently operations of this type are not carried out in a rigorous form in any case – for example in 2D it is common for a datum correction to be applied only to the end-points of a line segment, with the assumption that the line remains straight.

Conversely, the application of an approximation sweep through the database can be an advantage, in reducing the space requirements for the storage of points.

12.5.7 Storage Requirements

This approach requires arithmetic operations to be available in the computational language that are effectively unlimited in size. That is to say, for integers A and B, it must always be possible to calculate $A+B$ and $A \times B$ exactly, and with no possibility of overflow. This is possible in several languages, including Java, but with the loss of some programming convenience. (In an language such as Java which does not permitting overloading of operators, the coding $A = B*C+D*E$; has to be replaced by A = B.multiply(C).plus(D.multiply(E)).

The disadvantage of this approach in that every operation is likely to increase the size requirements of the result. Thus after a long series of operations, the storage requirements of a representation increase (without bound), and processing time requirements for further operations increase (also without bound). Note that with numbers of the size found in spatial data – where 9 digits resolution is commonplace, the numerators and denominators can become very large indeed.

As a rough calculation, if the coordinates of points in 3D are stored as 32 bit integers, the formula for a plane which passes through them in the form $Ax + By + Cz + D = 0$, with A, B, C and D integers will in general require A, B, C to be at least 64 bit integers, and D to be 96 bits. Further estimates show that if three planes are intersected to define a point, and three such points used to define a plane, the storage requirements for this plane are ten times those of the original ones. (For example, if the original planes require 64 bits, the next generation of planes require 640 bits per parameter). Thus every operation on a three dimensional object defined by unrestricted rational number points potentially creates a very large (and multiplicative) rise in the precision requirements. Note, however, that this effect is not as extreme in the case of 2D spatial objects, where the multiplier effect is smaller.

It is clear that if a common factor can be found for the integers that define a point or a plane, the storage requirements can be reduced. Clearly:

$$\text{Point } (RX, RY, RZ, RW) = (X, Y, Z, W) \text{ and}$$
$$\text{Plane } (TA, TB, TC, TD) = (A, B, C, D).$$

If it exists, a common factor can be found using a fairly obvious modification of the Euclidean Algorithm [2], allowing simplification in some cases, but whether this is productive is not necessarily clear.

Given two random integers, the probability that they are relatively prime is given as $6/\pi^2$ [1]. This means that the probability that a common factor can be found is about 0.4. For four integers at random, the probability of any common factor drops significantly, but this does not really answer the question. There are many cases where the integers are not randomly distributed, and common factors can be found.

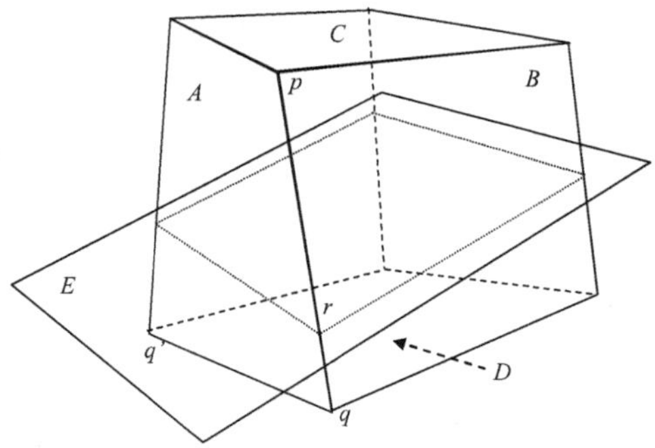

Fig. 12.8. Point of intersection of a line with a plane

For example, in Fig. 12.8, if *p* and *q* have each been defined as the intersection of three planes, (*ABC* and *ABD* respectively). If the planes require 64 and 96 bit integers for their parameters, then the points *p*, and *q* will require of the order of 224 bits for the *XYZ*, and 192 bits for the *Q* ordinates.

In general, the intersection of a plane using 64/96 bits (such as *E*) with the line joining two points (*p* and *q*) each requiring 224/192 bits would be expected to take a very large number of bits to represent the resultant point *r*. In fact, *r* is at the point of intersection of three planes, therefore it requires only 224/192 bits – the same as *p* and *q*. Thus the result of the calculation of *r* from *p* and *q* must be a 4-tuple of integers which have a common factor. If any calculation is made of a point or plane, it may be wise to test for common factors, even though this calculation is expensive.

Despite this case, and "accidental" cases of common factors, typically the precision requirements grow linearly (in terms of number of bits required in the result) with the number of operations executed.

As alluded to above, this type of database may benefit from an occasional sweep through all of the geometry (even if not needed to apply a datum change or other such action), approximating it back to reasonable sized coordinate values. It must be remembered that this action can introduce validation errors, and so must be accompanied by a validation sweep. If such an approximation operation is carried out at the completion of each operation, a result equivalent to the lazy evaluation approach of section 12.4 is obtained.

12.6 Round Only on Specific Operations

The regular polytope, based on domain-restricted rational (dr-rational) numbers [15] rounds at data load time, but also when some (rare) operations are carried out. It does, however provide rigorous evaluation of most of the important operations and predicates, including all those of topological and RCC theory.

Operations which round include:

- Datum shift or any point positional adjustment.

- Calculation of derived geometries such as convex hull.

Operations which are rigorous include:

- All RCC predicates
- All basic topological operations (union, intersection, inverse).

A regular polytope representation of spatial objects is defined as the union of a finite set of (possibly overlapping) "convex polytopes", which are in turn defined as the intersection of a finite set of half spaces (in 3D, half planes in 2D). These half spaces (planes) are defined by finite precision integer parameters (3 values in 2D, 4 in 3D etc). Although the definitions of the half spaces use integral coordinates, the points within them (and therefore the point sets defined by regular polytopes) are interpreted as domain-restricted rational points, and expressed in homogeneous coordinates.

12.6.1 Half Space Definition

In 3D a half space $H(A,B,C,D)$ is defined as the set of all points (in homogeneous coordinates) $p(X,Y,Z,Q)$: $Q > 0$, $-MQ \leq X,Y,Z < MQ$ for which computational evaluation of the following inequalities yields these results:

$$(AX + BY + CZ + DQ) > 0 \text{ or}$$
$$[(AX + BY + CZ + DQ) = 0 \text{ and } A > 0] \text{ or}$$
$$[(BY + CZ + DQ) = 0 \text{ and } A=0 \text{ and } B>0] \text{ or}$$
$$[(CZ + DQ) = 0 \text{ and } A=0, B=0 \text{ and } C>0],$$

where M is the limit of values allowed for point representations, and A, B, C and D are integers. We place the restrictions that:

$$-M < A, B, C < M, \quad -3M^2 < D < 3M^2 \text{ in 3D applications,}$$
$$-M < A, B < M, \quad -2M^2 < D < 2M^2 \text{ in 2D } (C \text{ is not required in 2D}).$$

Clearly each half space defines a plane in the same way as described in section 12.5.1 unless $A=B=C=0$, with the half space being the set of points to one side of that plane. $H(0,0,0,0)$ is not a permitted half space. Two special half spaces are defined:

$$H = H(0,0,0,-1) \text{ ('empty' i.e. points for which } -1 > 0).$$
$$H_\infty = H(0,0,0,1) \text{ ('everything' i.e. points for which } 1 > 0).$$

The complement of a half space is defined as:

$$\overline{H} = (-A,-B,-C,-D)\text{, where } H = (A,B,C,D)\,.$$

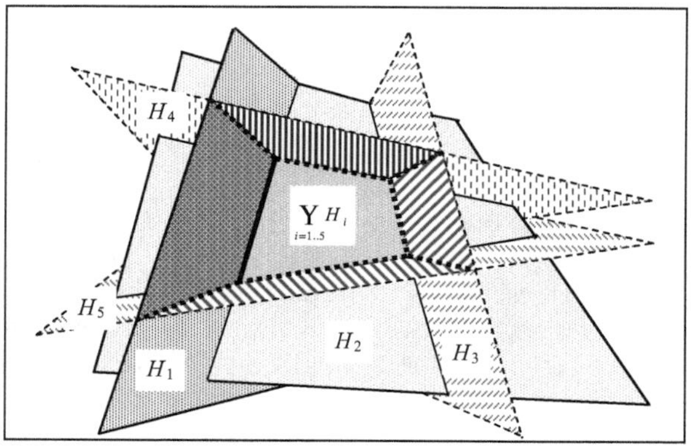

Fig. 12.9. A convex region defined by a set of half spaces (in 3D)

12.6.2 Regular Polytope Definition

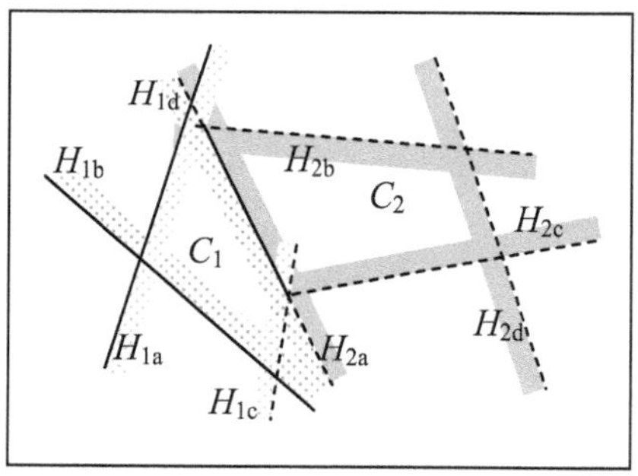

Fig. 12.10. Regular polytope O defined as the union of two convex polytopes C_1 and C_2[3].

A regular polytope O is defined as the union of a finite set of (possibly overlapping) non empty "convex polytopes" (C_1 and C_2 in Fig. 12.10), which are in turn defined as the intersection of a finite set of half spaces (H_{1a} to H_{1d} and H_{2a} to H_{2d}) [14]. This may be represented as $O = \bigcup_{i=1..m} C_i$, where $C_i = \bigcap_{j=1..n} H_{ij}$. Note – a regular polytope may consist of disconnected parts, and parts may overlap.

Points coincident with the boundaries, shown as dashed in Fig. 12.9 and Fig. 12.10 do not belong to the regions they define. In Fig. 12.10 $H_{2a} = \overline{H}_{1d}$, and points that lie along the common boundary are within C_2 but not C_1 (and therefore are within the regular polytope O).

[3] Note – Fig. 12.2 and many others are drawn as 2D cases for the ease of presentation where there is an obvious extension to 3D. By contrast, the text and mathematics are expressed in 3D.

The natural definitions of the union, intersection, and complement of regular polytopes is used, so that the meanings are exactly equivalent to the point set interpretations. E.g.:

$$\forall\, p: p \in O_1 \cup O_2 \Leftrightarrow p \in O_1 \vee p \in O_2$$

It is possible to define two forms of connectivity for the regular polytope, known as C_a (weak) and C_b (strong), and to show that for these definitions, the usual functions for a Region Connection Calculus apply. Thus it is guaranteed that all the functions and predicates based on union, intersection, overlap, connectivity, and complement will be rigorously calculated [15].

12.6.3 Validation of a Face as a Planar Object

This follows immediately from the fact that all objects are defined from the half space primitive which can only represent planar faces. Vertices are a secondary calculation from the half spaces, and are calculated exactly.

12.6.4 Validation of Polytopes

A regular polytope, as defined here, is similar to a polyhedron, but has some significant differences. It is not necessary that a regular polytope is fully bounded, nor does it need to be internally connected. In general, there is no such thing as an invalid regular polytope, but in certain applications, it may be useful to insist that a regular polytope should be:

non-empty – i.e. $\exists p: p \in O$.
bounded - $\forall p=(X,Y,Z,Q) \in O: -MQ < X,Y,Z < MQ$.
connected.

Whatever validity requirements are to be enforced, it is assured that, since integer arithmetic is mandated, the same results will be assured on any computing platform.

12.6.5 Union of Polytopes

The union of two regular polytopes $O_1 = \bigcup_{i=1..m_1} C_i$ and $O_2 = \bigcup_{j=1..m_2} C_j$ is defined as:

$$O_1 \cup O_2 = \left(\bigcup_{i=1..m_1} C_i\right) \bigcup \left(\bigcup_{j=1..m_2} C_j\right) \qquad \textbf{(8)}$$

This is simply the union of a set of convex polytopes, and is therefore a regular polytope. It clearly does not matter in what order this set is specified, and so the union operation is clearly commutative and associative. The cases for the intersection and complement operations, while more complex, are similar.

12.6.6 Equality of Volumetric Objects in Space

The functions and predicates, operating on regular polytopes, are defined in the following order, in order to implement them.

The Empty(C) predicate is defined first for the convex polytope – where Empty(C) is defined as $\forall p\colon p \notin C$. It is fairly simple to implement this while a convex polytope is being constructed. As each half space is added to the definition, the vertices of the convex polytope are calculated. If all vertices are outside or on the plane of the new half space, the resultant convex polytope is empty.

This is then used to define overlap: OV(C_1, C_2) as ¬Empty($C_1 \cap C_2$).

Which is used to define regular polytope overlap, OV(O_1, O_2) as $\exists C_i \in O_1, C_j \in O_2$: OV($C_i$, C_j).

This is then used to define part of: P(O_1, O_2) as ¬OV(O_1, $\overline{O_2}$).

Which defines equality: EQ(O_1, O_2) as P(O_1, O_2) and P(O_2, O_1).

Thus the definition is rigorous, and can be applied on any computational platform.

12.6.7 Operations Requiring Imprecise Calculations

All the operations that require rounding in the infinite precision rational approach clearly require rounding in this approach, but in addition some other calculations cannot maintain full precision. An example is the calculation of the convex hull. It is not possible to rigorously calculate a convex polytope which is the convex hull of a given regular polytope. The best that can be found is the "convex enclosure", which is a convex polytope, guaranteed to enclose the regular polytope, and is within one unit of resolution of the true convex hull.

Unlike the infinite precision rational number case, there is no need to run a periodic sweep through the database approximating the objects to reduce storage and processing requirements.

12.6.8 Storage Requirements

The storage requirements of this approach are constrained. The half planes are representable using conventional integers – for example, using 32 bit integers for A, B and C, and 64 bits for D, as used in the proof-of-concept coding described by [15]. (Note that by the definition of a half space (Section 12.6.1); A, B and C have a range of $\pm M$, but D has a larger range: $\pm 3M^2$). If it is desired to store vertices, they require extended precision representations, but these are of finite and pre-determined size. (In 3D, X, Y, Z require 128 bits, and Q requires 96 bits).

12.7 Conclusions

All spatial data that is represented in a computer is subject to imprecision and rounding errors. These may lead to failures of logic when they are applied, so it is advantageous to be aware of when this occurs. Different approaches have been investigated and compared, with the results tabulated in Table 12.1. No individual approach is optimum in all situations, but the regular polytope, based on dr-rational homogeneous coordinates has been shown to provide rigorous results in all of the most useful operations, without unconstrained growth in storage requirements or the need for periodical approximation sweeps of the database.

Table 12.1. Summary of Approaches

	Floating Point	Lazy Calculation	Unrestricted Rational	Domain Restricted Rational
Approximate on data capture	Yes	Yes	Yes	Yes
Approximate on adjustments	Yes	Yes	Yes	Yes
Approximate results to store in database	Yes	Yes	No	No
Approximation sweep through database	No	No	May be useful	No
RCC operations	Not rigorous	Rigorous, then approximated	Rigorously supported	Rigorously supported
Convex Hull	Yes	Yes	Yes	Approximate
Representation storage requirements	Fixed	Unlimited during operations	Unlimited	Fixed

References

1. Castellanos D (1988) The Ubiquitous pi (Part II). Mathematics Magazine Vol 61, 3: 148-161.
2. Courant R, Robbins H (1941) What is Mathematics? Oxford University Press, New York.
3. Coxeter HSM (1974) Projective Geometry. Springer-Verlag, New York.
4. Gaal SA (1964) Point Set Topology. Academic Press, New York.
5. Goldberg D (1991) What Every Computer Scientist Should Know About Floating-Point Arithmetic. ACM Computing Surveys, March 1991 http://docs.sun.com/source/806-3568/ncg_goldberg.html
6. IEEE (1985) IEEE Standard for Binary Floating-Point Arithmetic. IEEE Standards Department, New York.
7. ISO-TC211 (2001) Geographic Information - Spatial Schema. IS19107. International Organization for Standards, Geneva.
8. Kahan W (1996) Lecture notes on the Status of IEEE Standard 754 for Binary Floating-Point Arithmetic. From http://www.cs.berkeley.edu/~wkahan/.
9. Kazar BM, Kothuri R, van Oosterom P, Ravada S (2008) On Valid and Invalid Three-Dimensional Geometries. In: Advances in 3D Geoinformation Systems. Van Oosterom P, Penninga F, Zlatanova S, Fendel E (eds). Springer, Berlin.
10. Mehlhorn K, Näher S (1999) LEDA: A Platform for Combinatorial and Geometric Computing. Cambridge University Press.
11. Milenkovic VJ (1988) Verifiable implementations of geometric algorithms using finite precision arithmetic. Artificial Intelligence 37, 377-401.
12. Randell DA, Cui Z, Cohn AG (1992) A spatial logic based on regions and connection. 3rd International Conference on Principles of Knowledge Representation and Reasoning, Cambridge MA, USA, Morgan Kaufmann.
13. Sun (1993) Class Object. In: Java Documentation, from http://java.sun.com/j2se/1.3/docs/api/java/lang/Object.html.
14. Thompson RJ (2005) 3D Framework for Robust Digital Spatial Models. In: Large-Scale 3D Data Integration. Zlatanova S, Prosperi D (eds). Taylor & Francis, Boca Raton, FL.
15. Thompson RJ (2007) Towards a Rigorous Logic for Spatial Data Representation. Publications on Geodesy 65. Netherlands Geodetic Commission, Delft, The Netherlands.
16. Weisstein EW (2002) Hessian Normal Form. In: MathWorld -- A Wolfram Web Resource Retrieved June 2007, from http://mathworld.wolfram.com/HessianNormalForm.html.

Chapter 13
Automatic Digital Aerial Image Resection Controlled by LIDAR Data

Roosevelt De Lara Jr., Edson A. Mitishita, Thomas Vögtle and Hans-Peter Bähr

Abstract. During last fifteen years the Light Detection and Ranging (LIDAR) has been used as a very good option to terrestrial surface surveying. It can survey 3D positions over the terrestrial surface at speed beyond comparison to the Topography or Photogrammetry. The recent Airborne Laser Scanner (ALS) apparatus are made with add of digital cameras (photogrammetric or not) becoming easy the integration among LIDAR data and digital aerial images. In this article the authors desire to emphasize the use of the LIDAR data acquired at distinct time of the digital aerial images. The integration of data surveyed at different times is based in cross correlation among the intensity image from LIDAR and gray scale image from RGB digital aerial image. This procedure is not conventional considering the radiometric differences between the two different types of images. The methodology proposed is fully automatic, restricting the human interaction to the adoption of different processing criteria (inserted an only time at the beginning of the processing). The final results had demonstrated the efficiency of the developed methodology and suggest its application in other situations aiming the images orientation.

13.1 Introduction

Though traditionally the 3D mapping earth´s surface is carried out by photogrammeric methods, LIDAR (Light Detection and Ranging) has been coming into use. It can survey 3D points with incomparable speed to the topographic or photogrammetric methods. The ALS (Airborne Laser Scanner) apparatus is currently made with a digital camera, metric or not, targeting at complementing laser survey by integrating their data with the digital images. Such integration, being automatic or not, determines

Universidade Federal de Santa Maria, CESNORS-FW/RS
roosevelt@smail.ufsm.br
Universidade Federal do Paraná, Curso de Pós-Graduação em Ciências Geodésicas
mitishita@ufpr.br

some benefits to the two kinds of surveys, aiming cartographic products, such as: altimetric models, orthophotos and photogrammetric mapping. These benefits by integrating data may be summarized as the following: LIDAR data contribute towards the geometric features, and the digital images contribute towards the radiometric features. The (positional) geometric information of the local topography from LIDAR data is invariable in the time in significant proportions, so that the LIDAR mass data may be used as a database to reference new different kinds of surveys at different times. For instance, giving control points to photogrammeric survey. According to [8], four distinguished automation levels are considered as follow: a) Interactive Systems: Absolutely manually or with no automation kind; b) Semi-Automatic Systems: Automatic units are added to the process by more or less human interaction; c) Automatized Systems: The main tasks are done automatically; the human interaction is limited to processing criteria definition and acceptance of the final results; d) Independent System: Totally automatic, "Press the button and forget it", it calculates all necessary parameters to an independent process automatically. Such systems do not exist in photogrammetric applications [18]. To [12], the general problem about automatic control points extracting and consequent external orientation process automation is limited to corresponding establishment between a database from the object space and its representation on the image space. In [14], when there is a reliable database, a total automation, or a highly degree of it, the control points extraction and exterior orientation parameters determination are possible. Basically, it consists in: a) Defining and extracting of the primitive from database; b) Extracting of the primitive from image; c) Determination of the correspondence between the extracted primitives; d) Calculating the exterior orientation parameters. During the last five years the scientific community has been dedicating attention to LIDAR data and images integration. [9] integrated digital images from a digital camera (not photogrammetric) with a LIDAR intensity image to digital camera calibration. [10] carried out image´s exterior orientation parameters determination using features (free form) and LIDAR intensity image. The results indicated the possibility of good data integration. Thus, it regards to the semi-automatic process. [11] integrated LIDAR data and digital images in close range photogrammetry application. They used linear features built from plan intersection automatically detected. This method took manually stages. [4] carried out LIDAR data and aerial images (taken with a low cost digital camera). The control points were extracted manually from the LIDAR raw data. Two phototriangulation applications were made using the bundle method for a small block (two strips, 13 images), one of them conventionally controlled, and the other one used the control points, whose coordinates were extracted from the LIDAR raw data. The two adjustments evidenced compatible results between them. They suggested the great potential LIDAR data for photogrammetric controlling. [2] proposed linear features extracted from the DTM generated through the LIDAR data as control points to spatial resection of images taken from a digital camera. [19] presented a methodology to do automatic spatial resection of images based on the use of roads hypotheses. The IEKF (Iterated Extended Kalman Filtering) was used in estimation of the external orientation parameters; the mathematical model of equivalent plans [21] was used on a relationship between image spaces and an object. In this article the authors desire to emphasize the use of the LIDAR data acquired at distinct time of the digital aerial images. The integration of data surveyed at different times is based in cross correlation among the intensity image from LIDAR and gray scale image from RGB digital aerial image. This procedure is not conventional considering the radiometric differences between the two different types of images. The proposed methodology to images integration is fully automatic, restricting the human interaction to the adoption of different processing criteria (inserted an only time at the beginning of the processing). At present work the digital aerial images were taken from a low cost digital camera, considering the potential growth use of this type of camera in no conventional Photogrammetry.

13.2 Methodology

The developed methodology runs by three different stages shown in the Fig. 13.1.

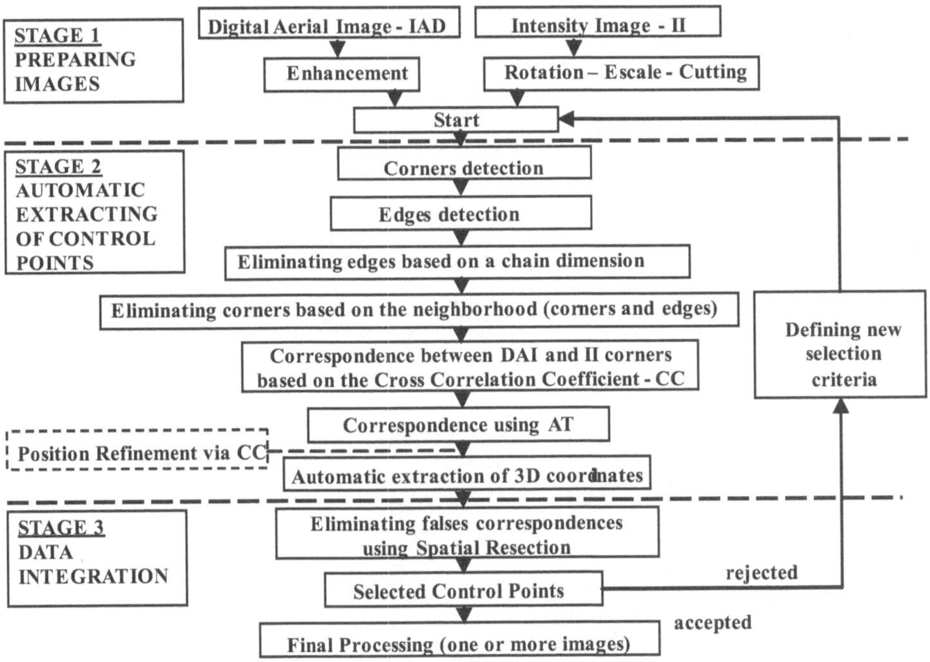

Fig. 13.1. Methodology stages

Digital aerial image and intensity image are prepared at the first stage, the preparation stage includes the interest area selection, scale, rotation and cutting. The second stage takes into account the automatic extraction from the 3D coordinates of the control points. The third stage corresponds to automatic integration of digital aerial images and LIDAR data (exterior orientation parameters determination, adjustment and evaluating the process).

13.2.1 Stage 1- Preparing the Images

The images preparing is based in information about photographic and LIDAR flights missions. The potential correspondence between areas of the intensity image and digital aerial image are selected in this stage. The key procedures are:

a. The area selection on the intensity image (total LIDAR surveyed area) is based on the approached coordinates of the digital aerial image centre and the respective flight direction line;
b. Scaling of the intensity image, so that two images have the same gsd (ground sample distance) in this work is maintained the gsd of the digital aerial image;
c. Rotating of the LIDAR intensity image (is based on the approached coordinates of the digital aerial image centre and the flight direction);
d. Cutting the interest area from total intensity image with the same dimensions (rows and columns) of the digital aerial image.

The Fig. 13.2 draws the main positional reference data used on the intensity image preparation.

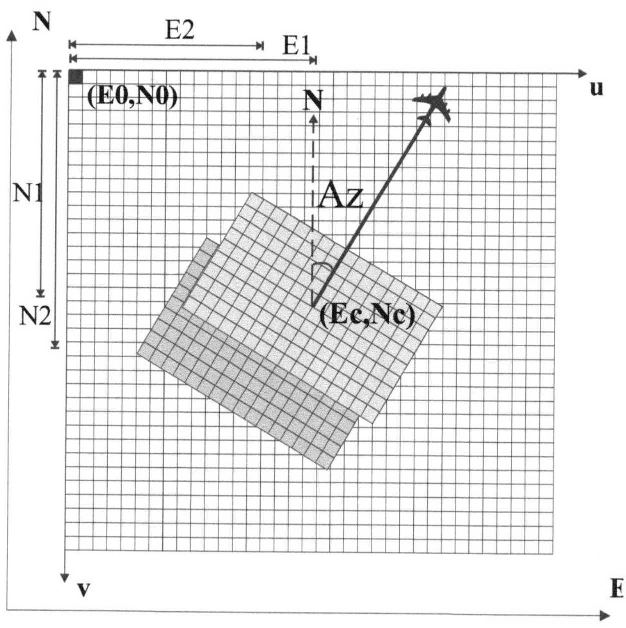

Fig. 13.2. Data for the image preparation

The Fig. 13.2 shows the greater grid (reference grid) performs the intensity image of the whole area by the LIDAR survey; the others grids on the greater grid then perform the digital aerial images; the flight line is designed through the line from approached coordinates of the digital aerial image centers; the systems (u, v) and (E, N) represent respectively the reference systems in the image and object spaces. Expected entry data: the digital aerial image and the intensity image (LIDAR); the gsd of the images; the digital aerial image perspective center approached coordinates (for two or more images); the approached flight line direction, and the approached coordinates of the image perspective center (in the case of an image); images dimensions (rows and columns), and the upper left corner coordinates of the intensity LIDAR image (the object space). The digital aerial image center coordinates in the object space coordinate system (the LIDAR intensity image) calculation is made as follow:

$$E_c = (u.gsd) + E_0 \quad (1)$$

$$N_c = N_0 - (v.gsd) \quad (2)$$

E_c, N_c, image center coordinates with reference to the object space;

E_0, N_0, located pixel coordinates on the upper left corner (1,1) of the intensity image with reference to the object space;

u, v, coordinates with reference to the image space (column, row)

gsd, related to the intensity image.

When working with only one digital aerial image, a primary condition to the process is having the former knowledge of the image center coordinates (with reference to the object space), and the flight line direction (κ), or the flight line azimuth (Az); these ones are related to each other through the (3) equation. Latter, they may determine the rotation angle (A_{rot}) in order to be applied for the LIDAR intensity image cutting through the (5) formula:

If working with two or more images, the process is limited to the knowledge of the approached coordinates of the digital aerial image centers. The azimuth calculation of the set up direction between two following image centers is made by the (4) formula. Latter, the rotation angle is calculated by (5) formula, with reference to an (n) image:

$$\kappa = 2\pi - Az \quad (3)$$

$$arctg(Az_{n,n+1}) = \frac{E_{n+1} - E_n}{N_{n+1} - N_n} \quad (4)$$

$$A_{rot} = -Az \quad (5)$$

κ, flight line direction;

Az, azimuth;

A_{rot}, rotation angle.

13.2.1.1 The digital aerial image enhancement

The objective of the digital aerial image enhancement is reduce the radiometric differences between two different types of images, digital aerial image (RGB) and the intensity image (256 gray scale obtained by the resample of the pulse intensity laser returning). The LIDAR survey (carried out through the superimposed flight strips) is represented by a cloudy of points. From this set of points, using interpolation, LIDAR intensity image is produced. The pulse intensity of a laser returning depends on the material and the object design which reflects it, causing considerable levels of noise to the intensity image produced. Therefore, a specific methodology was developed for the RGB digital aerial image enhancement. It was inspired by the follow works [7], [16, 17]. Working with RGB color images, [7, 16] presented a rate that allows the object classification in an aerial image into two classes: an artificial one (the human constructions) and a natural one (the vegetation, the shadow, the exposed topsoil, etc.). Such rate consists of getting an Artificial Degree of the scene features through the DoA (Degree of Artificiality), given by the following equation:

$$DoA = \frac{G-R}{G+R} \qquad (6)$$

R, G, B, colored image bands (RGB);

Trying to separate the artificial objects from those natural ones, through RGB band analysis, [17] developed the artificiality rate called NandA (Natural and Artificial), in order to enhance lighter gray tones for the vegetation, while darker gray tones for buildings and roads.

$$NandA = G - (R + B) \qquad (7)$$

In the present methodology an image treatment was developed to enable the correspondence between the RGB image and intensity image, without radiometric or geometric losses. The formula (9) develops the image enhancement pixel by pixel and gives positive values as a result (to the present case, between 0 and 255, to greater values than 255 it is applied the value of 255 as shown in Fig. 13.3). It is important stands out what [7, 16, 17] solutions can turn in negative values for the pixels of the image of intensity.

$$I = f(R, B, G) \qquad (8)$$

$$I' = -NandA = (B + R) - G \qquad (9)$$

I, earliest colored RGB image;
I', enhanced image;

Converting the RGB color image into an image of gray scale image (gs) comes from the (10) formula [6].

$$gs = 0.299R + 0.587G + 0.114B \quad \mathbf{(10)}$$

(a) (b) (c)

Fig. 13.3. Gray scale image (a), LIDAR intensity image (b), and Enhanced image (c)

13.2.2 Stage 2 – Automatic control points extraction

This stage implies in the location, selection, and positioning of the control points from the correspondence of corners detected in the image and in the LIDAR intensity image. The assumed processing criteria are shown in the Table 13.1. Such criteria have taken into account the experiments by [5].

Table 13.1. Assumed process criteria

Element	Criterion	Unit
Searching by radius through points for correlation	80	pixel
Sample matrix for correlation	31x31	$pixel^2$
Search matrix for correlation	91x91	$pixel^2$
Search matrix for correlation	41x41	$pixel^2$
GSD – LIDAR Intensity Image	25	cm
Standard deviation for Gaussian filtering (Harris e Canny)	2	pixel
Searching by radius for non-maxima-suppression (Harris)	2	pixel
Searching by radius for non-maxima-suppression (Canny)	1	pixel
Top threshold for hysteresis (Canny)	0.1	
Neighborhood filtering	5x5	$pixel^2$
Least number for connected pixels	60	pixel
Module of the cross correlation coefficient	0.7	
Exclusion criterion in affine transformation filtering	5	pixel

13.2.2.1 Corners detection

According to [22], it is advisable that a corner detector should meet the following requirements:

a. It detects all corners which really are there on the image;
b. It does not detect false corners;
c. The corner location should have the highest fidelity;
d. The corner detector should keep its efficiency under different situations;
e. It needs to be robust against noise;
f. It needs to be computationally efficient.

In the present methodology is used Harris corner detector [13], the Fig. 13.4 shows an example of the Harris corner detection.

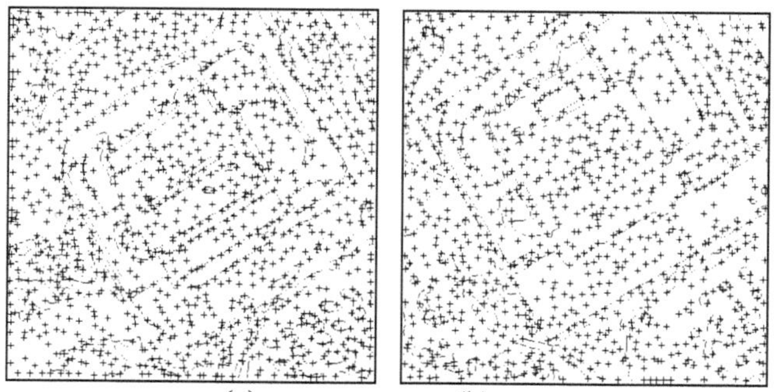

(a) (b)

Fig. 13.4. Detected corners in: enhanced RGB image (a) and intensity image (b)

13.2.2.2 Edge detection

According to [15], it is important that the edge detector ought to have the following features:

a. Good detection: it needs to have the ability to locate and to mark all existing edges;
b. Fine locating: it needs to lessen the distance between the detected edge and the real edge;
c. Good return: for each edge there would be just a return.

At present work is used the Canny edge detector [1], the Fig. 13.5 shows an application of the Canny edge detector.

(a) (b)

Fig. 13.5. Detected edges in: enhanced image (a), intensity image (b)

13.2.2.3 Corner filtering based on the chain dimension

On the binary images, resulting from the edge detection, the small open chains (segments of lines) and the small closed chains (islands) are eliminated, which are a significant part of undesirable segments, on each of the two images independently. The elimination criterion is based on counting the number of pixels connected (neighborhood of 8) on those islands, and on those segments of undesirable edges, compared to a pre-definite criterion, as shown on the Fig. 13.6 example.

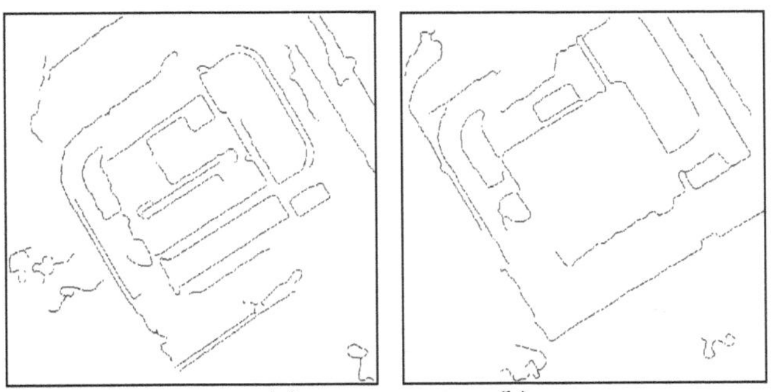

(a) (b)

Fig. 13.6. Filtering based on the chain dimension, enhanced image (a), intensity image (b)

13.2.2.4 Filtering edges based on the corners and edges neighborhood

On the binary resultant images of the detection of edges and of the filtering based on the chain dimension, carried out in the previous section, a test of neighborhood is applied in the positions of each selected corner. The test mask is a square matrix with odd order. And its elements are all zero centered on the concern corner. Then, adding the masks as well as the filtered images is the procedure on all over positions where corners were detected: when there is an element which is not invalid in the matrix, as a result of the test, then the corner is selected. Otherwise, it will be eliminated, as it is shown on the Fig. 13.7. This proceeding result to the enhanced and the LIDAR intensity images is shown on the Fig. 13.8.

1	0	0	0	0	0	0
0	1	0	+	0	0	0
0	0	1	0	0	0	0
0	0	0	1	1	+	0
0	+	0	0	0	1	0
0	0	0	0	0	0	1
0	1	0	0	0	0	1

(a)

0	0	0
0	+	0
0	0	0

(b)

1	0	0	0	0	0	0
0	1	0	+	0	0	0
0	0	1	0	0	0	0
0	0	0	1	1	+	0
0	0	0	0	0	1	0
0	0	0	0	0	0	1
0	1	0	0	0	0	1

(c)

Fig. 13.7. Earliest image (a), Mask (b), Filtered image (c)

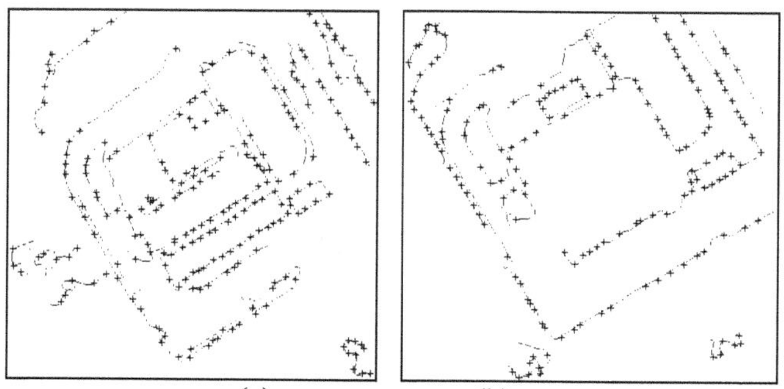

(a) (b)

Fig. 13.8. Filtering based on the neighborhood: enhanced image (a), intensity image (b)

13.2.2.5 The correspondence based on the cross correlation coefficient

The first level of correspondences between the detected corners on the digital aerial image and its respective in the intensity image is carried out taking as a basis the Cross Correlation Coefficient – CC, details on [20]. This first level of correspondences has as the main interest to take equivalent dimension of the detected corners sets on the mentioned images. Using the sets of corners selected (a set for the digital aerial image and another for the respective intensity image), the selection of corners potentially correspondent is implemented between the two sets of corners independently at each other.

The correspondence will be based on the correlation coefficient between the candidate corners, limited to their somewhat neighborhood, with focus on not turning the process slow. After this first trying to establish the correspondences, the two sets of corners come to have the same number of elements.

It takes into the account that the most part of these correspondences are false, easily noticed by the Fig. 13.9, on the central region of the (a) and (b) images. Comparing to the Fig. 13.8, it is also realized the increasing capacity for eliminating the spurious corners.

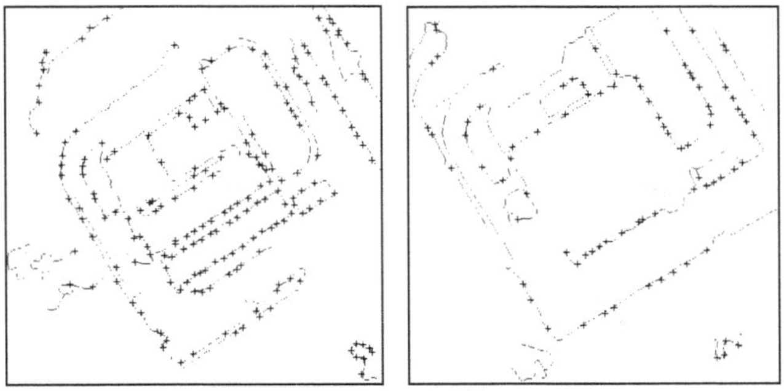

(a) (b)

Fig. 13.9. Correspondence based on CC, Enhanced image (a), intensity image (b)

13.2.2.6 The correspondence based on the Affine Transformation

The second level of correspondences between two detected corners on the digital aerial image and its respective in the intensity image is carried out based on the Affine Transformation – AT, focusing the improvement of the correspondences gotten in the last section. Now, this proceeding tries to eliminate the most number of false correspondences within a selective procedure of adjustment, in which are eliminated point by point, in reference to the points that exceed the rejection positional criterion. The mathematical model is the AT (the linear form, the equations 11). The adjustment model is the Parametric LSM. Putting great emphasis on the case of images (the representation on the tri-dimensional space plan, and the distortion of the camera lens), the AT is just a nearly mathematical model, which is relegating the two images to the inherent considerations for the bi-dimensional space. In addition to that, while the digital aerial image has a central perspective, the intensity image features an ortho image.

$$
\begin{aligned}
u_2 &= a_1 u_1 + a_2 v_1 + a_3 \\
v_2 &= a_4 u_1 + a_5 v_1 + a_6
\end{aligned}
\qquad \textbf{(11)}
$$

u_1, v_1, the enhanced image coordinates;

u_2, v_2, the intensity image coordinates;

$a_1, a_2, a_3, a_4, a_5, a_6$, AT coefficients.

The point elimination comes from the identification of the larger module element in the residual vector. The process repeats itself until the larger module element of the residual vector reach the smaller module, or the equivalent to the rejection criterion. This rejection criterion assumed here was determined empirically with the value of 5 pixels. The Fig. 13.10 (a) and (b) shows the ultimate configuration of detecting corners which may, or not, be used as control points during the stage of the image integration (see the white crosses and its correspondents black crosses).

(a) (b)

Fig. 13.10. Correspondence based on the AT, gray scale image (a), intensity image

13.2.2.7 Improving the position through the correlation coefficient

To the correspondent corners determined over the last section, it is applied the correlation again focusing for the improvement in identifying the positions of the potential control points. It is important to place emphasis on the previous stages which always kept the coordinates determined by detecting corners. Now, the responsibility of this stage is only to try the position improvement of the detected corners (with the changing consequence of the corner position to a local of a greater coefficient correlation). Yearning for the improvement in some applications may be carried out by applying the correspondence with the least squares method. The earlier coordinates, determined by the detection of the corners, are used as approached coordinates in the adjustment process [20]. Undertaking the position improvement is not a necessary stage through this suggested methodology. It may be quit according to the kind of the application.

13.2.2.8 Determining the planimetric coordinates

The Fig. 13.11 shows the general plan, and the implied elements, when transforming the planimetric coordinates in the image system into the planimetric coordinates with reference to the object space, through the application of the Orthogonal Transformation – OT, as it follows:

$$\begin{bmatrix} E \\ N \end{bmatrix} = \begin{bmatrix} \cos(-A_{rot}) & -sen(-A_{rot}) \\ sen(-A_{rot}) & \cos(-A_{rot}) \end{bmatrix} \begin{bmatrix} du.gsd \\ dv.gsd \end{bmatrix} + \begin{bmatrix} E_c \\ N_c \end{bmatrix} \quad \textbf{(12)}$$

$$du = [u - (nc/2)] \quad \textbf{(13)}$$

$$dv = [v - (nl/2)] \quad \textbf{(14)}$$

nc, nl, number of the digital aerial image columns and rows;

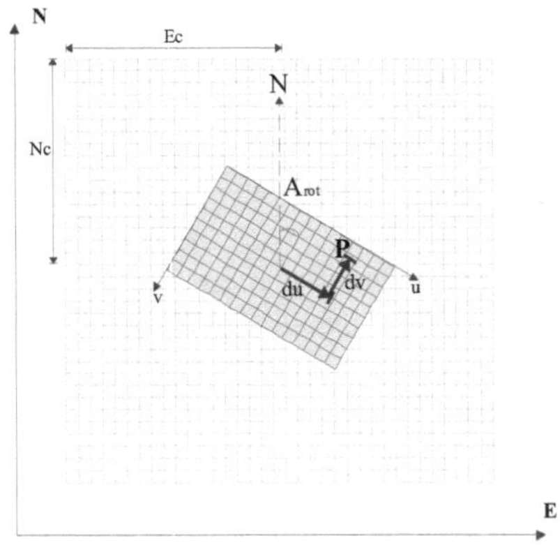

Fig. 13.11. Elements for determining the planimetric coordinates via OT

13.2.2.9 Extracting the altimetric coordinate

The planimetric coordinates determined in the last section will be used as an argument about in the search within the raw file of the generated coordinates by the LIDAR. The raw data point, which lends its height to the argument point, needs to follow two criteria simultaneously:

a. The distance between the argument points and the searched ones in the LIDAR raw data needs to be short (the nearest neighbor);
b. When there are two points at the same distance from the argument point, the height chosen needs to be higher (the highest point). Such criterion was determined because aerial photographs are being dealt. Then, the located points on the object tops have more probability to turn up on the images (the digital aerial and the intensity ones).

It is helpful to take into account that the LIDAR raw data are usually substantial (its content has million points). So, it is necessary to establish a strategy to manage such files. For this reason in this work LIDAR raw data file was divided into smaller files (cells), in which all integrating points are known and numbered.

The Fig. 13.12 shows a plan for managing the LIDAR raw files. The LIDAR raw file is a kind of text file, which has, in the case of the systems that record the returning pulsed laser intensity, at least, four pieces of information: three planialmetric coordinates (E,N,h), and the pulse laser intensity information. The raw file set up consists of, primarily, determining the arranged file extremes (the upper left corner coordinates and the lower right corner coordinates), and, latter, determining the dimensions of the arranged file cells. Two auxiliary files are created: the first one needs to get pieces of information for arranging each point in each cell (the points in each cell have a number of a sequence determined as the allotment of the cells); the second file needs to get pieces of information for positioning the cells (limits of the lines and the columns of each cell), and the starting and the ending number of the sequence of the points in the cell. From the planimetric coordinates of the argument point and the search by radius decision, the survey limits are determined in terms of the cells.

The Fig. 13.13 (a) shows the fixed position through the developed methodology in this work. Its planimetric position determined by the OT application, according to the 13.2.2.8 section description (a light tone on the building corner), and the nearly point (a dark tone), whose position was determined by description over this section, and, on its turn, the altimetric coordinate was established to the argument point. Also, it is noticed that the other points of the LIDAR raw data (really measured) get a darker tone. To the left region also is seen the concentration of points, as a result of superposing the LIDAR survey bands. There is, on the Fig. 13.13 (b), the hypsometric representation of the same neighborhood shown by Fig. 13.13 (a). The hypsometry displays the existence of two distinguished levels, one of them is shorter (dark), and the other one is higher (light). Such levels are due to a small building presence on the local place. The setting analysis of the Fig. 13.13 (a) and (b) suggests a view confirmation as an efficient methodology, which was developed to focus for determining the planimetric positions, as well as its association and extension to complement the 3D coordinates of the identified point (corner), when the LIDAR raw data are used for it.

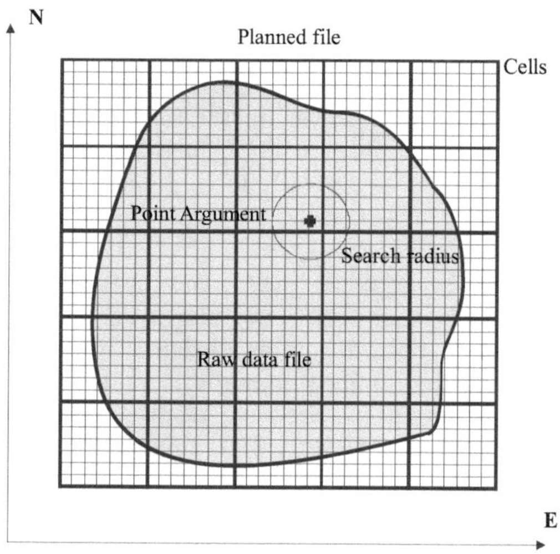

Fig. 13.12. Plan for surveying coordinates in the LIDAR data file

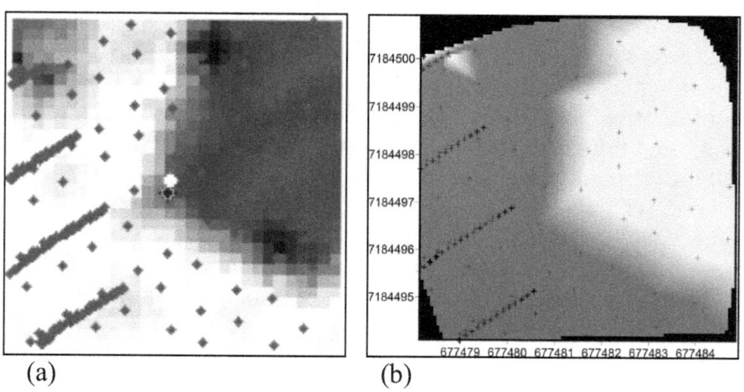

Fig. 13.13. LIDAR raw data over the intensity image (a), and over hypsometric representation (b)

13.2.2.10 False correspondences eliminating via Spatial Resection

The present step focus on eliminating (last) false correspondences when using, hitherto assumed, control points in the spatial resection process (SR). The SR performance over this step has its adjusting criteria less strict (it is admitted that the linear differences on the image space, up to 4 pixels, and on the object space, up to 1 m, the angular convergence criterion is 1") than those applied in the photogrammetric process, as a result of its focus on detecting and eliminating the false correspondences remaining in the process of automatic extraction of the control points. Summarizing, those rejected points on the adjusting period are eliminated, and those ones approved are actually considered as photogrammetric control points.

13.2.3 Stage 3 – Integrating Data – The Photogrammetric Process

When it is determined the photogrammetric control points, its application in the phototriangulation by bundle adjustment may be carried out at any photogrametric unit, that is, a spatial resection, a model, a strip or a block.

13.3 Experiment

The experiment consists of carrying out the automatic spatial resection (SR) of a digital aerial image, controlled by the extracted points automatically from the LIDAR database. The divergence as a result of checking the points up means the difference between its coordinates measured manually and its similarities, calculated by the collinear equations on an inverted state, from the exterior orientation parameters, gotten through the manual control points extraction (MCPE), and the automatic one (ACPE). Such procedure was chosen because the number and the allotment of the manual and automatic control points are not identical ones.

13.3.1 Presenting data

The digital aerial image (n. 216) belongs to a group of 13 RGB color aerial images. It was taken with a digital camera Sony DSC-F717 (CCD 2560x1920 pixel^2), which came from the A AGRITEC company. The images (at the moment of the acquisition) were recorded through specific JPEG format files. The photographic flight took place on 27 June 2003, which undertook two strips, through the nearly direction NE-SW (a flight strip to NE-SW, with a sequential number from 194 to 199, and the other flight band to SW-NE, with a sequential number from 212 to 218), covering about 2 km^2, around the Centro Politécnico, Campus III, Universidade Federal do Paraná - UFPR. The estimated flight height of this photographic mission was 750 m, and the estimated gsd was 25 cm. The LIDAR data were taken from the Optech ALTM 2050 system, which belongs to the LACTEC. It includes 2 text format files (first and last pulse returning), whose contents of the geodesic coordinates UTM-SAD69 (E, N), the orthometric height (h) and intensity image (with gray scale value from 0 to 255). The

LIDAR mission flight height was 1000 m, which took place on 09 May 2003, expecting the following accuracy: 50 cm for the planimetry, and 15 cm for the altimetry (data and all information provided by LACTEC, which was responsible for carrying out and processing the LIDAR mission). The cloud of LIDAR points produced had 8.593.331 points, and covered an area of 3.55 km² (the estimated density of points was 2.42 points/m²).

The intensity image process was carried out through the QTmodeler software, showing the gray scale via the bilinear interpolation again. The resulting image was recorded on the GeoTIFF format. The studied data may be seen on the Fig. 13.14. Through this paper, the computer programs, which are necessary to carry out all integrating process of the developed methodology, were performed (by the authors) in the Matlab environment.

13.3.2 Reference systems and units

The reference system to the object space, which was chosen here, regarding the small area used in the experiment, was a 3D, hybrid, and simplified system. The planimetric coordinates (E, N), on the UTM (Universal Transverse Mercator) projection, referred to the SAD69 (South American Datum), were equivalent to their own plane coordinates (X, Y). The orthometric height (h), assumed as equivalent to the Z altimetric coordinate. The measure unit used for the object space was the meter, and for the image space was the pixel.

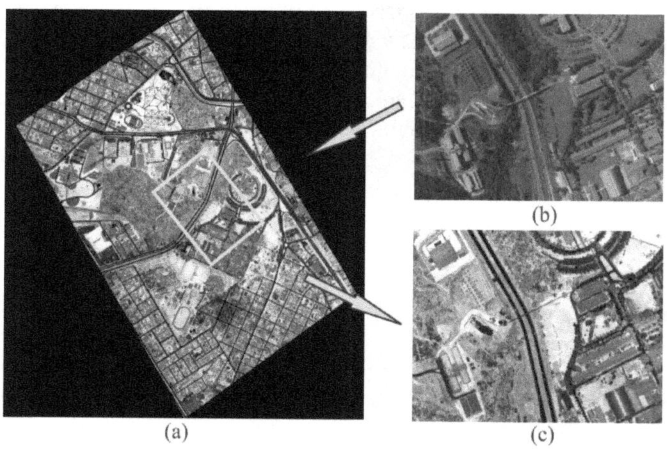

Fig. 13.14. LIDAR intensity image (a), Digital aerial image 216 (b), Intensity image cutting equivalent to the digital aerial image 216 (c)

13.3.3 Statistics of the developed methodology application

The methodology suggested may be faced as a process of filtering or cleaning (prospecting) of the data (corners), taking benefits of control points in focus. The 13.2

shows the indicated numbers for the detected corners on the two EI images (enhanced image), and II (the intensity image), the resulting numbers after filtering based on the neighborhood of corners and edges, the correspondence effect based on the cross correlation coefficient (pairing and dimensioning), the residual points after the correspondence application controlled by AT (affine transformation), and finally the residual points after carrying out the spatial resection (SR). Examining this table allows to find out the method powered by getting or carrying out the identification and the measurement of the control points automatically. Another examination is about the method powered by selecting the real correspondences, considering the residual points, which amount to about 0.14 % (20 points) of the early total (about 14000 points).

Table 13.2. Statistics of the developed methodology application

Image	EI, II Corners		EI, II Filtering		CC	AT	SR
216	14435	13698	3239	3870	951	38	20
%	100.00		25.27		6.76	0.27	0.14

13.3.4 The Spatial Resection of the Image 216

The Table 13.3 shows the parameters of Sony DSC-F717 camera inner guide taken with calibrating the camera [3]. The parameters were used on the spatial resections (with manual and automatic control point extraction) of the image 216.

Table 13.3. Parameters of interior orientation of Sony DSC-F717 digital camera

Parameter	Calibration	Standard Deviation	Unit
c	2931.722	1.172	pixel
x0	-71.648	2.012	pixel
y0	-40.965	1.953	pixel
k1	-2.640E-08	6.626E-10	pixel^{-2}
k2	3.242E-15	6.916E-16	pixel^{-4}
k3	3.061E-22	2.103E-22	pixel^{-6}
P1	-4.130E-07	5.186E-08	pixel
P2	2.429E-07	5.118E-08	pixel
A	-1.346E-04	1.042E-04	pixel
B	-2.033E-05	1.055E-04	pixel

The Table 13.4 below shows the exterior orientation parameters by spatial resections with manual (MCPE) and automatic (ACPE) control points extraction and their own discrepancies among them (MCPE-ACPE).

Table 13.4. MCPE(manual) and ACPE(automatic)

	MCPE	Stand.Dev.	ACPE	Stand.Dev.	MCPE-ACPE	Unit
k	5.43635	0.00030	5.43676	0.00026	-0.00041	rad
f	0.01272	0.00141	0.01736	0.00172	-0.00464	rad
w	0.01163	0.00154	0.00663	0.00104	0.00500	rad
X0	677486.065	1.113	677490.129	1.328	-4.065	m
Y0	7184230.551	1.207	7184233.150	0.829	-2.599	m
Z0	1660.648	0.279	1659.241	0.214	1.407	m

The Table 13.5 shows discrepancies between the checkpoints coordinates (extracted manually and assumed as a reference) and their own being calculated from MCPE and ACPE, applied to collinearity equations into an inverted state. The M index refers to the manual process, and the A index to the automatic process; the R denotes the resultant discrepancies outcome between dE and dN. The Mean and the Standard Deviation values refer to discrepancies and their own results.

The Fig. 13.15 (a) indicates the geometrical distribution of the control points manually measured (squares), and the geometrical disposition of the checkpoints (circles); the Fig. 13.15 (b) shows the geometrical configuration of the control points automatically extracted (squares) and the checkpoints (circles).

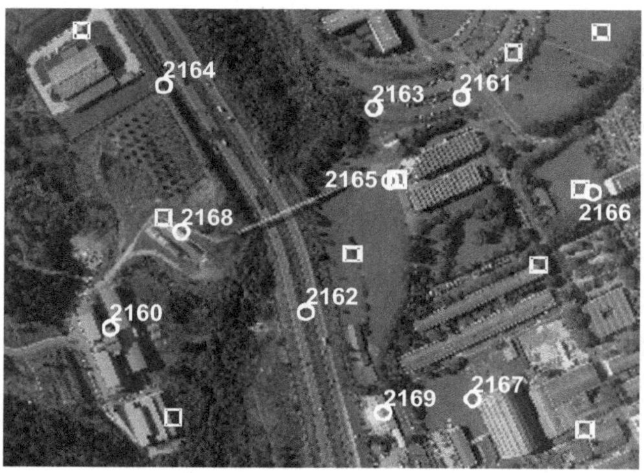

a)

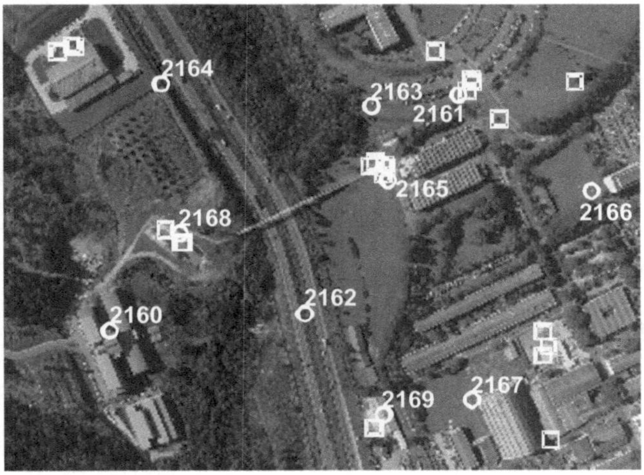

b)

Fig. 13.15. Image 216 control points extraction: manual (a), and automatic (b)

Table 13.5. Discrepancies in the image 216 checkpoints

Checkpoint	dEM(m)	dNM(m)	RM(m)	dEA(m)	dNA(m)	RA(m)
2160	-0.128	1.365	1.371	0.343	-0.374	0.507
2161	-0.101	0.081	0.129	-0.191	-0.534	0.567
2162	-0.394	0.120	0.412	-0.128	-0.591	0.605
2163	-0.045	0.399	0.401	0.000	-0.264	0.264
2164	-0.136	0.427	0.448	0.088	-0.664	0.670
2165	-0.228	0.074	0.240	-0.092	-0.590	0.597
2166	-0.200	-0.010	0.201	-0.148	-1.418	1.426
2167	-0.258	0.808	0.848	0.102	0.368	0.382
2168	0.296	0.175	0.344	0.818	-0.684	1.066
2169	-0.325	0.351	0.479	-0.008	-0.276	0.276
Mean	-0.152	0.379	0.487	0.078	-0.503	0.636
Stand.Dev.	0.190	0.422	0.368	0.303	0.447	0.360

13.3.4.1 Comments about the image 216 processing

a. The criteria for automatic process of digital images were efficient, because all process had neither failure nor interruption;
b. Twenty control points extracted via automatic; and their geometrical distribution fulfilled the phototriangulation by the bundle method; there is no control point at the left bottom of the image 216, though;
c. The exterior orientation parameters (determined through manual and automatic control points extraction) produced compatible coordinates to the checkpoints,

whose discrepancies related to the checking coordinates presented, comparable by themselves to positional evaluation (Table 13.5 – Mean and Standard Deviation). The differences between manual and automatic methods in this experiment got down under 1 pixel (or about 25 cm over the surface).

d. The automatic detection identified control points very nearly to each other, as it is indicated on the Fig. 13.15 (b); although not being desirable, this fact had a positive consequence by geometric compensation, according to the item c) report.
e. The experiment had a successful methodology to extract control points from the LIDAR database automatically, integrating them on the digital aerial image through the spatial resection performance.

13.4 Conclusions and Remarks

The methodology in focus on carrying out automatic integration of digital aerial images and LIDAR data – was efficient, according to the experiment outcomes.

About the enhanced image execution with focus on the correspondence based on the correlation coefficient, a treatment was developed to the whole image, which affords to systematize the acquired correspondences between these kinds of images.

All stages through the experiment had a continuous process, that is, neither interruption nor human interruption (limited to the process criteria definition), which got the methodology qualification as an automatic one.

As filter applications of borders based on the chain dimension as filter of corners based on neighborhood testing between corners and borders showed an excellent performance of filtering spurious corners, nearly 75% of the started total from the experiment carried out (Table 13.2, EI, II Filtering).

The correspondence based on the cross correlation coefficient, although applied between images captured by different sensors (correlation between digital aerial image and LIDAR intensity image) came up with a high performance of eliminating false correspondences (about 73% of eliminating the correspondences left (Table 13.2, CC) from the filter actions based on the chain and neighborhood dimension).

Selecting correspondences based on AT is essential in the developed method, because of its effective and proved capacity to select true correspondences (Table 13.2 – AT).

As long as the photogrammetric process is based on the phototriangulation by bundle method, the geometric distribution of the control points in the photogrammetric unit has a reiterated importance. The automatic process was less efficient than the manual through the experiment carried out, concerning the distribution of the control points.

The outcomes acquired by the tests carried out may come out a positive conclusion, giving an account of using the potential LIDAR database to bring into focus the control points, especially from this study, whose task assigns the automation of the extraction of the control points and their use in the automatic and semi-automatic Photogrammetry.

It is recommended the methodology might be applied its use to other situations, for instance, as integrating conventional (digitalized) aerial images and LIDAR data, as using in Close Range Photogrammetry for integrating digital images and terrestrial laser scanning system. It is suggested improvements and adjustments to be added throughout the developed methodology stages by choosing new algorithms, and by new developing technology.

13.5 Acknowledgments

The authors wish to thank the *Conselho Nacional de Desenvolvimento Científico e Tecnológico* – CNPq (Brazil), and the *Deutscher Akademischer Austauschdienst* – DAAD (Germany) for financial and logistic incentive of the fomentation programs.

References

1. Canny JA (1986) A computational approach to edge detection. IEEE Transactions on Pattern Analysis and Machine Intelligence 8:679-698
2. Dalmolin Q, Santos DR, Delara R, Bähr HP (2005) O uso de feições retas extraídas do MDT gerado pelo sistema de varredura laser como apoio para cálculo da ressecção espacial de imagens digitais. Boletim de Ciências Geodésicas CPGCG/UFPR 11:117-140
3. Delara R (2003) Calibração de câmeras digitais não métricas e de pequeno formato utilizando o pixel como unidade no espaço imagem. UFPR, Curitiba
4. Delara R, Mitishita EA, Habib A (2004) Bundle adjustment of images from non-metric CCD camera using LIDAR data as control points. International Archives of XXth ISPRS Congress 2004, Istanbul, Turkey, Comission III 13-19
5. Delara R (2007) Extração automática de pontos de apoio para integração de imagens aéreas digitais e dados de perfilamento laser aerotransportado. Curitiba. Thesis (Ph.D), Curso de Pós-Graduação em Ciências Geodésicas, UFPR
6. Gonzalez RC, Woods RE (1992) Digital Image Processing. Addison Wesley
7. Grün A (2000) Semi-automated approaches to site recording and modeling. International Archives of XIXth Congress ISPRS 2000, Amsterdam, v. XXXIII, WG V/3, part B5/1: 309-318
8. Gülch E (2000) Digital systems for automated cartographic feature extraction. International archives of Photogrammetry and Remote Sensing, Vol. XXXII, Part B2:.241-256
9. Habib AF, Ghanma MS, Morgan MF, Mitishita EA (2004a) Integration of laser and photogrammetric data for calibration purposes. International Archives of XXth ISPRS Congress, Istanbul, Turkey, Comission I
10. Habib AF, Ghanma MS, Kim CJ, Mitishita EA (2004b) Alternative approaches for utilizing LIDAR data as a source of control information for photogrammetric models. International Archives of XXth ISPRS Congress, Istanbul, Turkey, PS WG I/5 Platform and sensor Integration,
11. Habib AF, Ghanma MS, Tait M (2004c) Integration of LIDAR and photogrammetry for close range applications. International Archives of XXth ISPRS Congress, Istanbul, Turkey

12. Hahn M (1997) Automatic control point measurement. Photogrammetric Week '97. Fritsch, D.; Hobbie D. (eds), Heidelberg: Wichmann Verlag, 115-126
13. Harris C, Stephens M (1988) A combined corner and edge detector. Proceedings of Fourth Alvey Vision Conference. 189-192
14. Heipke C (1997) Automation of interior, relative, and absolute orientation. ISPRS Journal of Photogrammetry & Remote Sensing, 52:1-19
15. Jain R, Kasturi R, Schunck BG (1995) Machine Vision. McGraw-Hill, Inc, Cingapura
16. Niederöst M (2000) Reliable reconstruction of buildings for digital map revision. International Archives of XIXth Congress of the ISPRS 2000, Amsterdam, WGIII/4:635-642
17. Polidorio AM, Imai NN, Tommaselli AMG, Flores FC, Franco C (2003) Realce do grau de artificialidade de feições em imagens aéreas coloridas. Curitiba, CPGCG/UFPR: Série em Ciências Geodésicas, 3:277-291
18. Rottensteiner F (2001) Semi-automatic extraction of buildings based on hybrid adjustment using 3D surface models and management of buiding data in a TIS. Viena, Thesis (Ph.D), Wien TU, Geowissenschaftiche Mitteilungen, Institut fuer Photogrammetrie und Fenerkundung
19. Santos DR (2005) Automação da resseção espacial de imagens com uso de hipóteses de rodovias como apoio de campo derivadas do sistema de varredura laser. Curitiba. Thesis (Ph.D), Curso de Pós-Graduação em Ciências Geodésicas, UFPR
20. Schenk T (1999) Digital Photogrammetry - Volume I. TerraScience, Laurelville, Ohio
21. Tommaselli AMG (1993) Um método recursivo aplicado ao problema de localização em Visão de Máquina. Campinas, Thesis (Ph.D), Faculdade de Engenharia Elétrica da Universidade de Campinas
22. Wang W, Dony RD (2004) Evaluation of image corner detectors for hardware implementation. Proceedings of the 2004 Canadian Conference on Eletrical and Computer Engineering, Niagara Falls, Canadá

Chapter 14
Automatic Surface Patch Generation from a Video Image Sequence

Yixiang Tian, Markus Gerke, George Vosselman and Qing Zhu

Abstract. Automatic 3D building reconstruction is becoming increasingly important for a number of applications. This paper aims to present a new approach for an automatic pipeline to reconstruct surface patches from a video image sequence, which can deal with images acquired with an uncalibrated hand held camera. In order to skip the details to get the surface patches that present the building's main structure, the algorithm applies rules and a processing sequence that aim to reasonably group extracted sparse 3D points and 3D edges. Then the surface patches are estimated from these grouped points and edges. The final step is to define the surface patches' outline, which aim to both present the location and the actual shape of the surface patches. The results for two datasets show that the main surface patches for each dataset are successfully reconstructed.

14.1 Introduction

Acquiring 3D models of real world objects has been an interesting and challenging problem in computer vision and photogrammetry, and is beneficial to many applications such as urban planning, architectural design, civilian and military emergency response. 3D city models constructed from ground based data are becoming more useful as they present the realistic façades, which contain more details than the models constructed from aerial data. Traditionally, ground based building extraction has mainly relied on manual operations with the support of some modeling software, such as 3ds Max, Google Sketchup, ImageModeler, which still remains in an expensive and time-consuming process, especially when a large amount of data must be processed.

International Institute for Geo-Information Science and Earth Observation (ITC),
e-mail: (ytian, gerke, vosselman)@itc.nl
State Key Lab of Information Engineering in Surveying Mapping and Remote Sensing, Wuhan University,
e-mail:zhuq66@263.net

In recent years, there has been intensive research activity on the reconstruction of 3D objects and scenes from the image sequences or image sets, especially in the field of computer vision [2, 4, 9, 10]. The high overlapping of images within a video sequence leads to highly redundant information, which is exploited for tie point extraction, feature tracking and 3D object extraction. However, the short baseline between the images also may lead to a poor ray intersection geometry and thus to a weak geometry across the sequence. The current state of automation in reconstruction of buildings from video image sequences is still not flexible. Some algorithms and systems have been developed [3, 5, 8]. Most systems try to extract detailed 3D points from video streams using state-of-art dense matching algorithms, which incur high computational cost. Presenting the constructed 3D geometry from the result of triangulating the dense point clouds waste a lot of memory space, is largely affected by depth error and cannot directly represent a building's shape.

As a majority of buildings in existence nowadays satisfy the assumption that they can geometrically be modeled as an ensemble of planar polygonal surface patches using polyhedral models seem to be a relatively simple and efficient way to represent building structures [13]. Because surface patches are the main components of the polyhedral models, this paper aims to present a new approach for an automatic pipeline to reconstruct surface patches from a video image sequence based on sparse 3D points and 3D edges.

In section 14.2 we first describe the preprocessing steps on camera parameters, while point and edge feature extraction methods are explained in section 14.3. Section 14.4 is the main part of this paper. We describe the approach to group 3D points and edges, estimate plane parameters and define on the outline of surface patches. Results of the surface patch generation are shown and discussed in section 14.5. Finally some conclusions are given in section 14.6.

14.2 Preprocessing

Starting from a collection of images or a video sequence, the initial step consists in relating the different images to each other by feature tracking. Only the relative position of cameras can be recovered, the overall scale of the configuration can not be recovered solely from images without prior information in the scene [7]. We use the commercial software Boujou [1] to get camera projection information and through some visual assessment we assure that the results are correct. In the subsequence steps the 3D points, their corresponding 2D points and the individual projection matrices are used.

14.3 Feature Extraction

As edges can provide more constraints about objects' shape than points, both points and edges are considered as basic features for surface patch generation in our method.

Edges are first detected in each frame separately by using EDISON edge detector, which introduce a confidence measure into the Canny edge detector and results in more connected and smoothed edges [6]. Then Hough transformation is adopted to extract straight edges. After analyzing the quality of points obtained from Boujou, reliably matched points are used as guidance for edge matching, which means only edges near these good quality points are considered. 3D edge parameters are estimated from the matched 2D edges by using the Gauss-Markoff model with constraints. As the whole sequence is incorporated in this adjustment the aforementioned weakness of video sequences, namely the short frame-to-frame baseline does not have an effect here. End points of each 3D edge are found by analyzing the end points of the corresponding 2D edges. This method for 3D edge extraction has been explained in [11]. As the common way for taking video is to maintain an almost constant height of the camera during capture, the camera is moving horizontally. Vertical edges can be better estimated than horizontal edges. As the result, the vertical direction, which is important for the following steps, can be easily obtained after edge extraction.

To obtain anyhow horizontal edges we exploit the fact that most structures on manmade objects are orthogonal. So, if reliable points are placed on a line, perpendicular to a reliable vertical edge, we insert an additional horizontal edge by connecting those points.

Fig. 14.1. Feature extraction result, reliable points (dot), matched edges (finite line)

Fig. 14.1 shows extracted 3D points and edges projected on the first frame of a video image sequence. The video was taken by a moving car with a mounted camera. The camera was oriented sideways and captured the facades of buildings. 3D view for edge extraction result is shown in Fig. 14.2. As we only get sparse points and edges and some semantic edges are missing or not complete, we need to incorporate some object knowledge (model) to construct the patches, which is explained in the next sections.

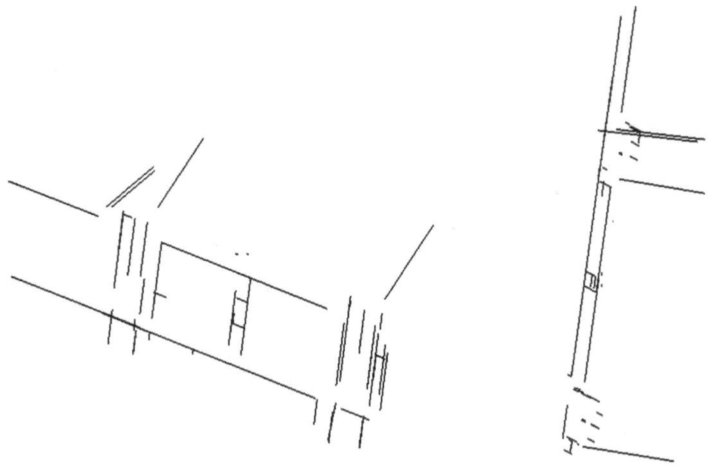

Fig. 14.2. 3D view of edge extraction result in Fig 14.1, side view (left), top view (right)

14.4 Surface Patch Generation

In this step, we group the estimated 3D points and 3D edges to more complex aggregates (surface patches). 3D edges can be aggregated to 3D corners based on the assumption that edges are coplanar and non-parallel. However, for the building façade, there may be only parallel edges on the same plane extracted, and some intersecting edges belong to detail objects, like windows, which are almost located in the same plane as the interesting wall. However, at this stage we focus on the main structure that let people recognize the building, such as walls and roofs.

Our method to get the surface patches can be divided into four steps. First, plane hypotheses are formulated from cues based on point-cloud segmentation and on some of the 3D edges derived beforehand. The hypotheses then are verified by incorporating unused 3D edges. Afterwards, plane parameters are obtained by all the edges and points in the plane. The one with the least residual RMS is chosen as the best fit. The final step is to define the surface patches' outline. The details are presented in the following.

14.4.1 Feature Grouping

To retrieve complete surface patches, coplanarity is the necessary but not sufficient condition for edge and point grouping. More conditions need to be satisfied to make sure that the surface patches being reconstructed from the features correspond to the realistic façade structure. Therefore, we consider cues from point cloud segmentation, intersecting edges and parallel edges and apply these cues in sequence. As some points and edges may lie on the boundary of planes, our grouping method allows an overlapping cluster result.

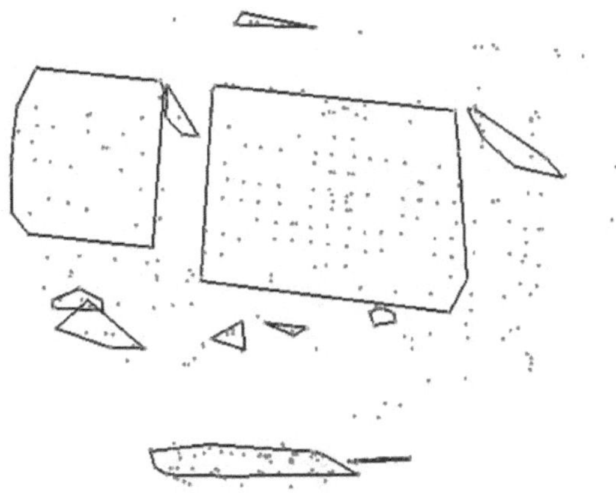

Fig. 14.3. Point cloud segmentation result of data in Fig. 14.1, reliable points (dot), point segments with convex outline (polygon)

1. Cue from point cloud segmentation. The purpose of the point cloud segmentation is to find the main building wall, which usually is the largest plane with some windows, doors and symbols to let people separate it from other buildings on the ground, so it contains more salient feature points then other parts of building. We adopt the planar surface-growing algorithm by [12], which consists of a seed surface detection followed by the actual growing of the seed surface. Because this method is mainly used for laser scanning data that is much denser then point cloud extracted from image sequence without using dense matching technique, we choose a large surface growing radius, which lead to some segments that do not exist in the real façade. 3D edges and the vertical direction are used to judge their reasonability. The rules will be explained in section 14.4.2. As shown in Fig. 14.3, two vertical walls can be reconstructed from the result of point cloud segment and small segments will be removed later.

2. Cue from two intersecting edges. If two 3D edges intersect, they must be in the same plane. But we do not want to group edges that are not corresponding to actual building façade. For example, edge 1 and edge 2 in Fig. 14.4 are intersected in the object space. Such plane hypothesis can be excluded by the relation between intersection point and two edges. Considering the units in object space, if the distance between the intersection point and any edge is lager than a threshold, the hypothesis is rejected. The distance between the intersection point and finite 3D edge is defined by the relationship between this intersection point and two endpoints of the edge. If the point is located between two endpoints, the distance value is zero, see the distance between the intersection point and edge 1 in Fig. 14.4. Otherwise, the distance is the minimal value of the distance between the point and endpoints, as the distance between the intersection point and edge 2 in Fig. 14.4.

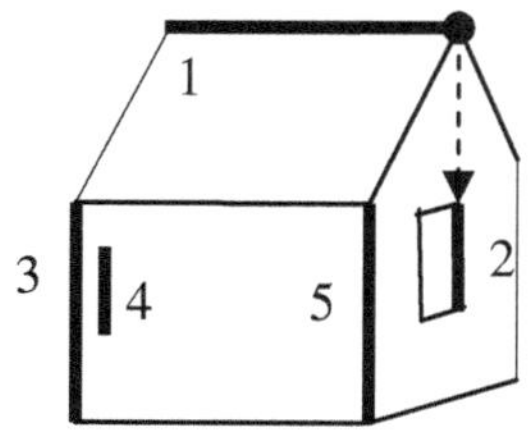

Fig. 14.4. A simple building with extracted edge 1-5

3. Cue from two parallel edges. As we known many parallel or almost parallel edges exist on the building façade. Furthermore, either too close or too far apart parallel edges are not reliable to make a surface patch hypothesis. We specify a minimal threshold and a maximal threshold between the two 3D edges, according to edge extraction result. There is one example in Fig. 14.4. If edge 4 is located a little in or out of the wall, edge 3 and edge 4 may result in a plane hypothesis. The minimal threshold is used to avoid such detail part and the maximal threshold is used to exclude edges that by chance lie on the plane formed by a cluster of closer edges. The sequence of edge processing within this kind of reasoning is not arbitrary. Therefore additional rules need to be applied. One is for each edge a hypothesis is first made from the parallel edge with a smallest distance above the minimal threshold. The other is if any one of the two edges was already grouped to a surface patch, we search for other parallel edges belonging to that patch and choose the two with smallest distance to make a hypothesis. For example, edge 2, edge 3 and edge 5 are parallel and recorded in that sequence. According to our rules, edge 2 and edge 5 form a surface patch before a hypothesis from edge 2 and edge 3. Meanwhile, the later one is modified by edge 2 and edge 5. So the plane hypothesis made from parallel edge 2 and edge 3 can be avoided.

14.4.2 Plane Verification

After defining the plane hypotheses, they are verified or modified by 3D edges that are not used for grouping so far. Two parameters must be given first to decide whether an edge belongs to a plane: A threshold determining the distance of endpoints from the plane and the maximum angle between the edge and the plane.

As we consider cues in sequence, the later cues are considered after verifying all hypotheses from the former one. However, the way to verify and implement cues is related to the particular cues, which are used to define the hypothesis. For the first type, i.e. the point cloud based cue, we segment all the points at one time. For each segment, a convex hull is calculated. Then the verification is done by testing whether there are some 3D edges belonging to that plane and located in that region. If there is no edge belonging to it and it is not parallel or orthogonal to the vertical direction, the respective plane hypothesis is rejected.

For the other two cue types, a new hypothesis is made after defining the outline of the former hypothesis. It doesn't need further verification. But additional edges that are belonging to it need to be included. Simply say this step is to grow a small surface with two edges to a bigger one by adding more edges to it. The way to judge whether an edge can be added to a hypothesis is the same as some rules used in making hypothesis from edges. If an edge is intersected with any edge belonging to the hypothesis with an intersection point near to it or parallel to any edge belonging to the hypothesis within the maximum distance, it is added to the hypothesis.

14.4.3 Plane Estimation

Plane parameters are obtained by all the edges and points in the plane. The one with the least residual RMS is chosen as the best fit. As some points and edges may lie on the boundary of planes, we label the edges inside the respective region after defining the outline of a plane. Therefore, the edges at the boundary can be grouped into more than one surface patch. This overlapping cluster method results in more surface patches, meanwhile requires effective outline reconstruction method that can present patches' shape and correctly judge edges at the boundary.

14.4.4 Outline Reconstruction

After estimating the plane, the normal direction of it can also be determined. In addition, according to the vertical direction, surface patches are separated to three types, vertical, horizontal and oblique surface patch. We assume that walls are vertical and most common surfaces have a rectangular shape, therefore in the absence of slanted edges within a particular patch we conclude it being rectangular. In addition, the horizontal part on the building is difficult to be observed from ground-based video, but there are usually some points and edges on the ground that is in front of the building. We can simply use a convex hull to present the ground.

By observing the general building structure and surface patches that can be seen from ground-based video, we make a conclusion in Table 14.1 about the relation between three kinds of surface patches, building structures and generalized shapes. Extrusion includes balcony, overhanging part on the wall or roof. The shapes in Table 14.1 can be rotated, stretched or have different length ratio between edges in object space. They are not complete to present all kinds of surface shapes. Nevertheless, they are commonly used shapes to simplify building facades. In order to exclude areas that are not belonging to a patch, the third and fourth shapes in the upper row are two different shapes, although the fourth shape is a combination of the third shape. All these knowledge about general building façade shapes are considered in this step, which is described in the following.

Table 14.1. The relation between surface type, building structure and generalized shape

	Building structure	Generalized Shape
Vertical	wall, window, door, extrusion	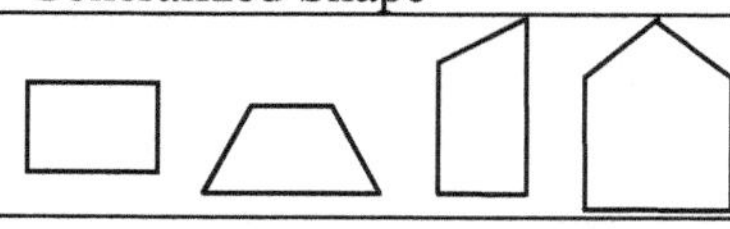
horizontal	eave ,extrusion	
Oblique	roof, extrusion	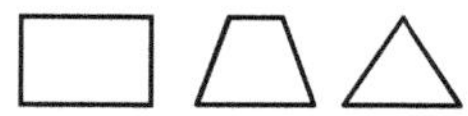

- Vertical surface patches
  If slanted edges are not on the side part, we assume there are two outline edges parallel to the vertical direction.
  If all the slanted edges are located in the upper part, we assume there is a straight edge parallel to the ground at the lower part, because those kinds of shapes mostly belong to wall, such as third and fourth shape in the upper row.
- Oblique surface patches
  If slanted edges are intersected and the intersecting point has a small distance to them, we use triangle shape to fit the surface patch outline, otherwise, the trapezoid is considered.

14.5 Results

Fig. 14.5 shows the surface patch generation result of the building façade in Fig. 14.1 and Fig. 14.2. All reconstructed surface patches are located in a correct position. From them, the basic structure for this part of construction can be recognized. However, outlines of some surface patches need to be modified. For example, there is a long 3D edge on the main wall plane (surface patch 1), which result in part of it occludes the other surface patch behind it (surface patch 2). As the edges of glass wall and door are in the same plane and near to each other, surface patch 2 includes all of them. In addition, one surface patch does not exist in real world (surface patch 3), because the intersection edge of surface patch 4 and surface patch 5 was not observed. Two parallel edges, which are at the boundary of those two surface patches, form surface patch 3.

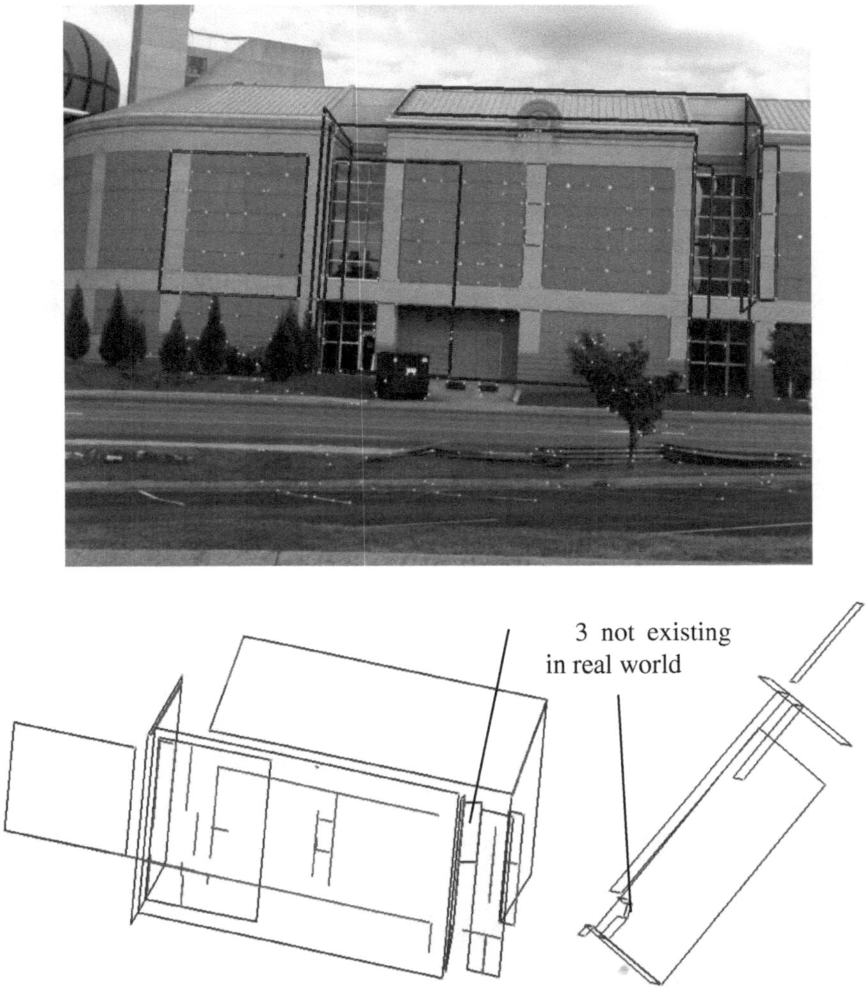

Fig. 14.5. Surface patch generation result of building façade in Fig 14.1 and Fig. 14.2, 2D view (top), 3D side view (down left), 3D top view (down right), reliable points (dot), extracted edges (finite line), surface patches (polygon)

Fig. 14.6 shows the first and last frame from another example. The video was captured by a hand-held SONY camera. The images have a resolution of 640×480 pixels and a frame rate of 30 frames per second. There are 189 frames in total in this case.

Fig. 14.6. Input image sequence with eight surfaces (each surface with one number), first frame (left), last frame (right)

Eight surfaces can be seen from the image sequence and seven of them are reconstructed as shown in Fig. 14.7, although most of them need to be extended. The real boundary can be reconstructed by intersecting adjacent patches. The only one lost is a small extrusion (surface 7), which cannot be easily seen from the image sequence, as shown inside the circle in Fig. 14.6.

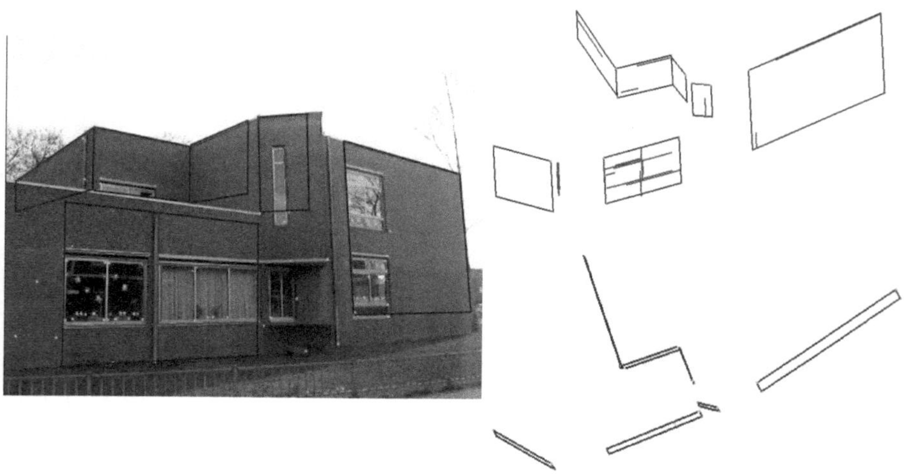

Fig. 14.7. Surface generation result of building façade in Fig 14.5, 2D view (left), 3D side view (right top), 3D top view (right down), reliable points (dot), extracted edges (finite line), surface patches (polygon)

14.6 Conclusions

Within this paper, an approach of surface patch reconstruction from a video image sequence was presented. The rules and processing steps can reasonably group extracted sparse 3D points and 3D edges. The final step is to reconstruct the surface patches' outline, which aim to both present the location and the basic shape of the surface patches. The shape information helps to restrict planes in the region corresponding to the actual case and make it easier to connect adjacent patches later. Up to now, only geometric information conveyed by the 3D points and 3D edges has been used for surface patch generation. Although the reconstructed surface patches present the building façade geometry, three aspects need to be mentioned. First, if there is no point or edge extracted from a surface patch, the pure geometric reconstruction fails to determine it. Secondly, small surface patches appear between surface patches which should be adjacent. Third, the outlines still need to be modified in the following steps.

In the near future, we will focus on how to set up relationship between these surface patches, extent or separate them, remove surface patches that do not exist and add new ones. Both geometric information and intensity information will be considered and some knowledge about building structures will be used. Finally, we want to build geometrically and topologically correct 3D models of the building facade.

14.7 Acknowledgements

Parts of the image data (Stadium, cf. Fig. 14.1) was made available by Matthias Heinrichs, Technical University of Berlin, Germany. We gratefully acknowledge his support. The work described in this paper is supported by National Natural Science Foundation of China (40871212 and 40671158), and National High Technology Research and Development Program of China (2006AA12Z224).

References

1. 2d3 Homepage for boujou software, http://www.2d3.com (accessed 18 August 2008)
2. Baltsavias, E.P.: Object extraction and revision by image analysis using existing geodata and knowledge: current status and steps towards operational systems. ISPRS Journal of Photogrammetry and Remote Sensing 58(3-4): 129-151 (2004)
3. Cornelis, N., Leibe, B., Cornelis, K., Van Gool, L.: 3D Urban Scene Modeling Integrating Recognition and Reconstruction. International Journal of Computer Vision 78(2): 121-141 (2008)
4. Hartley, R.I., Zisserman, A.: Multiple View Geometry in Computer Vision (second edition): Cambridge University Press, Cambridge (2000)

5. Mayer, H., Reznik, S.: Building facade interpretation from uncalibrated wide-baseline image sequences. ISPRS Journal of Photogrammetry and Remote Sensing 61(6): 371-380 (2007)
6. Meer, P., Georgescu, B.: Edge detection with embedded confidence. IEEE Transactions on Pattern Analysis and Machine Intelligence 23(12): 1351-1365 (2001)
7. Nistér, D.: An efficient solution to the five-point relative pose problem. IEEE Transactions on Pattern Analysis and Machine Intelligence 26(6): 756-770 (2004)
8. Pollefeys, M., Nistér, D., Frahm, J.M., Akbarzadeh, A., Mordohai, P., Clipp, B., Engels, C., Gallup, D., Kim, S.J., Merrell, P., Salmi, C., Sinha, S., Talton, B., Wang, L., Yang, Q., Stewénius, H., Yang, R., Welch, G., Towles, H.: Detailed Real-Time Urban 3D Reconstruction from Video. International Journal of Computer Vision 78(2): 143-167 (2008)
9. Pollefeys, M., Van Gool, L., Vergauwen, M., Verbiest, F., Cornelis, K., Tops, J., Koch, R.: Visual modelling with a hand-held camera. International Journal of Computer Vision 59(3): 207-232 (2004)
10. Remondino, F., EL-Hakim, S.: Image-Based 3D Modeling: A Review. The Photogrammetric Record 21(115): 269-291 (2006)
11. Tian, Y., Gerke, M., Vosselman, G., Zhu, Q.: Automatic Edge Matching Across an Image Sequence Based on Reliable Points. In: Chen, J., Jiang, J., Förstnner, W. (eds.) the International Archives of the Photogrammetry, Remote Sensing and Spatial Information Science Beijing, Vol.XXXVII, Part B3b, pp. 657-662 (2008)
12. Vosselman, G., Gorte, B., Sithole, G., Rabbani, T.: Recognising structure in laser scanning point clouds. In: Thies, M., Koch, B., Spiecker, H., Weinacher, H. (eds.) International Conference NATSCAN, Laser-Scanners for Fores and Landscpae Assessment - Instruments, Processing Methods and Applications, ISPRS working group VIII/2. Freiburg im Breisgau, Germany, pp. 33-38 (2004)
13. Werner, T., Zisserman, A.: New techniques for automated architectural reconstruction from photographs. In: Seventh European Conference on Computer Vision, Vol.II, pp. 541-555 (2002)

Chapter 15
Indoor 3D Modeling and Visualization with a 3D Terrestrial Laser Scanner

Jia Dongzhen, Tor Yam Khoon, Zhong Zheng and Zhou Qi

Abstract. For indoor wireless location, it is useful to have 3D models to document and to validate the results of the locations derived from wireless sensing equipment. Visualization using a 3D model also allows the user to generate a realistic mental model while navigating their virtual surroundings. While it is possible to build a 3D framework model from CAD drawings, the results do not feature the exact locations and models of indoor equipment, such as tables, fire sprinklers, fluorescent lamps, and so on. A 3D Terrestrial Laser Scanner (3DTLS) is a tool that can be used to record indoor 3D coordinates of real objects efficiently and effectively. It has remarkable advantages including high data acquisition rate, high accuracy and excellent spatial data density (Yusuf A. 2007). The resulting dense "point-cloud" of data makes it easy to model any scanned object. In this study, we summarize our experiences in scanning and modeling. Using visual C++ and OpenGL, we developed a 3D visualization platform that enables browsing and navigation of 3D models and other applications. A case study using 3DTLS at the NTU Position & Wireless Technology Center (PWTC) is also presented in this paper.

15.1 Introduction

Laser scanner technology was introduced in the 1990s [3]. The first generation terrestrial laser scanner was introduced in the surveying industry in 1998. 3D Terrestrial laser scanners make use of non-contact high-speed laser pulses for measuring the

---

Institute of Satellite Navigation & Spatial Information System, Hohai University, Nanjing
jiadongzhen@hhu.edu.cn
School of Civil and Environmental Engineering, Nanyang Technological University
cyktor@pmail.ntu.edu.sg
LIESMARS, Wuhan University
seven-laand@hotmail.com

surface profiles of objects. Scanners directly acquire distance and reflection intensity of object surfaces and then automatically store the data in a database. As a remote and rapid data acquisition tool, the technology has matured over the last few years through an impressive and steep development curve. The latest terrestrial laser scanning technology captures up to 500,000 points per second [8] and achieves point accuracy of ±6 mm or better [5]. Such systems are able to acquire immensely accurate 3D points within a short period of time. In view of advantages such as speedy long-range measurements [1], high spatial resolution, high speed data acquisition and high density point clouds, the technology opens new opportunities for data acquisition and modeling.

An entire scan collects a series of 3D points called "point clouds." These provide the basis for surface reconstruction or modeling. The point clouds in most cases are just an intermediate step to reach results and deliverables. The real value is in the information implicitly stored and contained in the point clouds, such as the size, shape, form, location and surface characteristics of objects in the real world [4]. The objectives of this study are to make use of point clouds and extract information from these point clouds. Subsequently, we use points, polygons, curves, etc. to reconstruct a surface or object and then simulate the real world.

PWTC, the studied region, has a total area of about 550-600 square meters. There are about 10 rooms, 100 fluorescent lamps, 80 fire sprinklers and 20 tables. In this paper we present the 3D structure of the whole building, especially the 3D elements mentioned above.

15.2 Data Acquisition and Processing

15.2.1 Accuracy and Resolution of Scanning

Generally, without taking into account the inherent measuring signal error of an instrument, scanning accuracy depends mainly on the size of the laser spot and the distance between two successive points [7]. The smaller the laser spot and the shorter the scanning distance, the higher will be the resulting accuracy. The Leica ScanStation 3DTLS used in this study has a point accuracy of 6 mm at 1 sigma as per specification. The accuracy of the modeled object is specified as 2 mm at 1 sigma. The size of the laser spot is 4 mm (FWHH - based) from 0 – 50 m and the point spacing is fully selectable horizontally and vertically; <1 mm minimum spacing, through the full range of 300 m.

The resolution of the scanning, i.e. the distance between the adjacent scanned points, depends on the incident angle and range of the laser beam. For instance, the resolution of the scanned points on the ceiling and the floor of a room is a function of the incident angle between the laser beam and the plane of the ceiling and the floor. The desired maximum resolution can be achieved by limiting the radius of scanning.

As shown in Fig. 15.1, where θ is the angle between two successive incident laser beams; L & S are the preset range & resolution; h is the distance between the horizontal axis of the instrument and the ceiling or floor; R is the limiting scanning radius; r

is the maximum resolution of the model; and the distances between laser scanner and two scanned points are x and y, respectively, and $x \bullet y$. θ can be evaluated as follows:

$$\theta = r / y \tag{1}$$

and,

$$\theta = S / L \tag{2}$$

so,

$$x \bullet y = L \bullet r / S \tag{3}$$

According to right-angle triangle geometry:

$$R = \sqrt{y^2 - h^2} = \sqrt{\left(\frac{L \cdot r}{S}\right)^2 - h^2} \tag{4}$$

Since L, S, h, r are known or preset/desired values, the maximum value of the scanning radius R can readily be computed. The number of instrument stations for an indoor scanning system can thereby be planned.

15.2.2 Data Acquisition

Following the above calculation, one must survey all the lights and fire sprinklers, before creating a plan outlining the location of each scanner. In order to combine the different scanning scenes into a single coordinate system, it is necessary to acquire certain common points, defined by either targets or distinct features, between two adjacent scanner stations.

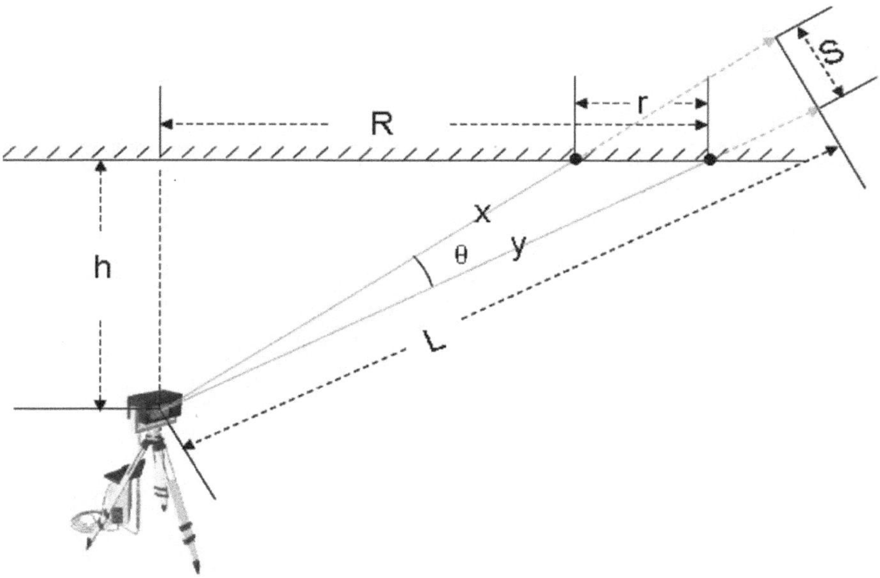

Fig. 15.2. Calculation of scanning radius

Although scanning the whole PWTC in high resolution with more scanning stations is feasible, it would not only take more time, but would also acquire superfluous point clouds that would be of little use for modeling. Level of detail (LOD) refers to the situation where one has a group of models at different scales, which are used to describe a single scene or its object [6]. Adopting the same concept as LOD, an initial prescan was taken at a low resolution of 5 mm to acquire panoramic data, followed by a higher resolution of 1 mm to scan key elements, such as the fluorescent lamps and fire sprinklers. This not only meets the accuracy requirements, but also substantially reduces redundant data. Ten sets of point clouds, one from each scanning station, were obtained and stored in a single database. The scanning process lasted one and a half days.

15.2.3 Data Processing

The ten sets of point clouds, each with its own coordinate system, must be aligned into one common coordinate system using a registration process. The process is depicted in Fig. 15.2.

Point clouds are composed of a number of data including some noise components that are derived from surface quality and laser dispersion [2]. Noise removal and smoothing should be performed after registration.

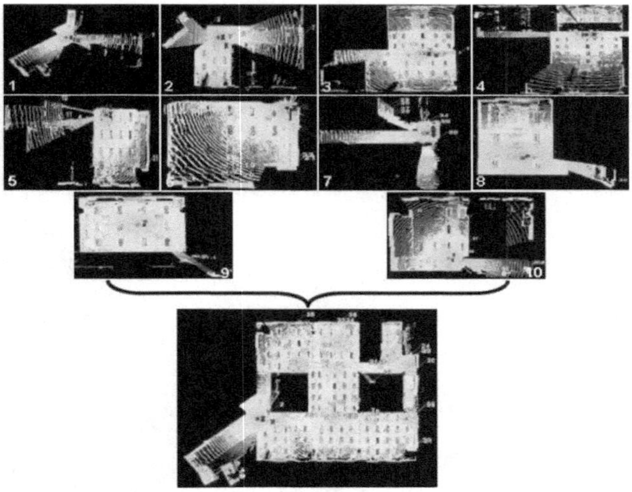

Fig. 15.2. Results of point cloud registration

15.2.4 Modeling

Point clouds acquired by 3DTLS comprise discrete vector points that themselves contain rich characteristic information. However, the points do not include any shape and topological relationships. Accordingly, modeling is usually performed by experienced operators. The process requires skill to produce a surface that has sufficiently high quality and precision [2].

The 3D object model comprises geometric and textural models. Since the most common representation of such objects in 3D models uses lines, surfaces and volumes, we used these basic geometric elements as the smallest units in our model. From the point clouds, the operator interactively constructs such geometric elements and integrates these elements to complete the model. The process comprises six main steps:

1. Remove noise, i.e. spurious point clouds, from the whole scene;
2. Segregate the point clouds defining an object from the whole scene and model the object as lines, surfaces and volumes;
3. Merge the modeled object into the original scene;
4. Repeat step 2 and step 3 till the whole model is completed;
5. Fine tune the whole model, for instance by intersecting and trimming surfaces, replicating objects etc;
6. Textually map the surfaces into the model.

There are five basic models in our representation of PWTC, namely wall models, floor and ceiling models, desk models, fire sprinkler models and fluorescent lamp models. The entire model is shown in Fig. 15.3. The process of modeling took about two days.

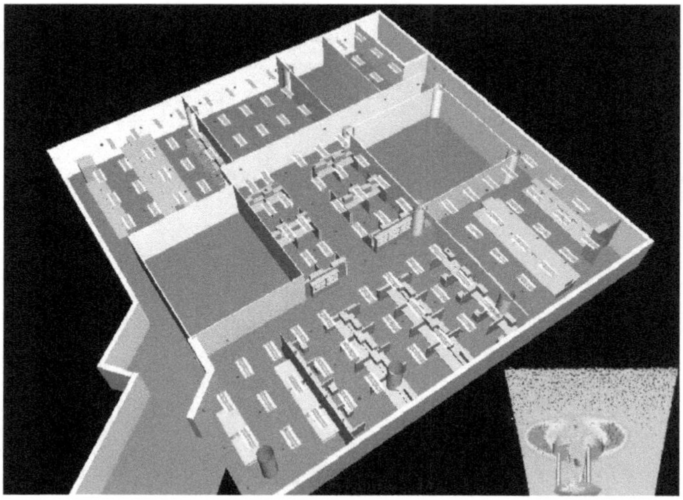

Fig. 15.3. Entire 3D model of PWTC

3 3D Visualization and Use of the Interactive Platform

Visualization is the basis of human-machine interaction. For the user, it provides both a window and a tool to interact with a real environment. It can immerse the user in a 3D scene and allow them to live in the model.

The work flow of the visualization processes consists of three parts, namely data preparation and preprocessing, establishment of the 3D model, 3D model management and display. The architecture of the system is shown in Fig. 15.4.

A data conversion system was developed with VC++ for 3D model creation, attribute data edition, management, etc. The 3D model of PWTC was converted to a 3D geography object library by the conversion system and this library was used in subsequent computation.

The visualization was performed using VC++ and OpenGL. The platform has the following features: multi-angle viewing, walkthrough, spatial query, distance measuring and so on. A 2D navigation map is provided for an intuitive browsing experience and to aid in user orientation. An interactive display of the 2D map and the associated 3D scene was implemented. The interfaces are as shown in Fig. 15.5 and Fig. 15.6. Crosses in Fig. 15.6 represent accurate positions of the fire sprinklers.

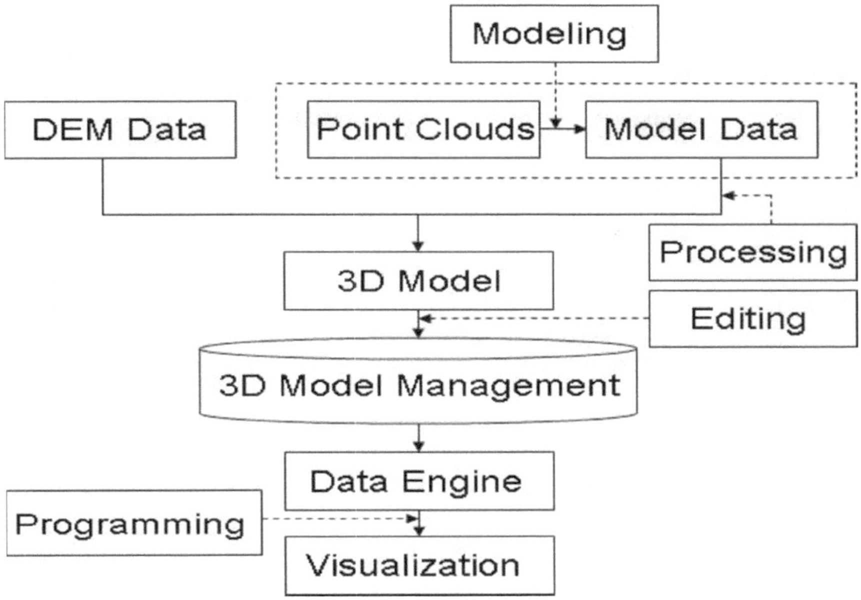

Fig. 15.4. Flowchart of the visualization process

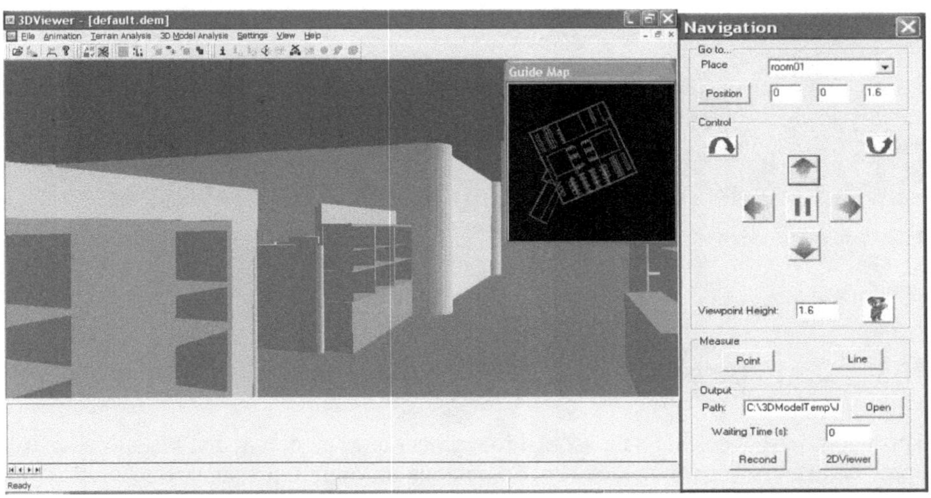

Fig. 15.5. Navigation of the 2D map and 3D scene

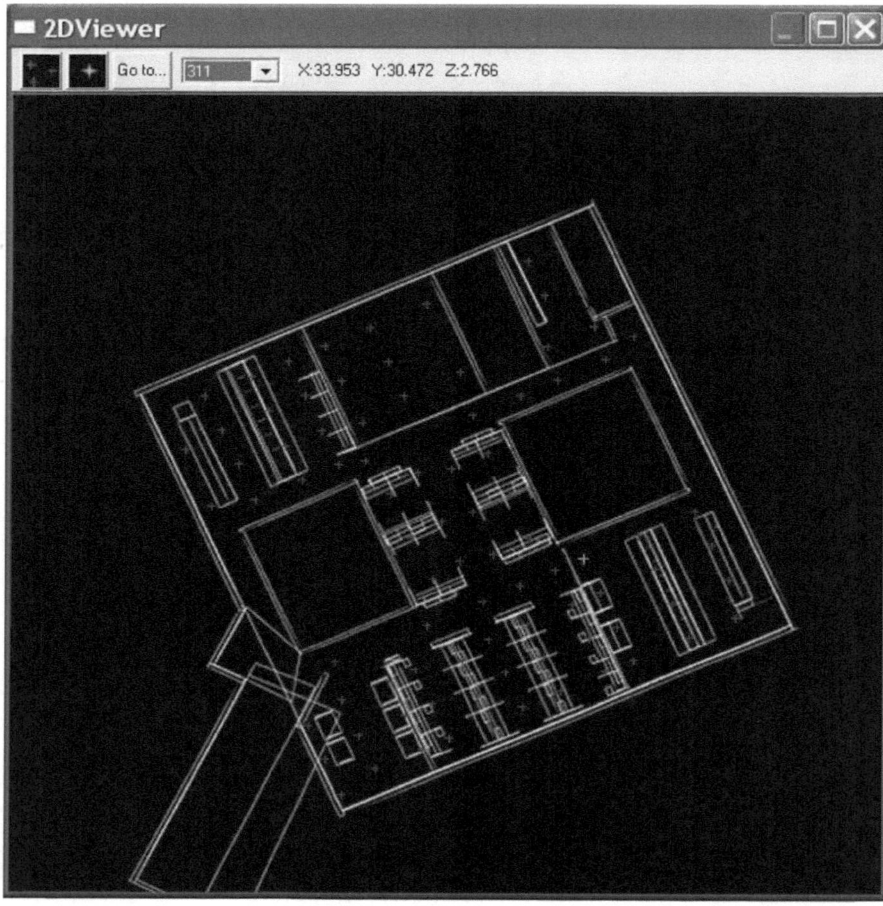

Fig. 15.6. Interface of the 2D viewer

15.4 Conclusions

This paper introduces 3D data acquisition, processing, modeling and visualization using a 3D Terrestrial Laser Scanner. To achieve an optimum scanning resolution for 3D point clouds, a detailed calculation method to select an appropriate laser scanning radius is presented and applied in this study. Due to different complexities of indoor objects and varying accuracy requirements, various scanning resolutions were used to scan the indoor objects. Our process helps reduce time and cost to meet the stipulated accuracy. A 3D application platform was developed for the 3D model of PWTC, which allowed for the dynamic display of 3D models and interactive user operation.

However, our modeling process is not fully automated and the accuracy of the model relies to a certain extent on the experience of the operator. With more than

100,000 surfaces in the model, it was very time consuming to textually map these elements. The modeling process can be further improved.

15.5 Acknowledgements

The work reported in this paper was supported by Professor Tor, Technician Mr Eric Low of the Spatial Information Lab, and other colleagues during the first author's stay at NTU. I would like to thank all of them for the opportunity to conduct this research.

References

1. A. Giussani and M. Scaioni (2004) Application of TLS to support landslides study: survey planning, operational issues and data processing. J International Archives of Photogrammetry, Remote Sensing and Spatial Information Sciences, Vol. XXXVI - 8/W2:318-323
2. Choi J, Hur M and Lee H (2002) Free-form Surface Generation from Measuring Points using Laser Scanner. J International Journal of the Korean Society of Precision Engineering, 3: 15-23
3. Dott. Geol. Lucio Amato, dott. Geol. Giovanni Antonucci, Arch. Barbara Belnato (2003) The three dimensional laser scanner system: the new frontier for surveying. J International Archives of the Photogrammetry, Remote Sensing and Spatial Information Sciences Vol. XXXIV-5/W12: 17-22
4. Frei E, Kung J, Bukowski R (2003) High-definition surveying (HDS): a new era in reality capture. J International Archives of the Photogrammetry, Remote Sensing and Spatial Information Sciences Vol. XXXVI-8/W2: 262-271
5. J. Roca, H. Lorenzo, P. Arias, J. Armesto (2008) From laser point clouds to surfaces: Statistical nonparametric methods for three-dimensional reconstruction. Computer-Aided Design 40 :646–652
6. Lu W, Chen W, Pan Z, Shi J (2001) Smooth transition between models at multiple levels of detail. J MINI-MICRO SYSTEMS, 22: 171-174
7. Luo D, Zhu G, Lu L, Liao L (2005) Whole Object Deformation Monitoring Based on 3D Laser Scanning Technology. J Bulletin of Surveying and Mapping, 7: 40-42
8. Milev I, Gruendig L (2007) New Tools for Terrestrial Laser Scanning Applied for Monitoring Rails and Buildings. In: FIG Working Week 2007, Hong Kong, pp 1294-1303.
9. Yusuf A, Y (2007) An approach for real world data modelling with the 3D terrestrial laser scanner for built environment. J Automation in Construction, 16 (6), pp. 816-829

Chapter 16
Automatic Image Mosaic-Building Algorithm for Generating Facade Textures

Yunsheng Zhang, Qing Zhu, Jie Yu and Yeting Zhang

Abstract. Our goal is to create seamless texture mosaics for photo-realistic modeling of building facades. This paper presents an automatic image mosaic algorithm that resolves both geometry and color in a wide variety of images. A geometrical preprocessing step based on vanishing geometry theory is first used to rectify the typical distortion of terrestrial imagery. Subsequently at least 4 evenly distributed corresponding points are extracted as tie points for solving the parameters of affine transformations between two adjoining images. A coarse-to-fine matching strategy with good reliability and efficiency is employed for the automatic extraction of tie points, where point features are extracted using the Harris operator at the top level of the image pyramid and by means of the SUSAN operator at the bottom level. For image auto-dodging, the color transfer algorithm in $l\alpha\beta$ color space is introduced. Experimental results prove that the presented method is feasible and is robust to modeling repeated patterns in imaging building facades.

16.1 Introduction

Increasingly, cities all over the world are being encouraged to build photorealistic 3D city models. Texturing these models provides a photo-realistic effect and conveys the visually detailed geometrical structure of a building. Such details are becoming more important for many simulation applications, such as urban planning, virtual tourism, digital documentation of historic landmarks, and so on. Currently feasible means of collecting building surface texture include analyzing roof texture and facade texture directly from satellite or aircraft imagery, and collecting facade texture data from terrestrial images. In the case of facade texture, aerial images always suffer from

State Key Laboratory of Information Engineering in Surveying, Mapping and Remote Sensing, Wuhan Universit
sheng_650@163.com, zhuq66@263.net, yujie2xw@126.com, zhangyeting@263.net

distortion, perspective changes and non-ideal resolutions, this paper focuses only on the terrestrial image mosaicking. Manually collecting and processing terrestrial images seriously limits the production efficiency and visualization effects of the 3D cityscape. The result is a bottleneck problem for large-scale 3D city modeling. In most cases, because of the small field of view of cameras and due to limiting environmental conditions, such as the physical boundaries of narrow streets, it is very hard to capture an image that covers a whole facade, and there is a need to mosaic multiple images to cover the whole facade. Terrestrial images are often shot with a large inclination angle. Consequently, the original texture images often demand a great deal of geometric rectification and mosaicking, which is a time-consuming and labor-intensive manual task. Automatic processing methods therefore seem promising.

Due to major distortions of terrestrial images, basic geometrical preprocessing is necessary. Based on the abundant line information in facade images, vanishing point geometry is often used to recover the camera's pose [5].

The purpose of mosaicking is to recover mapping relations among multiple images, ensuring the mosaic image is geometrically and color-wise seamless. For image mosaicking, an appropriate mathematical relationship is needed to map pixel coordinates from one image to another before image registration. A variety of such parametric models are available, including planar motion models such as affine transforms, projective transforms, etc; 3D transformations can also be used. However, this paper only focuses on mosaicking facade images, so planar motion models are most suitable. Projective transformation uses eight parameters, and can be readily achieved using parameter estimation. Thus, affine transform is chosen in this paper to describe the relationships between images.

Once a suitable model has been chosen to describe the relationships between a pair of images, the basic problem is to estimate the parameters. There are two kinds of geometrical image mosaic methods: direct and feature-based methods. The direct method is to shift or warp the images relative to each other and to look at how much the pixels agree. Pixel-to-pixel matching is often used in the direct method, such as the multiple projection merge method [9], equidistant matching (Kyung and Soon 1999) and similar. The direct method operates on practice pixels, and the result is generally inferior in terms of efficiency. Feature-based methods utilize corresponding points or corresponding lines to estimate the parameters to create the image mosaic, which is robust to illumination change. One example is the image mosaic method based on SIFT [1], which uses regional segmentation and matching steps to estimate the rotation, the translation and the scale factor between two successive images in the input sequence. Harris point matching is guided by the estimated parameters [16]. Many feature-based methods are robust to viewpoint change and illumination change but also have low reliability with repetitive patterns.

Color differences in terrestrial images are usually very obvious because of variable imaging conditions. Multiple-image dodging methods are therefore necessary to decrease differences, using the mean-variance method, histogram matching, and color matching based on Wallis filter [8]. The mean-variance method needs image registration and the processed image has local contrast that often decreases after dodging. If large homogeneous regions exist, color aberrances will result after histogram matching because gray tones get overlaid and concentrated. Color matching based on the Wallis filter is widely used to process remote sensing images, but its efficiency is lower.

Large inclination-angle and many similar transparent windows often exist in building surface images owing to the image capturing environment constraint which is a big problem for image alignment. In such cases the direct mosaic result is rarely ideal. This paper utilizes a method based on vanishing point theory and object space parallel geometric constraints to automatically rectify building surface images. We utilize different operators to extract feature points in different pyramid levels, adopting a coarse-to-fine matching strategy, thereby compromising match reliability and efficiency. We also adopt color transfer strategies to adjust the color among images and we blend the images to get a seamless mosaic. Section 2 provides details. Real image experimental results are given in section 3.

16.2 Image mosaic algorithm

As shown in Fig. 16.1, the automatic mosaicking algorithm includes two parts: one is for seamless geometric mosaicking and the other for image dodging.

For geometrical mosaicking, vanishing point theory and parallel geometric constraints are firstly introduced to automatically rectify the original facade images, and the images are transformed from a central perspective projection to an orthographic projection.

Second, to increase the reliability of feature point matching due to the repetitive pattern of facade image textures, a coarse-to-fine matching strategy and relaxation method are employed to disambiguate the corresponding points. Both the Harris operator and the SUSAN operator are introduced for extracting feature points at different pyramid levels. The Harris operator offers a higher repetition rate, in favor of image matching. In contrast, the SUSAN operator has a high efficiency in extracting corner point features [11, 12]. Accordingly, Harris feature points are extracted at the top level of the image pyramid, matched through correlation, relaxation and robust estimation. We then obtain a group of reliable corresponding points, which is used for calculating initial affine parameters between original images. At the bottom level of pyramid image, the SUSAN operator is used to extract feature points in the first image overlap region where grids are evenly divided. Subsequently we select feature points in each grid and utilize the obtained parameters to guide correlation matching on the original image. When each grid has one feature point successfully matched then we turn to the next grid, remove the gross error in the obtained corresponding points set, and calculate affine transformation parameters between images to guide least square matching of each grid feature point. Finally, we utilize the corresponding points to complete the image geometrical mosaic.

In order to remove color or luminance differences within a single image and between multiple images, color transfer in a $l\alpha\beta$ color space is presented for image dodging.

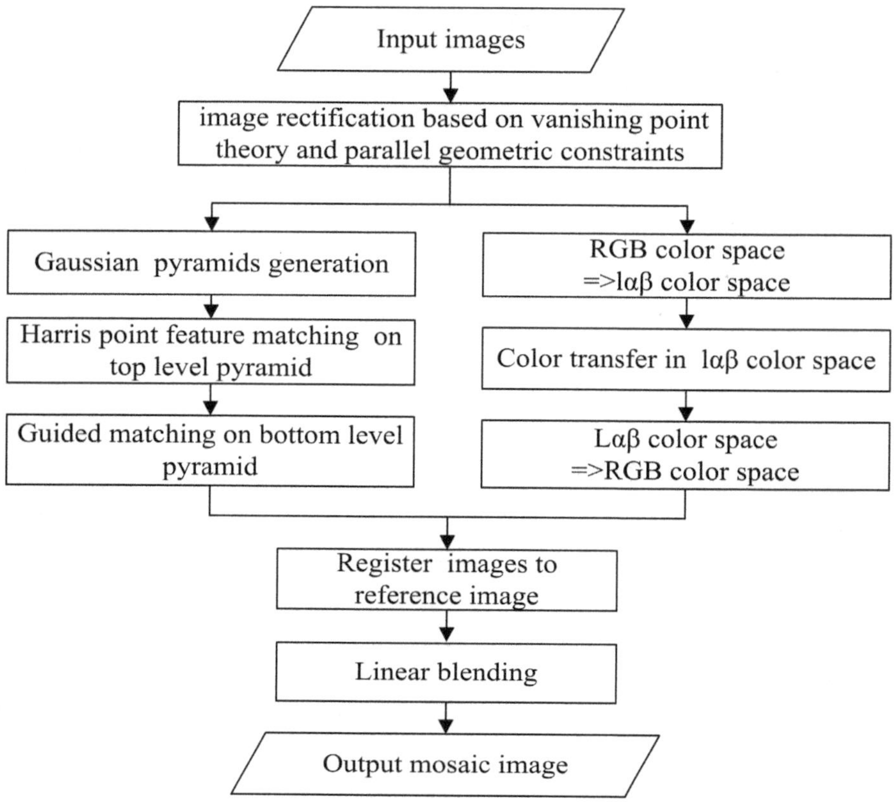

Fig. 16.1. Flowchart of image mosaicking

16.2.1 Rectification based on vanishing point theory and object space parallelism

The method based on vanishing point theory and object space parallel geometric constraints is considered useful for automatically rectifying facade texture images [7, 5]. According to vanishing point geometry, the interior and exterior orient elements of an image are a function of three vanishing points. Building facade images feature considerable amounts of parallel line information, which provides a basis for vanishing point theory. However, in most building facade images, only lines clustered parallel to the two coordinate axes can be extracted, and consequently only two vanishing points can be calculated through adjustment. Hence, a principal point is first fixed to the image center, then based on the rule that principal point lies in the orthocenter of a triangle composed by the three vanishing points we can deduce the third vanishing point. The image orientation elements can then be calculated according to the relationship between the vanishing point and orient elements.

One key to rectifying the images in this paper is the extraction of line information. One classical line detection method is the Hough Transform that is calculationally complex and has difficulty selecting proper parameters. For easier line extraction, the freely available software PtLens [2] is used to remove the radial distortion of the original images. The extracted line lacks local features and has low distinguishability, so the following improvements are proposed when utilizing the Hough Transform to detect facade image line features.

First, set the line initial orientation manually, and then only detect the lines which orientate near the initial orientation. Second, adopt a coarse-to-fine detection strategy, namely first adopt coarse quantized parameters to detect lines, and then adopt fine quantized parameters to detect the line again on the basis of the result from the earlier coarse detection step.

After line extraction, eliminate some outlier line features based on length and orientation, then divide the detected orthogonal line cluster into two sets according to orientation. In the object space these should be separately parallel. Establish an adjustment model based on line clustering of geometrical constraints that utilizes the object space parallel geometrical constraints. Use least square adjustment to solve image orientation elements iteratively. Finally, project the image onto a virtual plane with a focal length distance from the shot center to get a rectified image.

16.2.2 Reliable coarse-to-fine feature matching

Initial matching of the repetitive pattern of building facade images would result in one-to-many or many-to-one matching. If just the matching measure in (1) is used to constrain matching, a point in the first image may be paired to several points in the second image (which are called candidate matches), and vice versa. Relaxation techniques are commonly used to eliminate this kind of matching ambiguity [13], because most of the time facade images can simply be treated as a plane.

A correlation coefficient is commonly used to measure the symmetry between feature points and invariant to gray linear transformation, it is defined as in (1):

$$\rho(x,x') = \frac{\sum_{i=1}^{m}\sum_{j=1}^{n}(g_{i+u,j+v} - \bar{g})(g'_{i+u',j+v'} - \bar{g}')}{\sqrt{\sum_{i=1}^{m}\sum_{j=1}^{n}(g_{i+u,j+v} - \bar{g})^2 \bullet \sum_{i=1}^{m}\sum_{j=1}^{n}(g'_{i+u',j+v'} - \bar{g}')^2}} \tag{1}$$

Where

- $\bar{g} = \frac{1}{m \bullet n}\sum_{i=1}^{m}\sum_{j=1}^{n} g_{i+u,j+v}$ is the mean gray value of x=(u,v)
- $\bar{g}' = \frac{1}{m \bullet n}\sum_{i=1}^{m}\sum_{j=1}^{n} g'_{i+u',j+v'}$ is the mean gray value of x'=(u',v')
- $\rho(x,x')$ is the gray correlation between the n×m window with center x and x'

The relaxation technique allows the candidate matches to reorganize themselves by propagating certain constraints, such as continuity and uniqueness, through the neighborhood. This is justified because most of the façade image can be treated a plane.

If we define two sets N(x) and N(x') respectively in the neighbors of(x,x')with radius R, then the correct corresponding points should be supported by more corresponding points. If (y,y')is a pair of corresponding point, then the relation between (y,y')and corrected corresponding point (x, x') is that the angle between xy and x' y' is no larger than θ

The support strength of (x,x') and (y,y')is defined as:

$$Sm(x,x;y,y') = \frac{\rho(x,x') \bullet \rho(y,y') \bullet \delta(x,y;y,y')}{1 + dist(x,x';y,y')} \quad (2)$$

- Where $dist(x,x';y,y') = \frac{dist(x,x') + dist(y,y')}{2}$ is the mean Euclidean distance.

$$\delta(x,x';y,y') = \begin{cases} e^{-r/\varepsilon r} & if\ r < \varepsilon_r \\ 0 & otherwise \end{cases} \quad (3)$$

- Where $r = \frac{|dist(x,y) - dist(x',y')|}{dist(x,x';y,y')}$ is the difference between relative distances.

In this paper ε_r is defined as 0.3, θ is 45 degrees, R is 1/4 of the shortest length. Finally, search support strength across the neighbors, and accumulate them to get matching strength. Because one point can have many candidate corresponding points and each candidate has its own support strength value, $\max Sm[x,x';y,y']$ is considered for each neighbor. Considering the symmetry of support strength, the matching strength is defined as:

$$Sm(x,x') = \sum_{y \in N(x)} \max_{(y,y')} Sm[x,x';y,y'] + \sum_{y' \in N(x')} \max_{(y,y')} Sm[x,x';y,y'] \quad (4)$$

Due to the repetitive pattern, some feature points may have more than one candidate match which can confuse the matching algorithm very easily. If all the candidate matches are reserved without filtering, the relaxation time is often a little longer. In this paper, we use the candidate match number to filter out some feature points that can cause highly ambiguous matches before relaxation.

After relaxation, there are also often a number of false matches, and a robust technical approach must be adopted to reject false matches in the sets. The RANSAC method (Martin and Robert 1981) was applied in order to reject these outliers. By iteratively choosing three corresponding points to fit the affine transform as in (5), we can accumulate the number of corresponding points to support the motion model under a certain threshold. The model with the largest number is chosen as the "correct" model.

$$\begin{cases} x = a_0 + a_1 X + a_2 Y \\ y = b_0 + b_1 X + b_2 Y \end{cases} \tag{5}$$

Where

- x,y are the pixel coordinates of first image
- X,Y are the pixel coordinates of second image
- $a_0, a_1, a_2, b_0, b_1, b_2$ are affine transform parameters

Initial matching mainly includes the following steps.

1. Utilize the shortest side to determine the pyramid layer, then construct the Gauss pyramid image.
2. Extract the Harris feature point [4] on the top level pyramid in both images. For each feature point in the first image, calculate the correlation coefficient with all the feature points in the second image. Treat the points with correlation coefficient greater than 0.85 as candidate matches. Sort the feature points according to the number of candidate matches, and then remove the feature points which have more than 35 candidate matches. This step can reduce the relaxation time and the ambiguity.
3. Use (4) to calculate all the matching strengths of each corresponding point candidate.
4. Sort the feature points by matching strength, choose the top 60% to be the winning points. If the filtered number is no more than 1/3 of the original feature points number, then continue to (5),otherwise return to (3);
5. Based on (5), use RANSAC to filter the inliers, convert the corresponding points coordinates to the original image, and utilize a least square adjustment to calculate the affine transform parameters between original images.

Guided matching After obtaining affine transform parameters between images, the feature point extraction and matching on the bottom level of pyramid image are confined to overlapping areas. The SUSAN operator [12] does not require gradient operation, and operates quickly in corner point detection. However, the corner point can be only located to a pixel level [14]. Other researchers have analyzed the problem of how to pinpoint exact location features using image acquired by aircraft, and they proposed a kind of exact locationing methodology based on the Harris interest value. Using this approach, the theoretical precision reached 0.14 pixels. However, this method is based on the Harris interest value. Consequently, it cannot precisely locate some of the feature points that are extracted via the SUSAN operator.

A good feature algorithm for image matching should identify zones distinct from their neighborhoods. Information content is a measure of the point features' distinctiveness, and the more distinctive the feature is, the easier it is to create a successful match. Information content can be described in many ways and entropy is one commonly used method [14].

Utilizing the parameters obtained from an initial matching step to determine the overlap area between images, the Susan operator is used to extract feature points in the first overlap area, then calculate the entropy of all the feature points in a 11×11 window with the points located in the center. All the points are sorted according to entropy and the square grids are divided at the same time. For every point in each grid, we utilize correlation matching to search for the corresponding point in a 40×40 search window confined by the known parameters. If the correlation coefficient's extreme value is greater than 0.85 in the region of a given search area, the center of the window with the extreme value can be thought of as the corresponding point. In this case, we move on to the next grid until all the grids are processed.

The parameters obtained from the initial matching process are coarse parameters. The corresponding points selected by these parameters still include some gross errors. Iteration can be used to reject these gross errors until all points have residuals of less than 2 pixels.

The corresponding points obtained from correlation matching can only be located to pixel level, so the least square matching is adopted to increase matching accuracy. We utilize the corresponding points excluding gross error to calculate affine transformation parameters between images. Feature points in each grid are first located to the sub-pixel level, if the location correction value is more than one pixel, and then the feature point is abandoned. The points chosen via the affine transformation parameters are treated as initial points for least square matching. If the differentiation between the co-ordinates obtained from least square matching and forecasted coordinates is more than 3 pixels, the feature point is abandoned. Once all the grids are processed, an evenly distributed number of corresponding points can be obtained.

After extracting at least 4 evenly distributed corresponding points, affine transform parameters can be computed via least squares adjustment. According to the affine transformation, one image can be registered to another through bilinear resampling.

16.2.3 Modifying Color and Tone

Because of variable image capture parameters such as viewpoints and scaling, there are always color and tonal difference among images. It is hard to maintain complete color seamlessness if the blending is performed directly after geometrical mosaicking. Accordingly, the dodging method should be adopted to adjust whole color across all the images. Existing dodging methods, such as histogram matching and color matching based on the Wallis filter, are processed in the RGB color space. Color aberrance easily appears with these methods because there are correlations between the different channels' values in the RGB color space. A color space called $l\alpha\beta$ minimizes the correlation between channels for many natural scenes. (Reinhard et al. 2001) contributed research on borrowing one image's color characteristics from another based on this kind of color space. A similar method inspired by Reinhard's work is developed to modify the color of the images. One image is chosen as the standard image, and then its color is transferred to other images to make all the images have similar color characteristics which can make the final mosaic image smooth color-wise.

The RGB color space can be transformed to a $l\alpha\beta$ color space via (6)(7)(8):

$$\begin{bmatrix} L \\ M \\ S \end{bmatrix} = \begin{bmatrix} 0.3811 & 0.5783 & 0.0402 \\ 0.1967 & 0.7244 & 0.0782 \\ 0.0241 & 0.1288 & 0.8444 \end{bmatrix} \begin{bmatrix} R \\ G \\ B \end{bmatrix} \tag{6}$$

$$\begin{aligned} L &= \log L \\ M &= LogM \\ S &= LogS \end{aligned} \tag{7}$$

$$\begin{bmatrix} l \\ \alpha \\ \beta \end{bmatrix} = \begin{bmatrix} 1/\sqrt{3} & 0 & 0 \\ 0 & 1/\sqrt{6} & 0 \\ 0 & 0 & 1/\sqrt{2} \end{bmatrix} \begin{bmatrix} 1 & 1 & 1 \\ 1 & 1 & -2 \\ 1 & -1 & 0 \end{bmatrix} \begin{bmatrix} L \\ M \\ S \end{bmatrix} \tag{8}$$

Because there is little correlation between the axes in $l\alpha\beta$ color space, a linear method can be used to transfer one image's color to another. For each channel in $l\alpha\beta$ color space, we get the destination image value according to (9)：

$$l_{dest} = \frac{\sigma_{dest}}{\sigma_{sorce}} (l_{sorce} - M_{sorce}) + M_{dest} \tag{9}$$

Where

- M_{dest} , σ_{dest} is the mean value and variance of the standard images
- M_{sorce} , σ_{sorce} is the mean value and variance of the other images

After the three channels are processed, the resulting image is transformed back to the RGB color space to obtain the dodged image, and then all the images thus have almost the same color characteristic.

After color transferring there is still a slightly noticeable stitch line, which will affect the mosaic result if is it not processed. Linear blending and multi-band blending are common methods to solve this problem. The former method has a high efficiency but may cause some blur when the geometrical mosaic is not ideal, the latter approach can preserve more detail but requires more time. Because of our high accuracy of geometrical mosaicking, this paper adopts linear blending. In the overlap region we select the distance between a current pixel and the seam pixel as the power, and we take the weighted average gray value as the corresponding pixel value in the mosaic image. The formula is as in (10) :

$$G = P_L G_L + (1 - P_L) G_R, \quad P_L = \frac{D_{cur}}{2D_{all}} + 0.5 \tag{10}$$

Where

- P_L is the gray power of the corresponding point on the first image;
- G_L , G_R is the gray of the pixel;
- D_{cur} is the distance between current pixel and seam pixel,
- D_{all} is the distance between the edge pixel of this line and the seam pixel

After dodging and linear blending, the color of the mosaic image is very smooth and the result is more realistic.

16.3 Experimental results

To test the feasibility of the proposed algorithm, a VC++ program was implemented, and the experiments were conducted on a PC with a 512MB EMS memory. All the test images as shown in Fig. 16.2 are a series of photographs taken using a hand-held camera with autofocus from multiple viewpoints along a facade. The images shown in Fig. 16.2(a) were shot in multi-view mode to avoid occlusion from trees and were exposed using a longer timeframe to avoid occlusion from moving vehicles. The image, as shown in Fig. 16.2(a), has repetitive patterns which potentially make image matching more difficult.

The rectified images in Fig. 16.3 show many repetitive patterns in the facade image which can cause ambiguity of image matching. The parameter accuracy can be measured by the mean square error of corresponding points through matching and checks on other corresponding points. Mean square error (m_0) can be estimated using (11):

$$m_0 = \sqrt{\frac{v^2}{n}} \qquad v^2 = v_x^{\ 2} + v_y^{\ 2} \qquad \begin{cases} v_x = a_0 + a_1 X + a_2 Y - x \\ v_y = b_0 + b_1 X + b_2 Y - y \end{cases} \tag{11}$$

Where

- (X,Y) (x,y) are the pixel coordinates of corresponding points
- $a_0, a_1, a_2, b_0, b_1, b_2$ are affine transform parameters

This paper chooses a Gauss window width equal to 5, the variance is 3, and the non-maximum suppression threshold for fixing corners is 3000 when using the Harris operator to extract features. The grid number on the original image's long edge is fixed at 15.

The traditional image stitching method directly extracts Harris feature points from the original image for matching, then uses RANSAC to filter the inliers and estimate the transformation parameters. It is very easy to match the inliers to part of the overlap region and this can be performed with good efficiency. The proposed method ensures the corresponding points for mosaicking distribute evenly all over the overlap region and the m_0 of all stereo images is less than 2 pixels.

(a) Horizontal 1

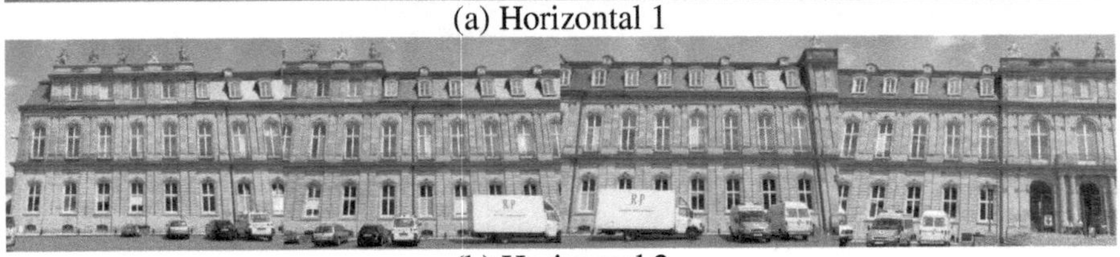

(b) Horizontal 2

Fig. 16.2. Original images

As shown in Fig. 16.4, the seamless mosaic image can be directly mapped onto a model surface. Since multi-view images are used, several occlusions such as pedestrians, trees or cars are avoided. Of course, because of the limited number of source images, there is still some occlusion as shown in Fig. 16.4(b)

Fig. 16.3. Part of rectified images from Fig. 16.2.

(a) Result of horizontal 1

(b) Result of horizontal 2

Fig. 16.4. Final mosaic result

Fig. 16.5 shows that the proposed algorithm is not limited to mosaicking images in the horizontal plane, but also can mosaic images vertically. Fig. 16.5(a) is composed of 2 images that represent the facade of a cathedral; Fig. 16.5(b) is composed of 4 images that represent the facade texture of a modern building which has many transparent windows that make the matching problem hard.

(a) Vertical 1

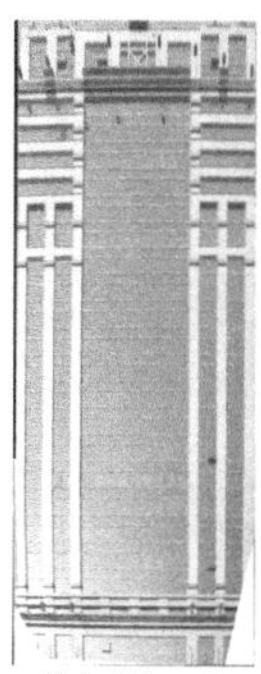

(b) Vertical 2

Fig. 16.5. Mosaic result from vertical images

Our experimental analysis results suggest the following conclusions: As a type of preprocessing, image rectification based on vanishing point theory is appropriate for subsequent geometrical mosaicking. By means of the coarse-to-fine matching strategy and the integrative application of different feature point operators at different level of the pyramid images, seamless geometric mosaicking is reliable and effective for complicated terrestrial images. Color transfer into the $l\alpha\beta$ color space is also useful to improve the image blending. This proposed algorithm provides an easy and efficient method for automatic generation of building facade textures. Future work will be focused on reducing occlusions by using image sequencing techniques.

16.4 Acknowledgements

The work described in this paper is supported by the National Natural Science Foundation of China (40871212 and 40671158) and the National High Technology Research and Development Program of China (2006AA12Z224)

References

1. Brown, M., Lowe,D.G.(2003). Recognising panoramas. In:International Conference on Computer Vision (ICCV'03) .pp:1218-1225
2. EPAPERPRESS 2008 http://epaperpress.com/ptlens/
3. Fischer, M.A., Bolles, R.C. (1981). Random sample consensus: A paradigm for model fitting with application to image analysis and automated cartography.J Communication of the ACM.24:381• 395.
4. Harris, C., Stephens, M.(1988).A combined corner and edge detector. In: Alvey Vision Conference, pp 147-151,.
5. Hu,J (2007) Integration complementary information for photorealistic representation of Large-scale environments Ph.D thesis. University of Southern California.
6. Jang, K.H., Jung, S.K., Lee, M.(1999).Constructing cylindrical Panoramic image using equidistant matching.J Electronics letters.35 (20):1715—1716
7. Kang, Z., Zhang, L.,Zlatanova, S. (2007).An automatic mosaicking for building facade texture mapping .In:International Archives of Photogrammetry, Remote Sensing and Spatial Information SciencesVolume XXXVI – 4/W45 covering contributions of the Joint Workshop "Visualization and Exploration of Geospatial Data" Stuttgart (Germany) pp: 1-9
8. Li, Z.(2005).Theory and practice on tone reproduction of color photos(in Chinese).Ph.D. thesis. Wuhan University
9. Peleg, S., Herman, J. (1997). Panoramic mosaic by manifold projection. In:IEEE Computer Society Conference on Computer Vision and Pattern Recognition (CVPR'97).pp 338-343
10. Reinhard, E.,Ashikhmin, M.,Gooch, B.,Shirley, P.(2001). Color transfer between images .J IEEE Computer Graphics and Applications.21: 34 – 41
11. Schmid, C., Mohr, R., Bauckhage, C. (2000).Evaluation of interest point detectors. J International Journal of Computer Vision 37(2):151-172
12. Smith, S.M., Brady, J.M. (1997). SUSAN-A new approach to low level image processing.J International Journal of Computer Vision.23: 45–78
13. Zhang, Z., Deriche, R., Faguaras, O., Luong, Q.(1995).A robust technique for matching two uncalibrated images through the recovery of the unknown epipolar geometry. J Artificial Intelligence 78:87-119
14. Zhu,Q.,Wu,B.,Wan,N.(2007). A sub-pixel location method for interest points by means of the Harris interest strength.J Photogrammetric Record 22(120): 321-335
15. Zhu,Q.,Wu,B.,Wan,N.(2007). A filtering strategy for interest point detecting to improve repeatability and information content.J Photogrammetric Engineering & Remote Sensing. 73(5): 547-553
16. Zagrouba, E.,Barhoumi, W., Amri, S. (2008). An efficient image-mosaicing method based on multifeature matching. J Machine Vision and Applications DOI:10.1007/s00138-007-0114-y.

Chapter 17
3D Continuous K-NN Query for a Landmark-based Wayfinding Location-based Service

Najmeh Samany, Mohmoud Reza Delavar, Sara Saeedi and Reza Aghataher

Abstract. Wayfinding in unfamiliar indoor and outdoor environments are particularly intricate problems of all the people's activities. The development of assistive technologies to aid wayfinding is hampered by the lack of reliable and efficient methods providing location information in the environment. Location-based services (LBS) are systems which support wayfinding task as process. They do not support landmark-based wayfinding although researchers agreed on its efficient role. In addition, one of the challenges of LBS is the continuous query which guarantees the performance of the system. In this paper, we propose an efficient method for 3D continuous k-NN query. Our method is based on three ideas, (1) The dynamic mobile objects have 3D coordinate space (x,y,t) and static landmark with 2D spaces (x,y) (2) selecting landmarks which are the nearest neighbor of one or more continuous queries, (3) indexing the queries rather than the landmarks. Through some experimental evaluation we demonstrate that our method is applicable on increasing runtime and decreasing power consumption of mobile devices.

17.1 Introduction

Wayfinding and orientation form important parts of people's daily lives. We have to find our routes through cities, through buildings, along streets and highways, etc. [18]. Location-based services are systems which present appropriate services for users

GIS Division, Dept. of Surveying and Geomatics Eng., College of Eng., University of Tehran
National Cartographic Center, Tehran, Iran
nsamani@ut.ac.ir, Najmehsamany@yahoo.com, Saeedi@ut.ac..ir
Center of Excellence in Geomatics Eng. and Disaster Management
Dept. of Surveying and Geomatics Eng., College of Eng., University of Tehran
mdelavar@ut.ac.ir
Head of GIS department in National Geographic Organization
Reza_aghtaher@yahoo.com

according to their locations. Location based services are gaining much importance for all kinds of human activities ranging from tourist navigation to support of rescue teams in disaster management. One of the fundamental processes theses systems proceed are navigation and wayfinding task. Although there are different approaches which provide guiding instructions, landmark-based wayfinding is known as a reliable and efficient method [19]. In landmark-based wayfinding, we have two main actors: human wayfinder which is a mobile actor and the landmarks which are used to guide the users and they are static in nature. We should select a suiatable landmark at decision points to navigate the user in the environment. Selection of these landmarks is based on three different parameters including: distance between the position of mobile object and landmark (also the visibility of it), orientation of mobile object and landmark and the saliency of landmark. In this context using the appropriate method for continuous query to monitor and track the wayfinder and giving the relevant instructions is an important factor in performance of these systems. There are a vast number of solutions for query on various types of data [10, 23, 26]. We distinguish between two different types of similarity queries: range queries and nearest neighbor queries.

For both types, the user selects an object, the query object which is the starting point of the search. For a range search, the user must additionally specify the query radius, i.e. a threshold for the maximally allowed distance from the query object. Since similarity measures are often not very intuitive, it may be difficult to specify such a query radius. Whereas, K-NN have a distance function to find the k nearest neighbor objects (landmarks at decision point). Therefore, in practice, the k-nearest neighbor query (k-NN) is more important, because the user only has to specify k, the number of objects that he wants to retrieve, and the system automatically retrieves the k most similar results. In our proposed algorithm we utilize K-NN method and considering three parameters introduced above. In this paper, we focus on k-NN queries but our technique can be extended to range queries in a straightforward way [5, 6, 23, 27].

In this research we are trying to apply and enhance a location-based service which supports landmark-based wayfinding by 3D continuous K-NN query. We address the problem of how to perform continuous queries on a LBS supporting wayfinding in an efficient way. In this paper, we propose an efficient method for 3D continuous k-NN query. Our method is based on three ideas, (1) The dynamic mobile objects have 3D coordinate space (x,y,t) and static landmark with 2D spaces (x,y) (2) selecting specific landmarks which are the nearest neighbor of one or more continuous queries, (3) indexing the queries rather than the landmarks. To reduce the large data volume of trajectory we assume the route of the wayfinder as a polyline, a sequence of line segments with each segment connected to the next.

The paper is structured as follows: The existing methods are briefly introduced. The notion of 3D continuous query, wayfinding and landmark is explained in the third section. Our methodology to design a 3D continuous K-NN query for landmark-based wayfinding is described in section four. Section five investigates the implementation of the proposed approach and presents the experimental results. Section six provides the conclusions and future research.

17.2 Literature Reviews

Mobile data model and management have been received increased attention in the past few years and methods to model and index moving object databases have been developed. Sistla et al. [21] proposed the MOST data model. They classified the queries in three categories as:

- Instantaneous query. In this case the query is responded immediately after the query sending and the answer is transmitted to the user;
- Continuous query. It needs to be evaluated at every time instant in order to ensure the correctness and validity of the query answer;
- Persistent query. It needs to be evaluated at every time instant as well, but the query assessment must involve all previous mobile object database states starting from the time of query issuance (database history).

Based on the mobility of query (client) and the objects queried by the clients, such queries can be further grouped into three different classes [23]:

- Mobile clients query about static objects (e.g. tourist services, m-commerce);
- Static clients query about mobile objects (e.g. fleet management, traffic control and management);
- Mobile clients query about mobile objects (e.g. tourist services, digital battlefield, mobile games).

Continuous spatio-temporal query processing in an LBS environment for simple range and nearest neighbor queries has also received much of interest. In this case, the queries are ranges in space and time. The index is used to find the objects that are contained inside the queried space during the specified time interval. The time interval may refer to the past (historical queries) or to the future (predictive queries). We are encountered with the mobile objects (mobile users) searching static objects (landmarks) to navigate the environment which it needs to be evaluated at every time instant in order to assess the correctness and validity of the query answer, so we are focused on "continuous K-NN query" according to the defined categories [21, 23].

For the historical queries, a number of indexing approaches based on conventional spatial access methods have been proposed. R-tree is the very first access methods created by [7]. Then some well-known index structures extended from this are R*-Tree, packed R-Tree, CR-Tree, and R+-tree, etc. [9, 22]. Grid file is another earlier indexing technique developed by 15]. Also Quad-tree has been developed into multiple new approaches such as Point quad-tree, pyramid, octree, PR-tree, MX-CIF quadtree, PMR quad-tree, etc. [25]. With deeper research carried out and more needs stems from spatial database usage, more and more new indexes are innovated for different specific kind of queries. Q+R-tree tempted to combine benefits from quad-tree and R-tree so updating and searching performance are more balanced [26]. Some efforts is put on the querying demand of historical, present and future positions of moving objects to better support the tracking services, where BBx-tree, Bx-tree, HR-tree, STR-tree and TB-tree are concerned and studied [11, 16, 20]. One significant variant is 3D trees. One example is trying to introduce 3D in order to benefit interval queries with paying some storage caapcity, implemented by modifying MVR-tree to MV3R-trees [24].

Each basic technique has its advantages and drawbacks. For example, quad-tree from the partitioning root has advantage of updating but suffers from the searching process. So the innovative way of preserving the advantage and making up for the disadvantage is the key to successful new solutions. In this paper we utilize R-tree index method for indexing landmark (static object in the environment) as it is the most popular indexes for Euclidean query processing due to its simplicity and efficiency.

The regular K nearest neighbor queries has been intensively studied and for which numerous algorithms proposed. A majority of the algorithms are aimed at m-dimensional objects in Euclidean spaces, and are based on utilizing one of the variations of multidimensional index structures. A K-NN query on static databases is a well studied problem for which many index structures have been introduced [5, 8]. Hjaltason et al. [8] propose an incremental nearest neighbor algorithm that is based on using an index structure and a priority queue. Their approach is optimal due to the structure of the spatial index but not with respect to the nearest neighbor problem. [17] introduced two elementary methods called Query Indexing (Q-index) and Velocity Constrained Indexing (VCI) and also proposed the important concept of safe regions. The Q-index approach is based on static range queries over mobile objects. The queries are indexed by an R-tree and mobile objects search the index structure to recognize the queries whose response they may involve. By indexing queries, Q-index approach avoids consequent updates of the index structure. In addition, it follows the concept of safe regions which force the mobile object to issue an update only if such mobile object passes the region and thus participate in answering some other queries. An improved technique based on generic framework was developed [8] to handle continuous queries by leveraging the concept of safe regions through which the location updates from mobile clients can be further reduced.

Although there are a number of continuous query methods which could support wayfinding task, there is no comprehensive method reported for continuous query supporting landmark-based wayfinding which is a more reliable approach and could enhance the performance of services.

17.3 3D Continuous Queries to Support Wayfinding Task

In this section we discuss wayfinding task and its important component called "landmark" which follows the navigation systems to support landmark-based wayfinding. Then we explain 3D continues query as a method for tracking and giving wayfinding instruction for mobile users.

17.3.1. Landmark-based Wayfinding

Wayfinding i.e. getting from some origins to a destination is one of the fundamental everyday problems human encounters [1]. It has been defined as "purposeful, directed, and motivated movement from an origin to a specific distant destination, which can not be directly perceived by the traveler" [1]. Such behavior involves relations between the traveler and the space. Human wayfinding takes place in large-scale environment. Such spaces cannot be perceived from a single viewpoint; therefore, people have to navigate through large-scale spaces to traverse them. People use various spatial, cognitive and behavioral capabilities to find their ways. These abilities are a necessary requirement to use environmental information or representations of spatial knowledge about the space [19]. Landmarks act as significant features in the wayfinding task, and work as nodes for organizing other spatial information into a layout. They may be mentioned or remembered because of dominance of visible form, peculiarity of shape or structure, or because of socio-cultural significance. Paths are static linear structures in an environment; they may be streets, footpaths, pavements, canals, rivers or railway tracks. Landmarks play an important role when humans experience foreign environments. For example, trying to find the way is much easier if the navigator can rely on a description of the route based on well-recognizable objects in the space, instead of navigating only based on the street names and metric directions. Landmark-based navigation applies knowledge about salience objects in the environment to guide travelers through unknown areas [4, 12].

Among the different meanings of landmark, it is an object or structure that shows a locality and is used as a reference point [14]. The concept is limited to the prominence or distinctiveness of a feature in a large-scale environment or landscape. Thus the landmark saliency of a feature does not depend on its individual attributes but on the distinction to attributes of near features, being a landmark is a relative property. Studies show that landmarks are selected for route directions usually at decision points. Another study has shown that routes enriched with landmarks at decision points make better guidance, or less wayfinding errors, than routes without landmarks. [19].

Lynch [13] introduced landmarks as "external points of reference—points that are not part of a route like the nodes in a travel network. He characterizes the quality of a landmark by its singularity, where singularity is bound to a clear form, contrast to the background, and a prominent location. The prominent factor is the figure-background contrast. The contrast can be produced by any property, such as uniqueness in form or function in the local or global neighborhood. Landmarks are essential parts of wayfinding directions and any communication about space [19].

Today's car navigation systems provide driving instructions in the form of maps, pictograms, and spoken language. However, they are so far not able to support landmark-based navigation, as the most natural navigation concept is for humans and also plays an important role for upcoming personal navigation systems [3].

However, the primary concept of delivering the instructions has not been modified very much. Still, spoken language instructions use a relatively small set of commands (like 'turn right now'), which only refer to properties of the street network. This is not optimal, since i) features of the street network usually are invisible from a greater distance because of the low driver position and small observing angle, and ii) the most natural form of navigation for humans is the navigation through landmarks, i.e. the provision of a number of recognizable and memorizable views along the route. Obviously, the definition of buildings as landmarks together with corresponding spoken instructions (such as 'turn right after the tower') would be a step towards a more natural navigation [2, 3]. Fig. 17.1 shows the wayfinding procedure supporting by landmarks.

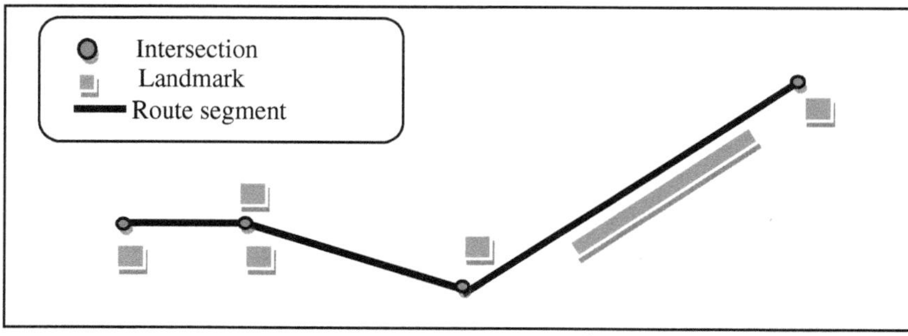

Fig. 17.1. Landmark-based wayfinding

Landmarks can be classified into three categories: visual, cognitive and structural landmarks. The more of these categories cooperate on constructing particular object, the more it qualifies as a landmark. This concept is used by Raubal and Winter to provide measures to determine formally the landmark saliency of buildings: the strength or attractiveness of landmarks is specified by the components visual attraction (e.g. consisting of facade area, shape, color, visibility), semantic attraction (cultural and historical importance, explicit marks, e.g. shop signs) and structural attraction (nodes (important intersection), boundaries (parting elements like rail tracks or rivers), regions (building blocks)). The combination of the property values results in a numerical estimation of the landmark's saliency. Hypothesis testing was used to select the most significant landmark at each decision point for integration in the wayfinding instruction [19]. Brenner and Elias proposed a method to extracting landmarks for car navigation system using existing GIS databases and laser scanning. They utilize data mining approach to achieve this goal [3].

17.3.2 3D Continuous K-nearest Neighbor Query

Continuous nearest neighbor queries are introduced as determining the K nearest neighbors of any object on a given path. An example of this type of query is shown in Fig. 17.2 where a moving object (e.g., a car) is traveling along the path *(A, B, C, D)* (specified by the dashed lines) and we are trying to find the first 3 closets restaurants (restaurants are specified in the figure by (r_1, ..., r_8) to the object at any given point on the path. "The result of a continuous KNN query is a set of *split points* and their associated KNNs. The split points specify the locations on the path where the KNNs of the object change "[10]. In the other words, the KNNs of any object on the segment (or interval) between two adjacent split points is the same as the KNNs of the split points.

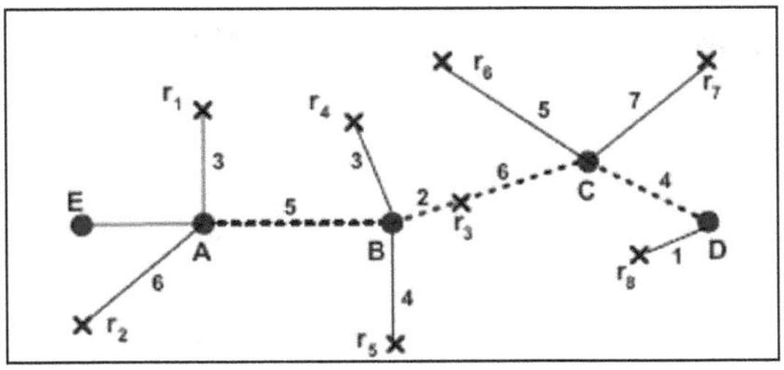

Fig. 17.2. Example of continuous K nearest neighbor query [10]

The current studies on continuous NN queries are focused on spaces where the distance function between two objects is one of the Minkowski distance metrics (e.g., Euclidean) [10].

17.4 Proposed Methodology

We base our approach on the wayfinding application scenario appropriate in LBS for monitoring and tracking mobile objects and giving instruction information through landmarks. We assume that the users have wireless devices (e.g., mobile phones or PDAs) that are online via some form of wireless communication network. We assume that users can obtain their positions using global positioning system (GPS) technology. We have setting in which a central database at the LBS server stores a representation of each mobile object's current position. Also we assume that the landmark-based wayfinding is a reliable and efficient method for wayfinding. So, it is not necessary for our enhanced query to search all of the objects surrounding the mobile clients, only finding the suitable landmark is sufficient to complete the wayfinding task.

We consider the route of mobile client as three-dimensional (3D) polyline trajectories (Fig. 17.3). To store the exact trajectory of a moving object would require storage of the location for each time instant during the moving object's lifespan. This would produce a potentially huge volume of data. Hence, in many applications, each object trajectory is approximated as a polyline, a sequence of line segments with each segment connected to the next.

We define the trajectory of moving objects as a set of line segments. Each line segment represents part of an object's trajectory for some time period. When the object's velocity (speed and direction) changes beyond a threshold, a new line segment is used.

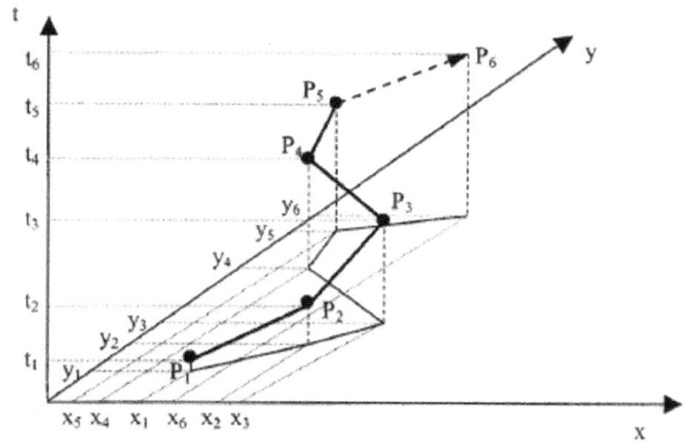

Fig. 17.3. Mobile point in a 3*D* space (2*D* space, time) [23]

Each mobile client "mo" in the dataset has the following properties: (moid; po_1; po_2; …; po_n). The trajectory is described by a sequence of 3-D points, where each po_i corresponds to (x, y, t). Also for each current mobile user we have some other parameters which receive on-line and then disappear. These parameters are speed and direction of the user. Also, the distance and orientation of the mobile clients are calculated with respect to nearest landmarks. According to these assumptions, we propose a methodology to enhance a LBS supporting wayfinding task through K-nearest neighbor query which is constrained on landmarks at decision points.

K-NN query retrieves *k* objects from the static dataset that are nearest to a query point according to Euclidean distance. Our algorithm manages the position of mobile client in spatial database by indexing them through a spatial access method. In this context we utilize R-tree indexing method and some branch-and-bound approach to restrict the search space. We indexed landmarks and network (consist of edges and nodes) using R-tree method and query about them by mobile user with KNN approach.

The procedure of our methodology is as follows:

- The mobile object (client) introduces the start point.
- The K-NN algorithm finds the nearest landmark/(s) to the query point and checks it/their orientation and the saliency.
- By selecting the appropriate landmark, the instruction guides the client to the next decision point (Places where the wayfinder need guiding instructions. These points usually are the intersections).
- Before the user arrives to the next decision point the K-NN query finds the appropriate landmark and the guiding process continues.
- The procedure continues until the user reaches the destination.

The described algorithm is illustrated in Fig. 17.4. By the selection of appropriate landmarksdhk, the wayfinding process continues according to the instructions presented at Fig. 17.5.

Algorithm 1: k-NN Search (GI, LT, q)
Input: Grid Index GI, Landmark Table LT, k-NN Query q; i^{th} landmark L (i)
Output: The appropriate Landmark for the guiding
Landmark list = Ø; /*Initialize k-NN*/
i = 0; /*Initialize cell level*/
Landmark.K.NN-dist = •; /*Initialize KNN-dist */
/*Construct initial k-NN candidates*/
For each K-nearest neighbor i=1 to k
If ((w_1*distance+w_2*orientation+w_3*saliency) (L (i)))> all of the K-NNs
Give instruction according to *Algorithm 2*

Fig. 17.4. The algorithm of K-NN searching suitable landmark

Algorithm 2: Give instruction based-on landmarks
Input: The selected landmark, the position of the user at time t (x, y, t)
Output: The appropriate instructions for the user
Start instruction: [When landmark_1 is (in front | back| right| left) of you, Start, Move Straight]
Stop instruction: [At landmark_k Stop]
Move instructions:
For i=2 to k-1
[At | Towards | Along landmark_i] +
[Turn Left | Right | Veer Left | Veer Right| Move Straight]+
{ONTO street name}
{(PASSING | CROSSING) LANDMARK J} 0 … n+
[UNTIL | TOWARDS | ALONG landmark_i+1]
Next for

Fig. 17.5. Wayfinding instructions based on landmarks

Location-dependent application is characterized by a large number of objects and a large number of continuous queries (or users). Most users require the system answer their queries as soon as possible or even process their query in real time. We expect using this approach could reduce the complexity and so the computational volume. Selection of landmarks at decision points are based on three different parameters including: distance between the position of mobile object and landmark, orientation of mobile object and landmark and the saliency of landmark. All of these parameters have an importance magnitude in choosing the suitable landmark for wayfinding which is considered as weights of these parameters (w_1, w_2 and w_3). The weight of distance (w_1) is measured based on the visibility of landmarks. The weight of orientation (w_2) is determined according to Eq.1F:

$$\text{Wori}=\begin{cases} -45° < \text{orientation angle} < 45° & ;1 \\ -90° < \text{orientation angle} < -45° & ;2 \\ 45° < \text{orientation angle} < 90° & ;3 \end{cases} \tag{1}$$

The orientation angle is the angle between the current direction of mobile object and the landmark. The saliency of the landmark is considered using Rauble's method [19].We proposed a model to measure the saliency of effective landmarks relying on attraction landmark criteria. Using these criteria, we could extract landmarks automatically [19]. Hypothesis testing was used to select the most significant landmarks at each decision point for inclusion in the wayfinding instruction.

17.5 Experimental Results

We perform experiments to obtain the performance of the proposed algorithm in the query processing methodology. Data structures have been implemented using Microsoft Visual Studio C sharp (C#).We tested the algorithm on 100, 200, 300, 400, 500 and 1000 mobile objects in a network route with 1023 segments. We measure the runtime parameter as a criterion for the performance of the system. The implementation showed that using this method could increase the speed of runtime of the system (Fig. 17.6). Increasing the temporal performance could result in the reduction of wayfinding process. Also, using this method reduces the volume of the computational process which could result in the reduction of energy consumption of the mobile system.

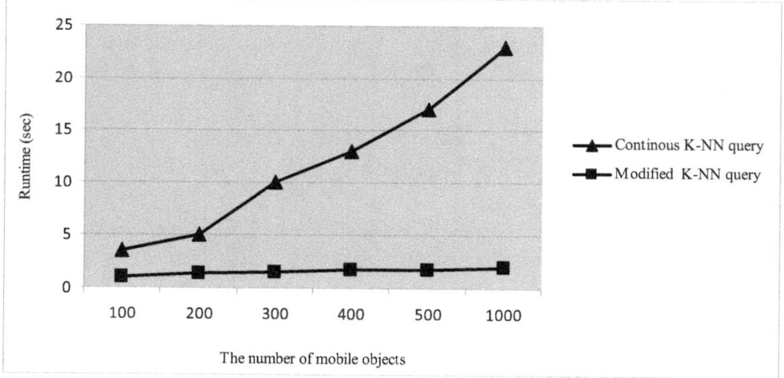

Fig. 17.6. Performance of the system based on the runtime.

17.6 Conclusions

Wayfinding and navigation are among the fundamental tasks supported by LBSs. There are different factors to provide a good means for human wayfinder. These parameters are: A reliable and efficient guiding instructions, the improvement of speed performance and response and as the users are mobile, the reduction of energy consumption. On the other hand, wayfinding based on suitable landmarks in traversing environment is known as a reliable method for guiding user. Current navigation systems apply various search and query approach to assist the wayfinder to find the destination such as KNN. These systems don't mention to the importance of landmarks in guiding the mobile users.

To overcome the explained challenges we utilized 3D continuous K-NN query method for finding appropriate landmarks guiding the user (we considered 3D as x,y,t). To achieve this goal we implemented a method to automatically extract local landmarks from datasets to be integrated in wayfinding instructions. Different individual properties for the attractiveness of a landmark were first defined and then put together to form a global measure of landmark saliency for each feature in a dataset. Then we indexed the static object or landmarks and network using R-tree. The K-NN query finds the appropriate landmark at decision points in spite of searching all mobile objects surrounding the users. The results showed that using this approach increases the speed of runtime of the process and decreases the energy consumption.

Our future research would be on the integration of different query approaches such as range query to enhance the process.

References

1. Allen, G.: Spatial abilities, cognitive maps, and wayfinding - bases for individual differences in spatial cognition and behavior. In: Golledge, R. (Ed.), Wayfinding Behavior Cognitive Mapping and Other Spatial Processes, Johns Hopkins University Press, Baltimore, pp. 46—80 (1999).
2. Baus, J. Krger, A., Wahlster, W.: A Resource-Adaptive Mobile Navigation System, International Conference on Intelligent User Interfaces IUI02, January 13-16, 2002, San Francisco. http://citeseer.ist.psu.edu/baus02resourceadaptive.html (2002).
3. Brenner, C. and Elias, B.: Extracting Landmarks for Car Navigation Systems using existing GIS databases and laser scanning. Proceeding ISPRS Workshop on Photogrammetric Image Analysis, Munchen, Germany, (2003).
4. Caduff, D., Timpf, S.: The Landmark Spider: Representing Landmark Knowledge for Wayfinding Tasks. In: Barkowsky T, Freksa C, Hegarty M, Lowe R (eds) Reasoning with Mental and External Diagrams: Computational Modeling and Spatial Assistance, AAAI Press, Stanford, CA USA, pp 30—35 (2005).
5. Cai, Y., Hua, K. and Cao, G.: Processing Range-Monitoring Queries on Heterogeneous Mobile Objects. In IEEE Int'l Conference on Mobile Data Management (MDM'04), pp. 27--38, January (2004).
6. Gedik, B. and Liu, L., MobiEyes: Distributed Processing of Continuously Moving Queries on Moving Objects in a Mobile System. In EDBT (2004).
7. Guttman, A. R-tree: A Dynamic Index Structure for spatial Searching. In SIGMOD '84, Proceedings of the ACM SIGMOD Conference. ACM Press (1984).
8. G. R. Hjaltason and H. Samet. Distance Browsing in Spatial Databases. TODS, vol. 24, No.2, pp. 265--318, June (1999).
9. Kim, K., Cha, S. K. and Kwon. K.: Optimizing Multidimensional Index Trees for Main Memory Access. In Proc. of SIGMOD (2000).

10. Kolahdouzan, M.R. and Shahabi, S.: Continuous K Nearest Neighbor Queries in Spatial Network Databases C. Proceedings of the Second Workshop on Spatio-Temporal Database Management (STDBM'04),Toronto, Canada, August (2004).
11. Lin, D., Jensen, C. S., Ooi, B. C. and Saltenis, S.: Efficient Indexing of the Historical, Present, and Future Positions of Moving Objects," in Proceedings of MDM (2005).
12. Luyten, K. Coninx, K.: ImogI: Take Control over a Context-aware Electronic Mobile Guide for Museums (2004).
13. Lynch, K.: The Image of the City. MIT Press, Cambridge (1960).
14. Merriam-Webster: Merriam-Webster's Collegiate Dictionary. Merriam-Webster, Inc (2001).
15. Nievergelt, J., Hinterberger, H., and Sevcik, K., The grid file: An adaptable, Symmetric Multikey File Structure. ACM Transactions on Database Systems, vol.9, pp.38--71 (1984).
16. Pfoser, D., Jensen, C.S and Theodoridis, Y.: Novel Approaches to the Indexing of Moving Object Trajectories. Proceedings of the 26th International Conference on Very Large Databases, Cairo, Egypt (2000).
17. Prabhakar, S., Xia, Y. Kalashnikov, D. V., Aref, W. G. and Hambrusch. S. E.: Query Indexing and Velocity Constrained Indexing: Scalable Techniques for Continuous Queries on Moving Objects. IEEE Trans. Computers, Vol.51, No.10, pp. 1124--1140 (2002).
18. Raubal, M.: Human Wayfinding in Unfamiliar Buildings: A Simulation with a Cognizing Agent, at the Department of Geography, University of Zurich, Switzerland (2001).
19. Raubal, M. and Winter, S.: Enriching wayfinding instructions with local landmarks. Egenhofer, M. and Mark, D. (Eds.), Geographic Information Science, Lecture Notes in Computer Science 2002, LNCS vol. 2478, pp. 243—259. Springer (2002).
20. Saltenis, S., Jensen, C. S. and Leutenegger, S. T. and Lopez, M. A.: Indexing the Positions of Continuously Moving Objects. In SIGMOD Conference (2000).
21. Sistla, P., Wolfson, O., Chamberlain, S, and Dao, S.: Modeling and Querying Moving objects, in: Proceedings 13th ICDE Conference, pp. 422–432 (1997).
22. Shekhar S. and Chawla, S.: Spatial Databases: A Tour, Prentice Hall (2003).
23. Stojanovic, D., Papadopoulos, A.N, Predic, B., Djordjevic-Kajan, S. and Nanopoulos, A.: Continuous Range Monitoring of Mobile Objects in Road Networks, J. Data & Knowledge Engineering, 64, pp. 77--100 (2008).
24. Sun, J., Papadias, D., Tao, Y. and Liu, B.: Querying about the Past, the Present, and the Future in Spatio-Temporal Databases. ICDE, pp. 202–213 (2004).
25. Thakkar, S. and Samet. H, Spatial Data Structures, http://infolab.usc.edu/csci599/Fall2001/note/SpatialDataStructures.ppt (2001).
26. Xia, Y. and Prabhakar, S.: Q+Rtree: Efficient Indexing for Moving Object Database. Database Systems for Advanced Application. Proceedings. Eighth International Conference on Volume, Issue, pp. 175 --182 (2003).
27. Zhang, W., Li, J. and Pan H.: Processing Continuous k -Nearest Neighbor Queries in Location- Dependent Application, J. Computer Science and Network Security, 6 No.3 (2006).

Chapter 18
3D Geo-Network for Agent-based Building Evacuation Simulation

Jinmu Choi and Jiyeong Lee

Abstract. This paper discusses 3D geometric network extraction for building evacuation simulation with an agent-based model. 3D geometric network represents the internal structure of a building, which provides agent-based models with the shortest path for evacuation. 3D geometric network of a building can be built from computer-aided design (CAD) file (vector) and scanned blueprint (raster) through wall extraction and 3D topology construction. We test two wall extraction methods: vector-based medial-axis transformation (MAT) and raster-based thinning. For vector-based MAT, straight MAT is used to extract wall structure from wall polygon generated from CAD data. For raster-based thinning, boundary peeling thinning is used to extract wall structure from scanned blueprint. The extracted 3D geometric network is then used in an agent-based model for building evacuation simulation. In the evacuation simulation, human beings are considered to be the only moving agents. To model human behavior, we adopt a social force model to consider human-to-human and human-to-wall interactions during evacuation. We test simple evacuation scenario in a situation of jam by enforcing different numbers of people in three rooms. The results show that the average velocity increases continuously before jams, decreases during jams at doorways and outer exits, and eventually increases again as individuals escape the jams.

18.1 Introduction

As geographic information system (GIS) continues to mature, three-dimensional (3D) modeling has become a tool for GIS analysis with varying levels of success [34]. There is increasing demand for 3D geometric network model in various disciplines including emergency services and cadastre management [2, 6, 16, 33, 34]. If there is

Department of Geosciences, Mississippi State University
jc778@msstate.edu
University of Seoul, Department of Geoinformatics, 13 Siripdae-gil, Dongdaemun-gu
Seoul 130-743, Korea, jlee@uos.ac.kr

an emergency situation in a building, 3D geometric network can provide fast way to egress the building.

Emergency services rely on both macro- and micro-level GIS [16]. The ultimate emergency management would function on macro scales such as urban areas and on a micro scale such as individual buildings. On the macro-level, an emergency response GIS in a large city might route responders to a building, and then, on the micro-level, route them to the emergency room through the shortest path once they arrive at the building's entrance [20]. Therefore, emergency response on micro-level features such as buildings depends upon 3D geometric network regarding the internal structure of a building.

Most current commercial GIS, however, do not provide tools to model 3D geometric network representing the internal structure of a building. They only include surface-based 3D representation methods. ESRI 3D Analyst provides tools for surface generation, volume calculation, and viewshed analysis. ESRI ArcScene emphasizes visualization through texture mapping and fly-through. ERDAS Imagine VirtualGIS and Intergraph GeoMedia also provide 3D fly-through tools. While these commercial GIS systems are primarily concerned with 3D visualization, they do not provide any tool for 3D geometric network representing and analyzing the internal structure of buildings [35].

On the other hand, simulation on the emergency situation requires a model that is a simplified representation of reality [21]. A model can be dynamic if the output represents a later point in time and represents time steps in the operation of a dynamic process [10, 22]. Dynamic models are used to assess different scenarios by attempting to project quantifiable impacts into the future [4]. The possible results of simulation can be visualized and help decision-makers avoid disastrous situations. As a dynamic modeling tool, agent-based model has become one of the key computational approaches to simulate collective outcomes of complex geographic phenomena based on individual agents' states and behaviors. In particular, human behavior with respect to the evacuation situation can be simulated on the 3D geometric network in a building.

The purpose of this paper is to provide a method to build 3D geometric network data and to use the 3D network data to simulate a building evacuation. 3D geometric network data can be produced from computer-aided design (CAD) files and scanned blueprints. Human behavior in the evacuation process is then simulated using agent-based model on the 3D network data.

In Section 18.2, we describe the framework of building evacuation simulation system that utilizes both agent-based model and 3D geometric network in GIS environment. Section 18.3 and 18.4 detail the procedure to extract 3D geometric network from blueprints that are either CAD file and scanned paper designs. In Section 18.3, 3D geometric network data are derived from CAD file using straight medial-axis transformation that extracts line wall structure from polygon wall. In Section 18.4, scanned blue print data are converted to 3D geometric network. In Section 18.5, building evacuation is simulated using agent-based model based on the internal structure of building and 3D geometric network information. Key ideas are summarized in the conclusion.

18.2 Framework of Building Evacuation Simulation System

An integration of agent-based modeling and GIS offers improvements to understand complex spatial phenomena. GIS provides agent-based model with spatiotemporal GIS data to simulate the distributed nature of actions and reactions at the agent (individual) level. Agent-based model provides GIS with the representation of feature dynamics such as temporal changes in spatial patterns. Therefore, in this study, a building evacuation system is designed in GIS environment (Fig. 18.1). For evacuation simulation, we utilize Agent Analyst[1] as agent-based modeling tool and 3D GeoNet for input data generation.

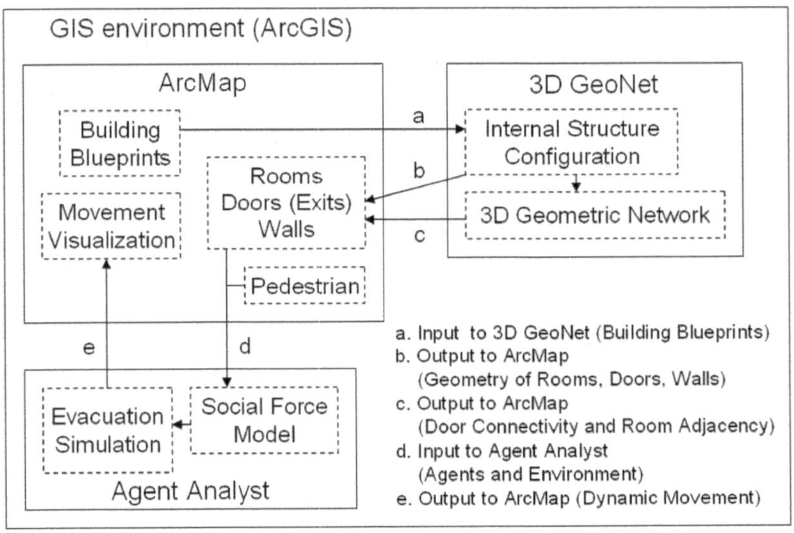

Fig. 18.1. Building Evacuation Simulation System Architecture

For agent movement, we adopt and implement Helbing et al.'s social force model [13] (see details in section 18.5.2) in Agent Analyst. Agent Analyst takes building structure data (Fig. 18.1d) and runs a social force model that incorporates a mixture of socio-psychological and physical forces of human-to-human and human-to-wall interactions for pedestrian movement. The simulation is iterative and the results in each step are visualized dynamically in ArcMap (Fig. 18.1e). So, we can explore agents' evacuation process in a building.

Current commercial software for pedestrian evacuation simulation such as ASERI[2] and Legion[3] utilizes CAD file for the physical environment of a moving agent in a building. To build shortest paths from CAD file, users should manually model agents'

---

[1] Agent Analyst is an extension of ArcGIS, software and materials available at http://www.institute.redlands.edu/agentanalyst/AgentAnalyst.html.

[2] ASERI (Advance Simulation of Evacuation of Real Individuals) is available at http://www.ist-net.de

[3] Legion is available at http://www.legion.com

routes from each room to destination, which is labor intensive process. Instead of manual modeling, we have developed 3D GeoNet that automatically extracts the internal structure of a building such as rooms, walls, and doors (and exits) and 3D geometric network from building blueprint data. The internal structures of a building from 3D GeoNet provides agent-base model with geometric configuration (Fig. 18.1b) for agents' egress movement and the 3D geometric network provides with connectivity and adjacency of rooms for agents' egress routes (Fig. 18. 1c). These data are used as input of Agent Analyst though ArcMap.

Two common data used as blueprint of a building (Fig. 18. 1a) are CAD files and paper designs that we can utilize in order to extract the internal structure of a building. Following two sections describes how to build 3D geometric network data from both CAD and scanned blueprint data.

18.3 3D Geometric Network Extraction from CAD Data

CAD data are commonly used as digital blueprint of a building [24]. The extraction of 3D geometric network of a building from CAD data requires the conversion of a room to a node and a shared wall to a link based on dual graph theory [18]. The process starts by extracting wall structure of a building since walls in CAD data are stored as closed double lines that need to be converted to single lines. From the wall structure, 3D geometric network can be generated through 3D topological structure. Details on individual steps are explained in the following sub sections.

18.3.1 Straight Medial-Axis Transformation

Walls in CAD blueprint of a building are represented as closed double lines. To build 3D network data, the double line walls need to be converted to single lines after the conversion of the closed double lines to polygons using ArcInfo's Build command. Then, lines are extracted from the wall polygons. There are several approaches to medial-axis transformation (MAT) for extracting a line from a polygon. A skeleton of a polygon can be extracted by the Delaunay triangulation based transformation [25]. One or two edges of the Delaunay triangles are inside the input polygon. Connecting the midpoints of the inside edges of the triangles produces the skeleton of the polygon (Fig. 18. 2a). The skeleton approximates a medial-axis of the polygon. This method has been implemented to extract morphological structures of natural features such as the human body [25].

A true medial-axis can be derived from the Voronoi-Diagram based MAT [5, 17]. Lee [17] has developed an O(n log n) algorithm for polygons with concave corners. His algorithm is based on the divide-and-conquer method, which requires four steps. First, the Voronoi edges for each vertex of a polygon are generated. Second, polygon edges are grouped into chains at the concave vertexes. Third, Voronoi edges are generated for each chain and merged with each successive chain. Finally, the medial-axis is created by merging the final two chains and by removing the Voronoi edges incident at the vertexes of the polygon (Fig. 18. 2b). In Chin et al.'s algorithm [5], the medial-axis is the connection of all centers of inner circles of a polygon. The inner circles touch the polygon in two or more points. Both MAT methods create second order lines at the 'T' intersections of skeleton lines. It is more difficult to calculate the interaction between walls and humans using a second order line rather than a straight line.

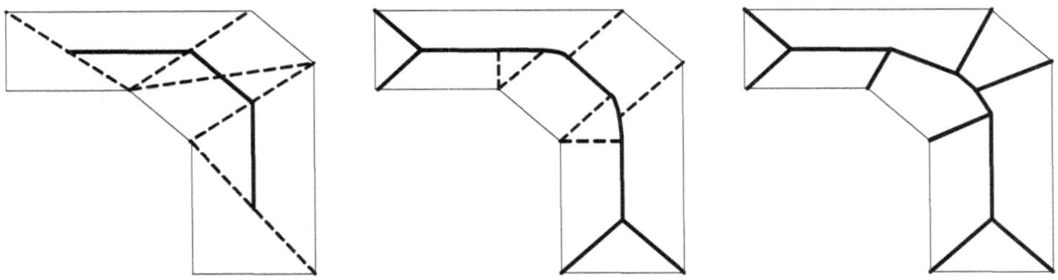

a. Skeleton based on Delaunay Triangulation [25]
b. Medial-Axis Transformation [17]
c. Straight Medial-Axis Transformation [7]

Fig. 18.2. Medial-Axis Transformation Algorithms

The straight MAT is another method to extract a medial-axis [7]. The straight medial-axis is constructed by a shrinking process in which the edges of a polygon move inward with the same speed (Fig. 18. 2c). All edges are reduced to three event points eventually: edge, split, and vertex events. An edge event occurs when an edge collapses down to a point (Fig. 18. 3a). If neighboring edges of the edge event still have nonzero length, they become adjacent. A split event occurs when a reflex vertex collides with and splits an edge (Fig. 18. 3b). A split event divides a component of the shrinking polygon into two smaller components. A vertex event occurs when two or more reflex vertices are collapsed to the same point (Fig. 18. 3c). A vertex event can introduce a new reflex vertex into the shrinking polygon, which eventually create another split event. Once the component of the shrinking polygon becomes a triangle, the straight medial-axis is completed by connecting three vertices of the triangle to its center. Connecting these reduced points forms a straightened skeleton. The algorithm is thus applied to construct a roof on a given ground plan [12]. To extract wall structure from wall polygon data in this paper, Eppstein and Erickson's straight MAT algorithm [7] is used because the algorithm can avoid the second order lines in order to calculate the interaction between walls and humans in a building evacuation model. The straight medial-axis transformation can also avoid the extraction of "Y" shape medial axis from "T" polygon.

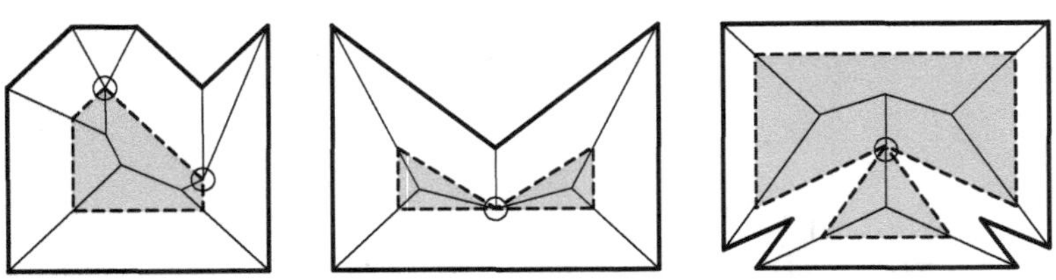

a. Edge Events
b. Split Event
c. Vertex Event

Fig. 18.3. Event Points of Straight Medial-Axis Transformation [7]

18.3.2 Topological Data Structure and 3D Geometric Network

Geometric network for 3D internal structure of buildings can be created through 3D topological data extraction. Boundary-based representations (B-Rep) constitute the most popular model to store 3D topological data. The Urban Data Model (UDM) [6] is based on the B-Rep model and uses boundaries: faces, edges, and nodes. UDM allows us to query topological relationships between rooms. However, topological relationships in UDM are implicit, which requires extra processing power for interpretation. Billen and Zlatanova [3] have proposed the Dimensional Model (DM) in which objects are composed of "dimensional elements." DM allows for topographical relationships between any combination of one, two, and three-dimensional objects. However, DM would also be resource intensive when applied to an entire building.

Since B-Rep models require a large volume of data and considerable power to process, Lee and Kwan [20] have proposed a simple 3D topological data structure called Combinatorial Data Model (CDM). CDM consists of node-relation structure (NRS) and hierarchical network structure (HNS) to represent topological relationships between objects. In NRS, a node represents a room and an edge represents a wall shared by two rooms. HNS, a subset of NRS, stores edges that connect two nodes through a door, an elevator, or a stair. Therefore, NRS represents adjacency information and HNS connectivity. This paper adopts CDM to build 3D topological structure because it allows for fast processing of 3D analysis with the concise 3D topological data structure. Based on the wall structure extracted from CAD data using the straight MAT, the NRS algorithm builds a 3D topological structure, in which a node represents a room and an edge represents a topological relationship (adjacency and connectivity) (Fig. 18. 4).

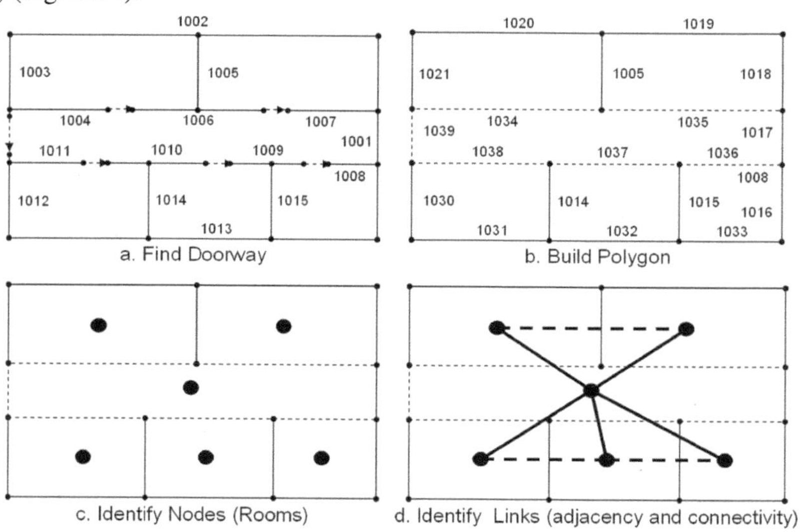

Fig. 18.4. 3D Topological Data Construction from Wall Structure

The algorithm to generate NRS data needs four steps. First, on the wall structure, doorways of individual rooms are easily extracted by extending direction vector of each end node to the nearest end node (Fig. 18. 4a). The doorway data are stored separately for later use to build connectivity. Second, walls and exits are then used to build room polygons using ArcGIS Build operator that split a line segment shared by three rooms (Fig. 18. 4b). Third, nodes for NRS are identified for individual rooms (Fig. 18. 4c). Finally, shared walls are converted to links that represent adjacency and doorways are converted to links that represent connectivity in 3D topological structure (Fig. 18. 4d). All edges represent connectivity is a subset of the edges that represent adjacency (Fig. 18. 5). Once horizontal NRS data is generated for each floor, HNS data links NRS of each floor by vertical connectivity through an elevator or a stairway [20, 30]. Each node (a room) is vertically adjacent to above and below nodes (rooms).

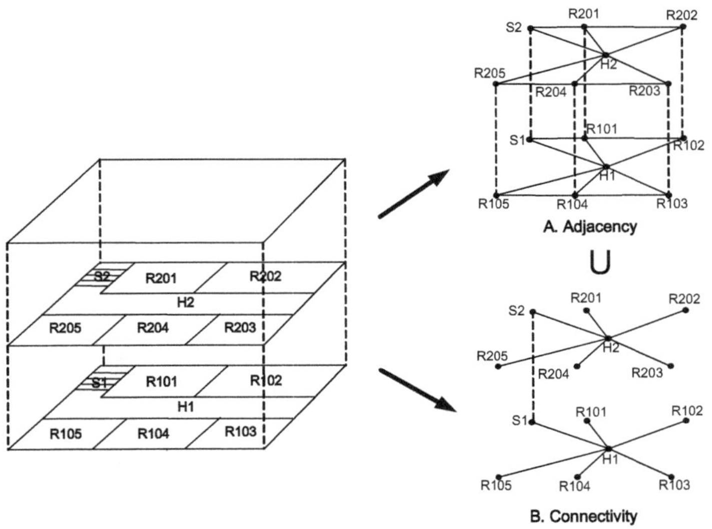

Fig. 18.5. 3D Topological Data based on Combinatorial Data Model [30]

Three dimensional topological data should be converted to 3D geometric network structure in order to model human behavior in building evacuation simulation. Hallways are represented by nodes in 3D topological data but they should be recognized as edges that connect rooms for navigation. The process to build 3D geometric network from 3D topological data requires two steps [19] (Fig. 18. 6). First, nodes that represent hallways are converted to edges. Hallway nodes are removed and hallway polygons are converted to edges using the MAT algorithm. Second, all edges from the other nodes need to be adjusted perpendicularly to the hallway edges in order to maintain shortest connectivity in 3D topological data.

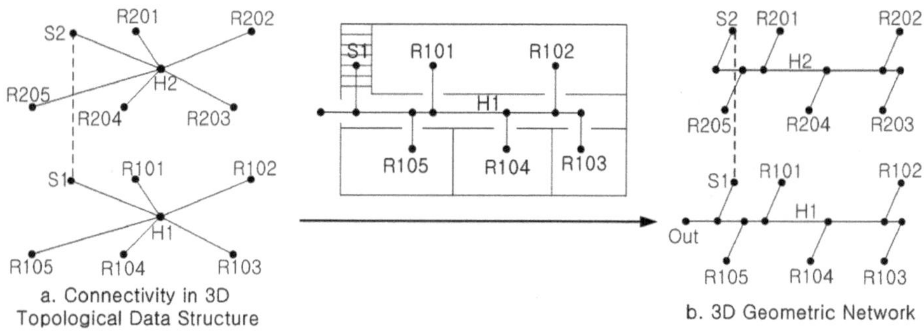

Fig. 18.6. 3D Geometric Network from 3D Topological Data Structure [19]

18.4 Conversion from Scanned Blueprint to 3D Geometric Network

The wall structure of a building can be extracted from scanned blueprint using three fundamental algorithms: Voronoi diagram-based thinning, mathematical operator-based thinning, and the boundary peeling method [1] (Fig. 18.7). The Voronoi diagram-based thinning algorithm is the vector-based method, which starts by collecting sample points of a region's boundary from raster data [29]. The, the incremental algorithm [11] computes Delaunay triangulations of the sample points. A Voronoi edge is a part of the skeleton if its corresponding dual Delaunay edge lies completely within the region's boundary (Fig. 18.7a). If a skeleton has parasitic branches, the skeleton can be simplified by retracting the leaf nodes of the skeleton to their parent nodes [9]. Finally, resulting vector skeleton is converted to raster skeleton. While the Voronoi diagram-based thinning algorithm may preserve the morphological shape of complex region, it does not extract straight line even from simple rectangular area without an extreme number of sample points (Fig. 18.7a).

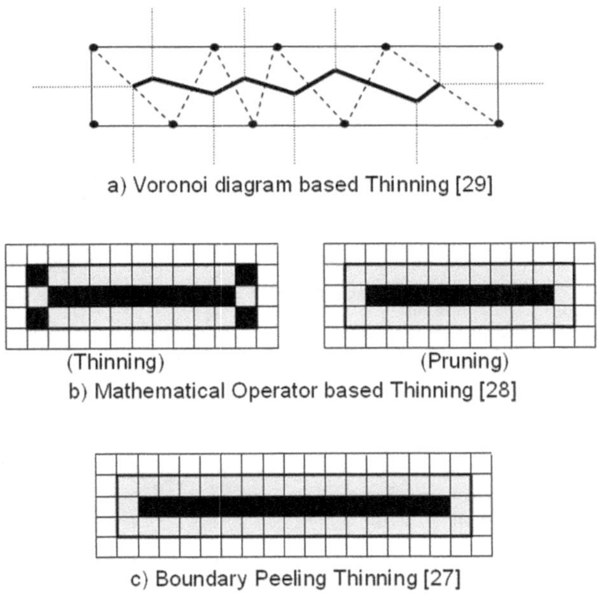

Fig. 18.7. Skeleton Extraction using Thinning Algorithm for Scanned Blueprint Data

The mathematical operator extracts skeleton of a region using two steps: thinning and pruning [28] (Fig. 18.7b). The thinning derives a general skeleton of an area feature and pruning removes parasitic branches of the skeleton [31]. Thinning process requires predefined sequential window filters (3x3 windows) that preserve eight directions from the center of the window. These sequential filters are applied one by one to thin an area feature iteratively until no changes occur. The thinning process produces the parasitic branches mostly at the end part of the skeleton (left image in Fig. 18.7b), which are removed by pruning. The pruning process utilizes some of the eight directional filters of which direction is corresponding to the directions of the parasitic branches [31]. Pruning is also iterative process until no further changes occur and produces the final skeleton of a region (right image in Fig. 18.7b). While the mathematical operator better approximates medial-axis of a region than the Voronoi diagram-based thinning method, it requires interactive selection of directional filters for pruning. Since the result skeleton depends on the selection of pruning filters, the mathematical operator method is not robust.

The boundary peeling algorithm [27] does not need the sequential filters and the pruning process of the mathematical operator method. The boundary peeling algorithm starts by initializing the region to ON and the background to OFF. Every ON pixel is inspected by a 3x3 window, in which the center pixels are erased (set to OFF) if the pixels are not required for preserving connectivity, maintaining end lines, and preventing inward erosion. In the thinning process, ON-valued center pixels are erased (ERASED) if three conditions are satisfied. First, if the connectivity, defined as the number of chains of connected ON pixels in the neighborhood, is equal to 1, the center ON pixel can be erased (Fig. 18.8a). The erasure of the center pixel will not destroy connectivity within any ON chains in the neighborhood. If the connectivity is

2 or more, the connectivity will be broken. Second, if the maximum length of a chain of the connected ON pixels in the neighborhood is greater than 1, the center ON pixel can be erased (Fig. 18.8b). If the maximum length is 1, the center ON pixel is the end location of the end line that should be maintained. Third, if the maximum length of a chain of the connected OFF pixels in the neighborhood is greater than 1 and less than 7, the center ON pixel can be erased (Fig. 18.8c). If the maximum length is 1, the erasure of the center pixel can intrude further into ON regions. Finally, if all these three conditions are satisfied, the center ON pixel is set to ERASED. The ERASED-value pixels are treated as if they were ON values to prevent uneven erosion on a single iteration. At the end of individual iterations the ERASED pixels are set to OFF pixels. Iterations stop when pixels are no longer erased. The results are the skeleton of input area (see Fig. 18.7c).

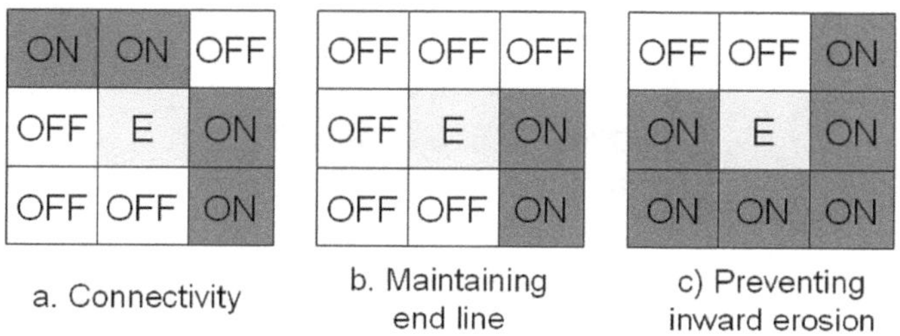

Fig. 18.8. Conditions to Erase Center ON-Value Pixel in Boundary Peeling Process

Skeletons from the boundary peeling method in raster domain can approximate the medial axis of a region better than those produced by the Voronoi diagram-based method. Further, the boundary peeling algorithm is robust than the mathematical operator-based method since it does not require pruning that uses interactive directional filters. Therefore, boundary peeling algorithm is implemented in this study and used to extract the skeleton of wall structure in order to build 3D geometric network data. For experiment, vector polygon walls (Fig. 18.9a) are converted to raster data (Fig. 18.9b) and used as input for the boundary peeling process. The output of the boundary peeling process is the skeleton of wall structure (Fig. 18.9c).

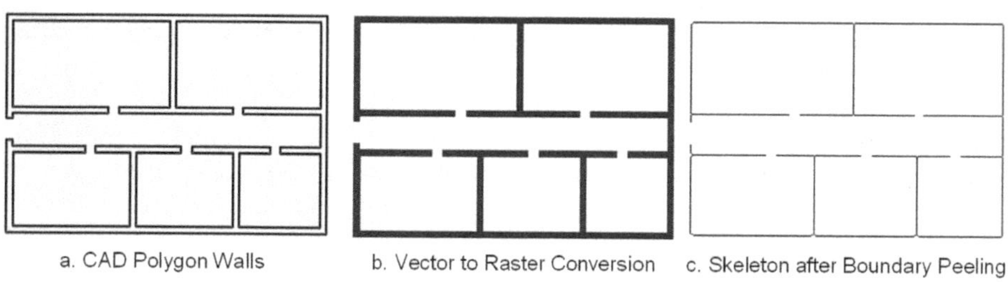

Fig. 18.9. Wall Skeleton Extraction using Boundary Peeling Algorithm

Although skeletons may not be medial-axis, they are enough to represent walls to identify rooms in a building. The result wall skeleton needs to be converted to vector line segments in order to build 3D geometric network through 3D topological structure.

The process to build 3D geometric network from wall skeleton is already explained in the previous section (see Section 18.3.2). In the next section, 3D geometric network information is used to generate input data for building evacuation simulation. The moving agents (human) in the evacuation simulation utilize the shortest path information from 3D geometric network to find the nearest outer exit from their respective locations.

18.5 Building Evacuation: Agent-Based Simulation

The key element of an agent-based model is agent. Agents are autonomous individuals who can make independent decision and influence on a simulation [8, 23, 32]. Agents have their own data and behaviors. Agent data are current states that include the agent's internal state and relationship with other agents. Agent behaviors sense their surrounding to solve complex problems, to communicate, to move, and to adapt their altered state [4]. The other elements of an agent-based model are relationships and environments. A relationship between agents is specified by linking agents within a system. Environments define the space in which agents interact with the environment and other agents.

18.5.1 Internal Structure of a Building and Agent's Data

Building evacuation simulation requires four input data: humans, rooms, walls, and doors (or exits). 3D geometric network is used to store connectivity in doors data and adjacency in rooms. Among these input data, humans and rooms are agents. Humans are moving agents that navigate to exit the building. Humans are located randomly in several rooms. Human agents have five data: LOCATION, SIZE, ROOMID, SITUATION, and SPEED. LOCATION stores current location of an agent for his movement. SIZE stores each agent's physical size for calculating the interaction with other agents. ROOMID stores agent's current room. SITUATION stores the awareness of an emergency situation in a building. SPEED is agent's desired speed. The direction of agent's movement is decided by the agent's location and the interaction with obstacles such as other agents and walls.

Rooms are modeled as static agents that update their own states in case a room has an emergency situation such as a fire. Rooms data are generated by ArcGIS Clean operator during 3D NRS construction process (see Fig. 18.4b and Section 18.3.2). Room agents have their own data: ROOMID, STATE, and NEIGHBOR. ROOMID is a room number. STATE stores the situation of the room. NEIGHBOR stores the adjacency information of each room, which is built from 3D geometric network.

Walls and doors are environments that provide humans with locational information to assess exit strategy in a building. Walls are extracted by straight MAT (see Section 18.3.1) and doors are generated by directional vector during 3D NRS construction process (see Fig. 18.4a and Section 18.3.2). Especially, geometric network data are used to build evacuation routes from each room to outer exit based on the connectivity of rooms. Doors data stores the connectivity of two rooms for human agent's moving direction at each door.

18.5.2 Human Behavior

Behaviors rule an agent's actions that change agents' location and update agents' status. In a building evacuation, the only moving agents are humans. The behaviors of human can be derived from the paradigms of pedestrian movements [14, 15, 26]. To simulate the crowd dynamics in a building, Helbing et al. [13] provide a social force model based on Newton's acceleration equation (Equation 1). The first term in Equation 1 represents socio-psychological force. The second and third terms represent human interaction (f_{ij}) (see Equation 2) and human to wall interaction (f_{iw}) (see Equation 3) forces.

$$f = m_i a = m_i \frac{dv_i}{dt} = m_i \frac{v^0{}_i(t)e^0{}_i(t) - v_i(t)}{\tau_i} + \sum_{j \neq i} f_{ij} + \sum_{w} f_{iw} \quad \textbf{(1)}$$

Here, each pedestrian (i) of weight (m_i) likes to move with a certain desired speed (v^0_i) in a certain desired direction (e^0_i) through time (t), and therefore tends to correspondingly adapt his or her actual velocity (v_i) with a certain reaction time (τ_i).

$$f_{ij} = \{A_i \exp[(r_{ij} - d_{ij}) / B_i] + kg(r_{ij} - d_{ij})\} n_{ij} + \kappa g(r_{ij} - d_{ij}) \Delta v^t{}_{ji} t_{ij} \quad \textbf{(2)}$$

In Equation 2, the first term $f_{ij} = A_i \exp[(r_{ij} - d_{ij}) / B_i] n_{ij}$ is a repulsive interaction force of two agents i and j to stay away from each other. A_i and B_i are constant that can reproduce the distance kept at normal desired velocities. Higher A_i produces greater repulsive forces with B_i range overlap between two agents. d_{ij} denotes the distance between center of two agents (H_i and H_j). r_{ij} is sum of radii r_i and r_j of two agents (H_i and H_j). $n_{ij} = (H_i\text{-}H_j)/\ d_{ij}$ is the normalized vector pointing from agent j to i. If two agents touch each other ($r_{ij} > d_{ij}$), body compression $kg(r_{ij} - d_{ij})n_{ij}$ and relative tangential motion $\kappa g(r_{ij} - d_{ij})\Delta v^t{}_{ji} t_{ij}$ are considered. If they are apart from each other ($r_{ij} < d_{ij}$), $g(x)$ is zero. k and κ determine the obstruction effect in cases of physical interactions.

$$f_{iw} = \{A_i \exp[(r_i - d_{iw}) / B_i] + kg(r_i - d_{iw})\} n_{iw} + \kappa g(r_i - d_{iw})(v_i t_{iw}) t_{iw} \quad \textbf{(3)}$$

Equation 3 captures the interaction between humans and walls. d_{iw} means the distance to wall W. n_{iw} denotes the direction perpendicular to W. t_{iw} is the direction tangential to W. Other terms are identical to those in Equation 2.

To implement Equation 1 for human movement in building evacuation simulation, six behaviors of human agents are used: MYSTATE, MYROOM, MYEXIT, MAINFORCE, P2PFORCE, P2WFORCE, MOVE, and UPDATE. The MYSTATE action checks an agent's states. MYROOM checks current room situation with room ID, which decides the agent's desired speed (v^0_i). MYEXIT checks current doorway to decide the desired direction (e^0_i) of the agent. MAINFORCE calculates the socio-psychological force (first term) in Equation 1. P2PFORCE calculates the human interaction force using Equation 2. P2WFORCE calculates the human to wall interaction force using Equation 3. MOVE changes the agent's location based on the result velocity of Equation 1 that is sum of MAINFORCE, P2PFORCE and P2WFORCE. Finally, UPDATE updates the agent's location and states.

18.5.3 Building Evacuation Simulation using Agent-Based Model

Among various agent-based modeling tools such as Swarm, MASON, Repast, and so on, Agent Analyst that is based on Repast[4] is used in this study since it is tightly coupled with ArcGIS for data management and visualization. There are two types of agents (vector and generic agents) in the Agent Analyst. A generic agent is a non-spatial agent such as emergency announcements. A vector agent is a spatial agent that is stored as features in a shapefile. In our evacuation simulation, all input data including humans, rooms, walls, and doorways are modeled as vector agents.

In building evacuation model, the only moving agent is human and therefore the simulation is based on the human behavior model (see Equation 1) and the characteristics of the crowd dynamics. We take first five characteristics from Helbing et al.'s [13] nine characteristics of crowd dynamics in order to simplify the situation: (1) people move or try to move considerably faster than normal; (2) individuals start pushing each other, and interactions among people become physical in nature; (3) moving and, in particular, passing through a bottleneck becomes uncoordinated; (4) at exits, arching and clogging are observed; (5) jams build up. We assume that all people in the building know its structure so that they know where the nearest exit is and how to get to it. Door data stores room connectivity from 3D geometric network, which provides the shortest path information.

To run the social force model (see Equation 1), we have specified the parameters as follows: a mass of m is average 60kg. Initial desired speed (v^0_i) is 5m/s; initial desired direction (e^0_i) is toward exit; distributed pedestrian diameters $2r_i$ in the interval [0.5m, 0.8m] considering shoulder widths; the constant A_i is 5000 and B_i is 0.08m; the parameter k is 750kg/s^{-2} and κ is 3000kg/ms. In the simulation, we consider 15 people (N) in three rooms to generate evacuation with and without a jam situation for each room.

The simulation shows the evacuation process through time (Fig. 18.10). People follow the shortest path to the outer exit through the door of their room. Figs. 10a to 18.10f show the evacuation situation in each room in the building. The number of moving people and the average speed are shown in Fig. 18.10g. When people start to jam in a room (Fig. 18.10b and Fig. 18.10c), the number of movements and the average speeds start to decrease (Fig. 18.10g). As people who are nearest to the door exit the room, the movements and the average velocity increase again. However, at the outer exit of the hallway, there is another jam because people out of different rooms meet at the exit, which decrease the average speed again (Fig. 18.10d and Fig. 18.10g). Therefore, if there are many people in a building and they want to evacuate

[4] Repast is an agent-based modeling tool and can download source codes and materials at repast.sourceforge.net.

at the same time, jams can occur at several places such as doorways of each room, stairways, and outer exits of the building. The jams make more distant people take even longer to evacuate and may result in many casualties in an emergency situation such as a fire.

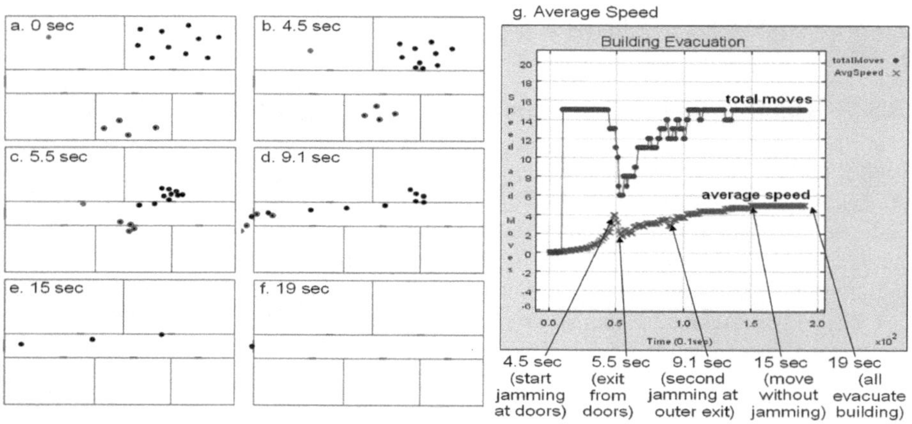

Fig. 18.10. Building Evacuation Simulation using Agent-based Model

In this study, we simulate egress movement simply in one story building in 2D display environment. We can easily extend it to multi-story building by the connection of multiple floors in 2D environment or may be directly visualized in 3D environment such as VRML. For evacuation modeling in multi-story building, we need to consider two aspects: vertical connection and agents' speed in vertical movement. For vertical connection of floors, 3D geometric network already identifies stairwells, elevators, and escalators and stores vertical connectivity information. For vertical egress movement, the agents' desired speed should be adjusted based on the type of space for vertical movement. In case of emergency, only stairwells may be considered for vertical movement and the desired speed should be reduced since agents could not move freely in stairwells with the same speed as in hallways.

18.6 Conclusion

An integration of agent-based modeling and GIS offers improvements to understand complex spatial phenomena. With agent-based models, GIS can represent feature dynamics such as temporal changes in spatial patterns for further time-series spatial analysis. However, current GIS are still nascent in the representation of 3D internal structure of buildings for micro-level spatial modeling such as building evacuation simulation.

In this study, we have discussed building evacuation simulation in GIS environment by integrating 3D geometric network construction procedure (3D GeoNet) with agent-based model (Agent Analyst). These tools are tightly coupled with ArcMap that is used for visualization. 3D GeoNet generates the internal structure data of a building and 3D geometric network data and provides input data for Agent Analyst. Agent Analyst produces sequential results of agents' egress movement. The sequential results of simulation are visualized in ArcMap.

Especially, the automated generation of 3D geometric network data is important for building evacuation in order to provide pedestrians' routing information with room connectivity. In this study, 3D geometric network has been created from building blueprint data through wall extraction and 3D topology construction. We tested two wall extraction methods: vector-based MAT and raster-based thinning. For vector-based MAT, we utilized straight MAT to extract wall structure from wall polygon from CAD data. For raster-based thinning, we tested boundary peeling algorithm to extract wall structure from scanned blueprint. With wall structure extracted using both extraction methods, doorways and rooms were identified. By 3D duality, rooms were converted to nodes and the walls shared by rooms were converted to links in 3D topological structure. 3D geometric network was built by converting nodes in 3D topological structure to links if the nodes represent hallways. In particular, the topological connectivity and adjacency information in 3D geometric network were stored doorway and room data, respectively, which were used in building evacuation simulation.

With spatial data including rooms, walls, doors, and humans, a building evacuation was simulated in GIS environment using Agent Analyst. Especially, human agents' behavior in building evacuation was modeled by adopting Helbing et al.'s social force model [13]. We tested simple evacuation with a jam situation by enforcing different numbers of people in three rooms. As expected, the results showed that the average velocity increased continuously before jams, decreased during jams at doorways and outer exits, and eventually increased again as individuals escaped from jams.

As we emphasized in the paper, 3D geometric network provides detailed sub-unit structure of a feature such as a building for observing and understanding micro-level navigation. Especially, the integration of 3D geometric network and an agent-based model provides current GIS with new methodology that allows a simulation for dynamic spatial process in a micro-level environment such as a building. Future building evacuation simulation will involve enhanced human behavior model and more complex building structure under disastrous situation such as fires.

18.7 Acknowledgments

This research was supported by a grant from 'Seoul R&BD Program (10592)' funded by the City of Seoul, South Korea

References

1. Asante, K., Maidment, D.: Creating a River Network from the Arcs in the Digital Chart of the World. Available at http://www.ce.utexas.edu/prof/maidment/grad/asante/dcw/rivernet.htm (1999)
2. Benhamu, M., Doytsher, Y.: Toward a spatial 3D cadastre in Israel. Computers, Environments and Urban Systems 27:359-374 (2003)

3. Billen, R., Zlatanova, S.: 3D spatial relationships model: a useful concept for 3D cadastre? Computers, Environment and Urban Systems 27:411-425 (2003)
4. Castle, C.J.E., Crooks, A.T.: Principles and Concepts of Agent-Based Modeling for Developing Geospatial simulations. UCL Working Papers Series: Paper 110. Available at http://www.casa.ucl.ac.uk/working_papers/paper110.pdf (2006)
5. Chin, F., Snoeyink, J., Wang, C.A.: Finding the Medial Axis of a Simple Polygon in Linear Time. In: International Symposium on Algorithms and Computation, pp 382-391 (1995)
6. Coors, V.: 3D-GIS in networking environments. Computers, Environment and Urban Systems 27:345-357 (2003)
7. Eppstein, D., Erickson, J.: Raising Roofs, Crashing Cycles, and Playing Pool: Applications of a Data Structure for Finding Pairwise Interactions. In: 14th Annual ACM Symposium on Computational Geometry, Minneapolis, Minnesota, pp 58-67 (1999)
8. Epstein, J.M.: Agent-Based Computational Models and Generative Social Science. Complexity 4(5):41-60 (1999)
9. Gold, C.M., Thibault, D.: Map generalization by skeleton retraction. In: the 20th International Cartographic Conference (ICC), Beijing, China, pp 2072-2081 (2001)
10. Goodchild, M.F.: GIS, Spatial Analysis, and Modeling Overview. In: Maguire DJ, Batty M, Goodchild MF (eds.) GIS, Spatial Analysis and Modeling, ESRI Press, Redlands, California, pp 1-17 (2005)
11. Green, P., Sibson, R.: Computing dirichlet tessellations in the plane. The Computer Journal 21:168-173 (1978)
12. Haunert, J.H., Sester, M.: Using the Straight Skeleton for Generalization in a Multiple Representation Environment. In: ICA Workshop on Generalization and Multiple Representation, Leicester (2004)
13. Helbing, D., Farkas, I.J., Vicsek, T.: Simulating dynamical features of escape panic. Nature 407:487-490 (2000)
14. Helbing, D, Farkas, I.J., Molnar, P.: Simulation of Pedestrian Crowds in Normal and Evacuation Situations. In: Waldau N, Gattermann P, Knoflacher H, Schreckenberg M (eds) Pedestrian and Evacuation Dynamics. Springer-Verlag, Berlin, pp 21-58 (2002)
15. Jiang, B.: SimPed: Simulating Pedestrian Flows in a Virtual Urban Environment. Journal of Geographic Information and Decision Analysis 3(1):21-30 (1999)
16. Kevany, M.J.: GIS in the World Trade Center attack-trial by fire. Computers, Environment and Urban Systems 27:571-583 (2003)
17. Lee, D.T.: Medial axis transformation of a planar shape. IEEE Transactions on Pattern Analysis and Machine Intelligence 4(4):363-369 (1982)
18. Lee, J.: 3D Data Model for Representing Topological Relations of Urban Features. In: ESRI International User Conference, San Diego, CA, Available at http://gis.esri.com/library/userconf/proc01/professional/papers/pap565/p565.htm. (2001)
19. Lee, J.: A Spatial Access-Oriented Implementation of a 3-D GIS Topological Data Model for Urban Entities. GeoInformatica 8(3): 93-113 (2004)
20. Lee, J., Kwan, M.: A Combinatorial Data Model for Representing Topological Relations among 3D Geographical Features in Micro-Spatial Environments. International Journal of Geographical Information Science 19(10):1039-1056 (2005)

21. Longley, P.A., Batty, M.: Advanced Spatial Analysis: Extending GIS. In: Longley PA, Batty M (eds) Advanced Spatial Analysis: The CASA Book of GIS. ESRI Press, Redlands, California, pp. 1-20 (2003)
22. Longley, P.A., Goodchild, M.F., Maguire, D.J., Rhind, D.W.: Geographical Information Systems and Science. John Wiley and Sons, Hoboken, NJ (2005)
23. Macal, C.M., North, M.J.: Tutorial on Agent-Based Modeling and Simulation. In: Euhl ME, Steiger NM, Amstrong FB, Joines JA (eds) The 2005 Winter Simulation Conference (2005)
24. Marion, B.: The History of CAD. MB Solutions. Available at http://mbinfo.mbdesign.net/CAD-History.htm (2004)
25. Prasad, L., Rao, R.: Morphological Analysis of Shapes. Available at http://computervision.lanl.gov/pub_001/pub_001.pdf (1998)
26. Schelhorn, T., O'Sullivan, D., Haklay, M., Thurstain-Goodwin, M.: STREETS: An agent-based pedestrian model. Working paper. Centre for Advanced Spatial Analysis UCL, London, UK. Available at http://www.casa.ucl.ac.uk/working_papers/paper9.pdf (1999)
27. Seul, M., O'Gorman, L., Sammon, M.J.: Practical Algorithms for Image Analysis: Description, Examples, and Code. Cambridge University Press, New York (2000)
28. Serra, J.: Image Processing and Mathematical Morphology. Academic Press, New York (1982)
29. Sharma, O., Mioc, D., Anton, F.: Voronoi Diagram Based Automated Skeleton Extraction from Colour Scanned Maps. In: Third International Symposium on Voronoi Diagrams in Science and Engineering (ISVD'06). Banff, Alberta, Canada. Available at http://ieeexplore.ieee.org/iel5/11131/35646/01691191.pdf (2006)
30. Stevens, M. and Choi, J.: CAD Data Conversion to a Node-Relation Structure for 3D Sub-Unit Topological Representation. Journal of the Korean Geographical Society 41(2):188-194 (2006)
31. Su, B., Li, Z., Lodwick, G.: Morphological Models for the Collapse of Area Features in Digital Map Generalization. GeoInformatica 2(4):359-382 (1998)
32. Torrens, P.M.: Simulating Sprawl: A Dynamic Entity-Based Approach to Modeling North American Suburban Sprawl Using Cellular Automata and Multi-Agent Systems. Ph.D. thesis, University College London, London (2004)
33. Zhou, Q., Zhang, W.: A preliminary Review on 3-dimensional City Model. In: Asia Geographic Information System Association 2003 Conference. Available at http://www.hku.hk/cupem/asiagis/fall03/Full_Paper/Zhou_Qiming.pdf (2003)
34. Zlatanova, S.: Advances in 3D GIS. Quarterly Review of Disegno Digitale e Design 1(4): 24-29 (2002)
35. Zlatanova, S., Rahman, A.A., Pilouk, M.: 3D GIS: Current Status and Perspectives. In: ISPRS, Ottawa, Canada. Available at http://www.gdmc.nl/zlatanova/thesis/html/refer/ps/SZ_AR_MP02.pdf (2002)

Chapter 19
Hierarchical Modelling of Multi-Geospatial Databases as Basis for Geo-Oriented 3D Analysis Capabilities

Sagi Dalyot and Yerach Doytsher

Abstract. Geospatial databases representing the natural environment are widely used and applicable for various mapping and engineering applications. Nowadays, the development of state-of-the-art measurement tools gives updated and more precise knowledge regarding the morphology of the acquired object, i.e., terrain relief. Geospatial analysis, such as comparison or integration, of two (or more) databases acquired on different epochs, while relying solely on their coordinate reference systems, will usually result in a wrong outcome. This occurs mainly because of inherent non-uniform topographic and topologic inconsistencies - as well as morphologic changes transpired - between the databases. Investigating and monitoring these factors prior to actual data analysis is essential. This paper suggests the division of the databases' mutual coverage area into homogeneous separate hierarchical working levels. 2-stage monitoring process involving spatial registration and matching is implemented. This enables to properly define, model and store the complete spatial local interrelations of the databases. Utilizing these values yields novel mathematical and analyzing geo-oriented capabilities, thus enabling visualization, simulation and integration processes.

19.1 Introduction

Nowadays, spatial topographic databases are amongst the main resources for a variety of terrain relief and natural phenomenon analysis and research applications, such as hydrogrpahy, urban mapping, risks and damages assessments. One of the more common spatial topographic computerized databases (model) is Digital Elevation Model - DEM (sometimes referred to as DTM - Digital Terrain Model). DEM databases store

---

Mapping and Geoinformation Engineering
Technion - Israel Institute of Technology
{dalyot, doytsher}@technion.ac.il

the spatial elevation data of the underlying surface or terrain. This data is usually stored as an equal-spaced rectangular grid and mostly presented in Cartesian or Geographic coordinate system (much like raster data). Common DEM databases usually store data in several meters resolution and have vertical accuracy around 1m. New data acquisition technologies, on the other hand, are now capable of collecting data in a much denser and more precise manner. For example, Light Detection And Ranging (LiDAR) technology, which have dispersal and irregular data structure, can acquire up to 16 points per $1m^2$ with position accuracy around 0.1m.

If two different databases acquired in different epochs are chosen for updating or morphological investigation task, it is quite obvious that they will present fundamental different data class, such as format, resolution, accuracy, density, datum, and more. Even though both databases are geographically geo-referenced to a certain coordinate reference system, a simple comparison and monitoring based on these reference systems might not suffice. Zonal topographic inconsistencies, as well as different data structures and formats, affect topographic ambiguity. These inconsistencies may occur due to natural phenomenon or human activities that took place during the data acquisition epochs, as well as having inherent errors occurring during the data acquisition or production (object modelling) stages [1]. These discrepancies can be categorized by two groups: global systematic ones that can be roughly monitored and modelled (even visually); and, local random ones, which reflect on different scales of geometric discrepancies, and can be quantified only by local rigorous modelling analysis [2].

As a result, a local thorough investigation of the relative spatial correlations exist between the databases is essential for achieving correct analysis capabilities that are based on the data stored in them. Moreover, this investigation will prevent distortions as well as an ambiguous and ill-defined modelling analysis. The chosen analysis mechanism has to address the height representation issue of the terrain, as well as its spatial characteristics - topology and morphological structures [3]. Fig. 19.1 shows an example of an up-to-date database superimposed ('inserted') on an outdated one while relying solely on their mutual reference coordinates systems. It is clear that there are irregularities in the topographic representation. For topographical analysis purposes these are to be solved antecedently via morphologic and accuracy adjustments set of rules to ensure continues and semantic modelling.

The structure of this paper is as follows: section 19.2 outlines the suggested concept and the contributions of the proposed automated hierarchical modelling; Section 19.3 outlines in further detail the algorithms developed and used; Section 19.4 describes geo-oriented experimental results derived from the discussed mathematical concept; Section 19.5 details relevant related work; and, the conclusions are given in section 19.6.

19.2 Proposed Approach

The analysis presented here can be implemented on any number and combination of geo-spatial databases. From here on, the concept is detailed in regard to two given databases (in the case where there are more than two databases, a sequence of the proposed approach will be implemented - resulting in a single outcome). A simultaneous mutual monitoring and modelling processes are essential for investigating the relative

existing spatial correlations. We suggest carrying out this task hierarchically, according to two main stages:

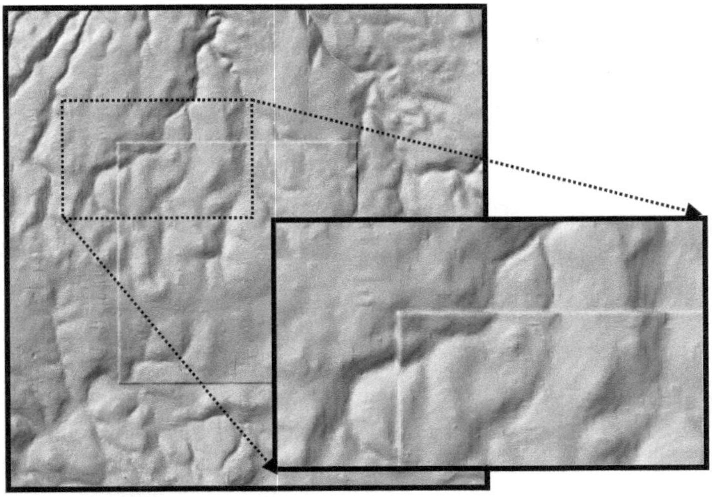

Fig. 19.1. An up-to-date database superimposed onto an existing outdated database showing distinctive discontinuities on the border of the mutual area: height gaps and truncated morphological features caused by spatial trends are visible in the zoomed area (on the right).

1. Registration (geo-referencing) - selecting a common mutual working schema (spatial reference frame) while relying on topologic relations between conjoint unique entities that exist in both databases.
2. Matching - spatial analysis process, which uses the registration knowledge, where a qualitative reciprocity is extracted for both databases. This enables precise and complete modelling of existing geometric conditions.

An independent hierarchical modelling process, which is based on these two stages, is implemented on several separate homogenous topographic areas (frames). This will enable geo-oriented analysis capabilities to utilize the quantitative matching and modelling knowledge (values). 2-stage hierarchical process is proposed, in which in each stage a smaller area, i.e., localized data, is utilized, while based on the previous stage modelling result: i) implementing a robust geo-referencing process on the databases - enabling global discrepancies monitoring; ii) utilizing this value for local spatial matching that is carried out on smaller data-portions, enabling precise monitoring of local random discrepancies that still exist. The complete 'from global to local process' validates that an analysis process utilizing the extracted databases' modelling will preserve the topology, geometry and morphology of features exist in both databases (terrain relief characteristics) - while preventing distortions.

Rough registration can be extracted by constraining an affine spatial transformation to selected unique homologous features representing the same real-world object that exists in both databases [4, 5, 6] showed that obtaining a-priori knowledge regarding the geometric spatial relations between geospatial databases is crucial for a geometric matching process, mainly to ensure a correct and non-biased (local minima) matching process. Rigid affine registration is achieved by various schemas, such as scene coherence [7], clustering approach [8] or invariant property [9].

Various geometric matching methods are cited in the literature, which are usually adapted to a certain data type in the selected test area. The matching algorithms mainly depend on the geometric types of the objects need to be matched, their topological relations, the databases' volumes, and the semantic attributes [10]. A robust and qualitative matching process suitable for the data characteristics at hand is the Iterative Closest Point (ICP) algorithm [11]. The ICP algorithm is designated for 3D point cloud matching process by nearest neighbor criteria, while using an iterative Least Squares Matching (LSM) process [12]. This algorithm utilizes a-priori registration value, which is basically needed to exclude local minima solution. We suggest dividing the entire mutual coverage area into a grid of homologous separate partially overlapping small frames. The ICP matching process is then implemented on each frame separately and independently. This enables a localized monitoring, thus categorizing more accurately local phenomena. This concept is in contrast to implementing a global matching process on the entire area, which consequently produces ill-defined and ambiguous results - mainly because it is not capable to model localized trends. The outcome of localized matching is the extraction of sets of accurate local spatial affine geo-registration (transformation) parameters - one for each homologues local frame. These sets of geo-registration values are stored in a novel grid of cells analogous to a 2D matrix - a DEM look-alike structure. Via the use of designated interpolation algorithms, utilizing this localized correct modelling enables the implementation of diverse geo-oriented analysis processes.

19.3 Algorithm Outline

19.3.1 Registration

19.3.1.1 Extracting Unique Entities.

Identifying distinctive topographic interest points is required for a robust registration of both databases. A novel extraction mechanism of surface-derived geomorphologic maxima features, such as local hill peaks, was implemented [13]. This mechanism is based on designated geometric, topologic, and topographic conditions. A generalization algorithm for interest area recognition is established by constructing four local second degree polynomials around each grid-point - one for each principal direction. These polynomials represent a generalized version of the point's topographic surroundings. According to a set of conditions and thresholds on the polynomials formulae the extraction of local maxima is feasible.

19.3.1.2 Extracting Registration Vector.

A coarse reference frame is achieved by registering (geo-referencing) selective homologous interest points identified in both databases while screening the conjoint ones. The process implemented here relies on the Hausdorff distance algorithm, in which no constraints or a-priori knowledge on the interest points' dispersal or their topologic relations is required. Given two sets of points $A = \{a_1,\ldots,a_m\}$ and $B = \{b_1,\ldots,b_n\}$, the forward Hausdorff distance - h - measures the degree of mismatch between the two sets, as defined in (1). This equation identifies point $a \in A$ that is farthest from any point in B, and then measures the distance from a to its nearest neighbor b in B. This value is later statistically evaluated by the correspondence it obtains between the remainder points in sets A and B. This Euclidean spatial distance gives an initial estimation of the global reciprocal working reference frame, i.e., systematic discrepancy modelling value. Here, a 3D translation vector was chosen: dx_0, dy_0, and dz_0. The output is a vector-set of preliminary registration values for the entire area chosen, which is used as a-priori data for the second working level - the matching process.

$$h(A,B) = \max_{a \in A} \min_{b \in B} \|a - b\| \tag{1}$$

19.3.2 Matching

The registration value enables the implementation of an adequate local matching process between the two databases. The ICP matching process is implemented on homologous corresponding local data frames that are divided from the complete mutual coverage area. It is obvious that by matching small frames more effective monitoring and modelling of local random discrepancies, i.e., inherent 'errors', exist between the databases is feasible. Monitoring - and then modelling - these errors is achieved by minimizing the target function, i.e., extracting the best possible correspondence between the databases. The geometric target function is defined by a spatial transformation model.

A geometrically constrained ICP algorithm between two homologous local frames is implemented (2). The constraints assure that the nearest neighbor search criterion is correctly achieved. In addition, they verify that each of the paired-up points - one in each database's frame - is the closest one exists, as well as having the same relative topography surroundings. This is achieved by implementing a perpendicular constraint on the first order derivative of the plane in both directions - X and Y - and the connecting line of the two pair-up points. It is worth noting that this algorithm is compatible in matching gridded databases - as well as non-gridded ones - by modifying the grid geometry into a Triangulated Irregular Network (TIN) one, such as in the case of LiDAR data.

$$Z_i^g = \frac{h_1}{D} \cdot X_i^g + \frac{h_3}{D} \cdot Y_i^g + \frac{h_4}{D^2} \cdot X_i^g \cdot Y_i^g \tag{2}$$

$$Z_i^g = -\frac{h_4 \cdot y_f^{'}}{D^2} \cdot X_i^g + \frac{h_3}{D} \cdot Y_i^g + \frac{h_4}{D^2} \cdot X_i^g \cdot Y_i^g + \left(z_f^{'} - \frac{h_3 \cdot y_f^{'}}{D} \right)$$

$$Z_i^g = \frac{h_1}{D} \cdot X_i^g - \frac{h_4 \cdot x_f^{'}}{D^2} \cdot Y_i^g + \frac{h_4}{D^2} \cdot X_i^g \cdot Y_i^g + \left(z_f^{'} - \frac{h_1 \cdot x_f^{'}}{D} \right)$$

where h_1 to h_4 are calculated from the height of local DEM grid cell corners: Z_1 to Z_4 ($h_1=Z_1-Z_0$, $h_2=Z_2-Z_0$, $h_3=Z_3-Z_o$, $h_4=h_2-h_1-h_3$); D denotes the DEM resolution; G and F denotes the DEMs' indexes; (X^g_i, Y^g_i, Z^g_i) denotes the paired-up nearest neighbor in G; and, ($x_f^{'}$, $y_f^{'}$, $z_f^{'}$) denotes the transformed point from dataset F.

The transformation model implemented here - depicted in (3) - was modelled according to 6 parameters - 3 translation parameters: dx, dy, and dz, and 3 rotation angles: φ, κ, and ω (a scale factor can easily be added if required). Because linearization is needed to solve this model, an approximated preliminary 6 transformation values per frame are required. The initial translation parameters used are the ones extracted in the registration stage (section 19.3.1.2), while the preliminary rotation angles values - φ_0, κ_0, and ω_0 - are given a value of 0 degrees (mainly for the fact that both databases are close to being orthogonal projections of the terrain due to different photogrammetric processes related to DEM production).

$$\begin{bmatrix} X_g - X^M{}_g \\ Y_g - Y^M{}_g \\ Z_g - Z^M{}_g \end{bmatrix} = R(\varphi, \kappa, \omega) \bullet \begin{bmatrix} X_f - X^M{}_f \\ Y_f - Y^M{}_f \\ Z_f - Z^M{}_f \end{bmatrix} + \begin{bmatrix} dx \\ dy \\ dz \end{bmatrix} \tag{3}$$

where, g and f denote databases indexes; (X, Y, and Z) denote coordinates, R denotes the rotation matrix, (dx, dy, and dz) denote three translation parameters, (φ, κ, and ω) denote three rotation angles, and M denotes center coordinates of homologues frame.

Implementing local matching processes on zonal data frames results in an output that consists of matching parameters sets, which are equivalent to the number of frames. Each set includes 6 geo-registration parameters that best describe the relative spatial geometry of the mutual homologous frames matched. Since this process yields better localized geo-registration definition, it ensures matching continuity on the entire area (as opposed to matching the entire data in a single matching process). These geo-registration sets can be described as elements stored in 2D matrix: each set is stored in the cell that corresponds spatially to the zonal area it covers in both databases, as depicted in Fig. 19.2. This matrix is actually analogues to a 'geo-registration DEM', where each grid node (cell) corresponds to the geometric center of local frames. This 2D matrix will comprise as the bases for implementing precise geo-oriented analysis capabilities between the matched databases (described in section 19.4).

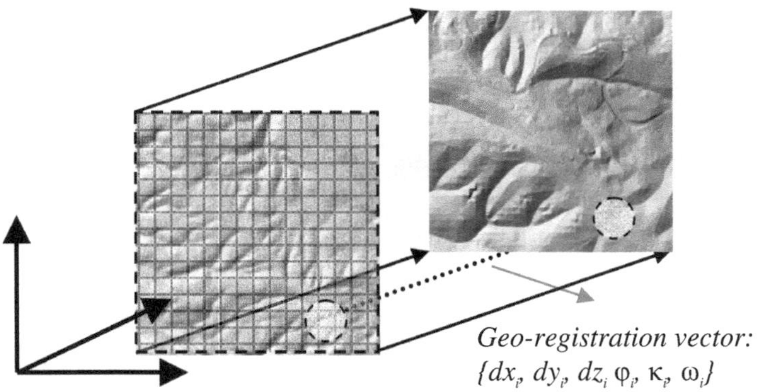

Fig. 19.2. Geo-registration matrix (superimposed on-top the source DEM) where each cell stores the geospatial relations, i.e., transformation, between mutual coverage frames.

3.3 Interpolation Concepts

3.3.1 Preface.

It is clear that the geo-oriented analysis procedures have to fit the databases' resolution in order to preserve actual precision. Still, the geo-registration modelling matrix resolution has significant different scale. As a result, an interpolation algorithm on the discrete data stored in the matrix is essential to ensure continuous interrelations representation - and not a truncated one that only rarely exist in the terrain relief. Interpolating between the vector sets of translation values is simple and straight forward because of their linear characteristics [14]. A problem arises when trying to interpolate between the orientation sets represented by three Euler angles. A naïve interpolation approach will produce a motion specified by a rotation-sequence that might be contracted, jerky at times, not continuous, and ill-specified. This occurs mainly for the fact that Euler angles ignore the interaction and dependency of the rolls about separate axes; hence there is not a unique path between every two orientations across different coordinate systems [15]. A resulting rotation depends on the order in which the three rolls are performed, which coincides with the fact that rotations in space do not generally commute.

The notation of Quaternions was first described by Sir William Rowan Hamilton in 1843, when he discovered a mathematical method to define a 3D number system by a 4D one. Four numbers describing rotation followed by scaling - just as complex numbers describe rotations in plane. Each Quaternion object contains four scalar variables: one real (W) and three imaginary (X, Y, and Z) - depicted in (4). Each of the imaginary dimensions has a unit value of different square root of -1, perpendicular to each other, known as i, j, and k (i.e.,: $i^2 = j^2 = k^2 = -1$, $i \bullet j = k$, $j \bullet i = -k$, $j \bullet k = i$, etc., with a cyclic permutation $i \bullet j \bullet k$). These Quaternion values describe rotation in a natural way: the vector part give the coordinates for the axis of rotation, while the scalar is determined by the angle rotated through. The main practical application of 4D Quaternions was found to be representing 3D rotations represented by a unit magnitude Quaternion.

$$q = W + X \cdot i + Y \cdot j + Z \cdot k \quad (4)$$

3.3.2 Translation values.

Due to the data structure of the 2D matrix an algorithm that will ensure continuous parameters representation among cells has to be chosen. Interpolating DEM heights using bi-directional third degree parabolic equations is described by [16]. This algorithm ensures a smooth interpolation within a grid-cell, as well as excluding surface representation discontinuities on cell borders while moving to the neighboring cell. This algorithm that utilizes 3rd degree parabolic equations is an improved version of the simple bi-linear interpolation that produces a jagged and unsmooth surface representation. The same concept is implemented on the geo-registration matrix; instead of interpolating height values we implement the suggested interpolation in three separate processes on the translation values stored in the cells: *dx*, *dy*, and *dz*. Hence, the precise quantification of the three translation parameters - describing spatial shift between the two databases - at a certain location within the matrix is feasible.

3.3.3 Rotation values.

The direct conversion from Euler angles (stored in the matrix cells) to Quaternions is not straight forward and involves mathematical considerations and constraints. Instead, converting Euler angles into a rotation matrix representation and then converting it into a unit-Quaternion is easier and more efficient [17].

Linear interpolation between rotations described by unit-Quaternions will result in a straight line, which ignores the natural geometry of rotation space, and would result in a motion that speeds up in the middle. The needed motion has to keep a constant velocity on the 4D hyper-sphere, which geometrically translates into a great arc drawn between two given Quaternions, i.e., a spline curve. using Spherical Linear Interpolation (SLERP) in unit-Quaternion space ensures a unique and correct path under all circumstances [18]. Consider 2 unit-Quaternion orientations to be 2 vectors in 2D: q_1 and q_2. The angle between the vectors is given by (dot product), which is normalized between the values 0 and 1, as depicted in Fig. 19.3. An intermediate third vector - q_i - creates angle t with q_1, and is derived from spherical linear interpolation between vectors q_1 and q_2, as shown in (5) (where λ_1 and λ_2 are the coefficients of a linear combination). The implementation of SLERP yields the unique shortest possible interpolation path between two unit-Quaternions on a unit sphere (a geodesic that follows the shortest great arc), which has constant angular velocity.

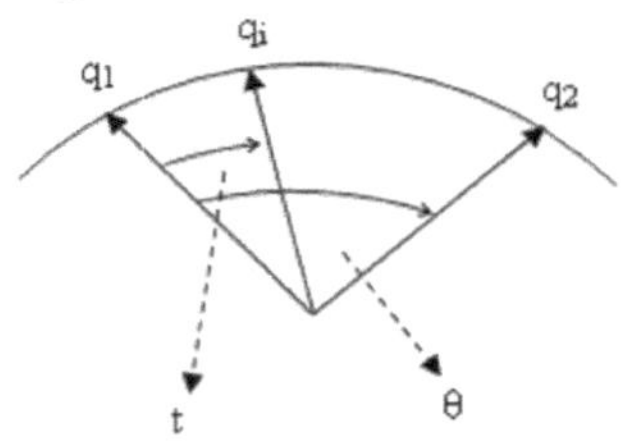

Fig. 19.3. SLERP 2D geometrical description.

$$
\begin{aligned}
q_i &= \lambda_1 \cdot q_1 + \lambda_2 \cdot q_2 \\
|q_i| &= slerp(q_1, q_2, t) = 1 \\
q_i &= \frac{\sin((1-t)\cdot\theta)}{\sin(\theta)} \cdot q_1 + \frac{\sin(t\cdot\theta)}{\sin(\theta)} \cdot q_2
\end{aligned}
\quad (5)
$$

Higher order of spherical interpolation within more than two orientations can be described as Bezier class curves [19], which draw arcs and proportions of arcs on unit sphere. This translates to calculating a 2-step SLERP sequence that will ensure a motion continuity described by the sequence of unit-Quaternions. Having a quadrangle grid of four corner-orientations: q_0, q_1, q_2, and q_3, the equivalent to Bezier curve with a spherical cubic interpolation - Spherical and Quadrangle (SQUAD) - is denoted in (6) [20]:

$$Squad(q_0, q_1, q_2, q_3, t) = slerp(slerp(q_0, q_3, t), slerp(q_1, q_2, t), 2t(1-t)) \quad (6)$$

(5) suits the given discrete data structure represented by the 2D matrix: given a grid of unit-Quaternion corners, the implementation of SQUAD enables the calculation of the required unit-Quaternion. It is worth noting that this process in non-commutative, hence the SLERP sequence is important. As indicated earlier, the rotations between two rectified and geo-referenced DEM databases are relatively small. This interprets to a geo-registration matrix with cells storing angle differences that do not exceed more than few degrees. As a result, any SLERP sequence chosen is adequate. This implies that implementing SQUAD interpolation mechanism on the orientations stored in the matrix ensure the precision needed.

19.4 Experimental Results

Several geo-oriented analysis processes were implemented to verify the effectiveness and advantages the complete hierarchical concept holds, while utilizing the 2D geo-registration matrix. It is worth noting that test examples concerning the modelling processes detailed in section 19.3 are not given here (though they were shown to be robust and accurate as can be found in [13]). Few geo-oriented analysis capabilities are presented here: i) Comparison, which can be designated for morphologic change detection among other uses; ii) Integration of grid databases (i.e., DEMs) as well as non-grid databases (such as LiDAR databases); iii) Visualization: in-between scene calculation and animating between topographic databases. It is clear that these capabilities are only few examples that make use of the concept and its outcome, as it can be utilized for other geo-oriented 3D analysis processes as well.

19.4.1 Comparison

Two DEM databases - depicted in Fig. 19.4 - acquired on different epochs and produced using different technologies were used to evaluate the effectiveness of the proposed concept in identifying morphological changes. One database presented 25m resolution and was produced by digitizing 1:50,000 contour maps. The other database presented 4m resolution and was produced from photogrammetric observations. The two databases covered approximately 2x2km and were produced 20 years apart from each other. Several unique entities were extracted in each database, enabling a robust registration process, which resulted in establishing a mutual correct working reference frame. The mutual coverage area was then divided into frames of 100x100m, in which the ICP matching process was implemented on each one separately and independently. The assumption is that all mutual frames should be matched accurately with approximately similar values, except for frames showing morphological changes.

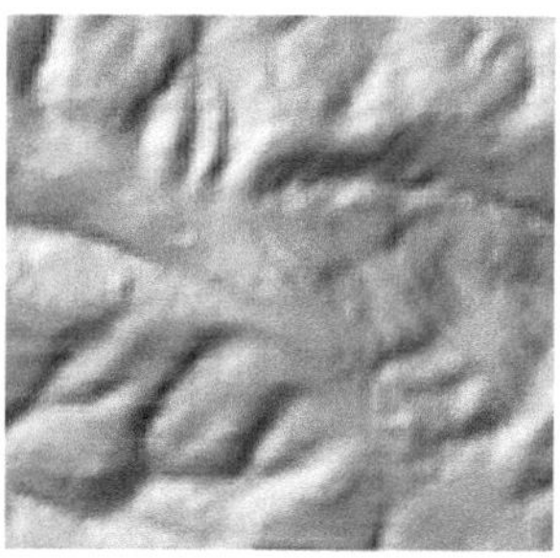

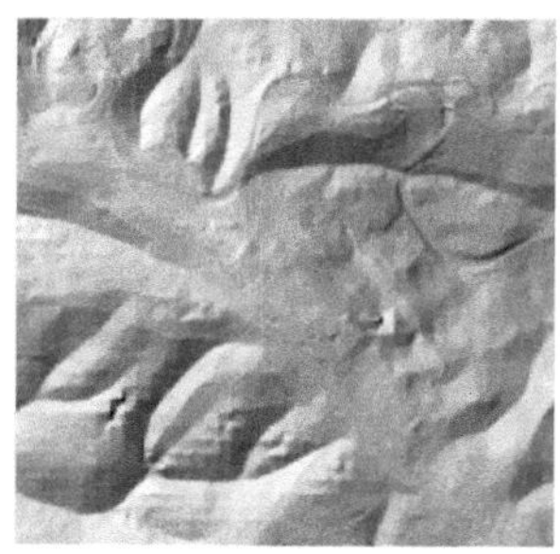

Fig. 19.4. Two databases utilized for the comparison process: digitized 25m - presented here in a 5m interpolation (left), and photogrammetric 4m (right).

Fig. 19.5 depicts the outcome of the proposed procedure (left) - as well as the outcome of the default mechanism that is based solely on the mutual coordinate reference frames (right). The figures depict a statistical evolution value - *rms-z* - for each frame. This value measures the correspondence between one database given heights and the heights calculated from transforming the other database via the geo-registration values calculated in the matching process. This calculation is carried out on each mutual frame separately. Roughly, this statistical test resembles the error vector that measures the matching quality of the frames, and as a result it embeds the stored 6 geo-registration parameters in its calculation. As can be seen, this statistical test identifies precisely areas with morphological changes denoted as frames with bright colors. It is clear that the *rms-z* values calculated via the proposed concept present better matching correspondence of both databases: around 1-2m (dark colors) with a maximum value of 5m, whereas in the default mechanism this value reaches 10m. Close inspection of Fig. 19.4 and Fig. 19.5 shows that the higher values in *rms-z* correspond to morphologic inconsistencies between the databases, whereas the default mechanism presents more noise with no exact correspondence to morphological changes. For example: the channel area on the low-left corner, where several channels appear only in the 4m database; or, the mountainous area in the upper-part of the area modified in the 4m resolution database as a result of a new paved road.

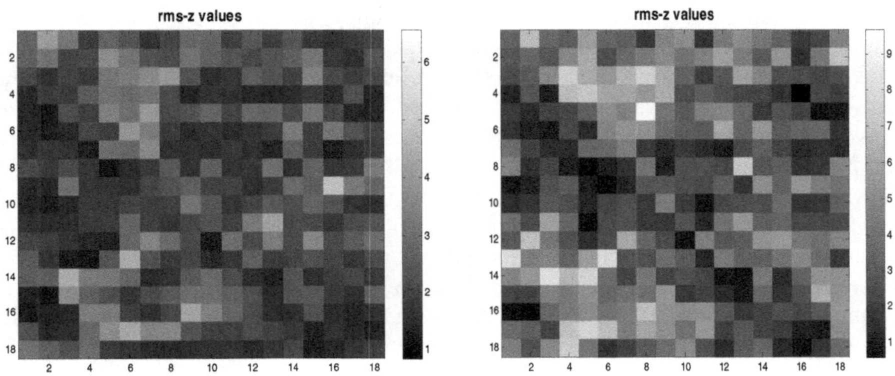

Fig. 19.5. Statistical value *rms-z* for each frame calculated via proposed concept (left), and via the default mechanism (right). Color bar values in meters.

19.4.2 Integration

As opposed to a comparison analysis, which merely utilizes the values stored in the 2D geo-registration matrix for further analysis, an integration process involves interpolation on these values as well. An integration process is designated to produce a new database that is derived from the data stored in the source databases as well as from their correlations. The integrated database has to preserve both databases morphology and topology. Compared to coercing single global transformation estimation, it is obvious that using localized interpolation yields more reliable and correct parameters calculation that describe both databases' local mutual correlations.

Conceptually, the integrated database is spatially located 'between' the databases used for its calculation. Thus, two interpolation 'directions' are implemented: i) Horizontal - among the matrix cells values. This calculation is needed to calculate the exact 6-parameters required for transforming each point stored in one database to its analogous location in the other. The horizontal interpolation utilizes SQUAD interpolation on the rotation values, and bi-directional 3rd degree parabolic interpolation on the translation values; ii) Vertical - the 6-paramerters calculated via the horizontal interpolation represent the correlation of a certain point in one database to its location in the other. As mentioned before, the integrated database 'lies' between the given databases. So, a weighted interpolation on these 6-parameters is therefore needed. The weight is derived from the mutual accuracy of both databases. This means that the integrated database will 'lay' closer to the more accurate database. The vertical interpolation utilizes SLERP interpolation on the rotation values, and linear interpolation on the translation values.

The adequacy and quality of an integrated database can be examined and evaluated by inspecting the preservation and continuity of morphologic entities exist in the terrain relief represented in both databases. Fig 19.6 depicts a topographic database that is the outcome of integrating the databases depicted in Fig. 19.1 according to the integration algorithm suggested here (including a 2nd degree polynomial fading seaming process on the borders of the mutual coverage area). As can be seen from this figure the spatial matching is correct and precise. The integrated database is unified, spatial continuous and free of gaps throughout the area. A careful examination of the morphologic structures shows a correct terrain relief representation with no discontinuities - including on the border of the mutual area.

Another integration using the 2D geo-registration matrix was carried out on DEM and LiDAR databases - with approximately 2x2km mutual coverage area - depicted in Fig. 19.7. The LiDAR database was acquired 25 years after the source DEM, and presented denser (approximately a ratio of 1:165 points DEM vs. LiDAR), and more accurate data. (It is worth noting that the LiDAR database had gone through a filtering process, as proposed by [21], so only the terrain relief was used in the process). As mentioned in section 19.3.2, the ICP matching process is not restricted to a specific data structure and data resolution - as implemented here: DEM vs. LiDAR. Fig. 19.8 depicts a 3D shaded relief representation of the integrated database showing continuous and unified database.

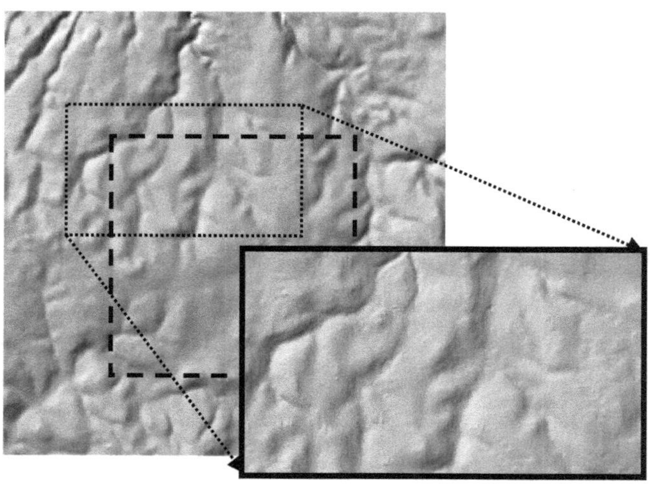

Fig. 19.6. Integration process showing continuous terrain relief and morphological representation (mutual coverage area is denoted by a dashed square).

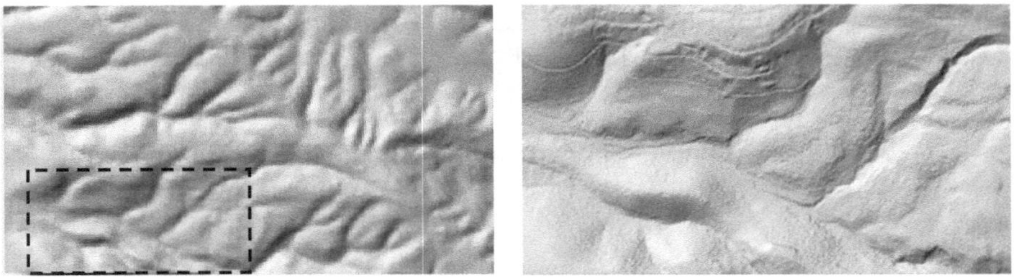

Fig. 19.7. DEM (left) and LiDAR (right) databases showing distinctive differences in level of detailing and accuracy (mutual coverage area is denoted by a dashed rectangle).

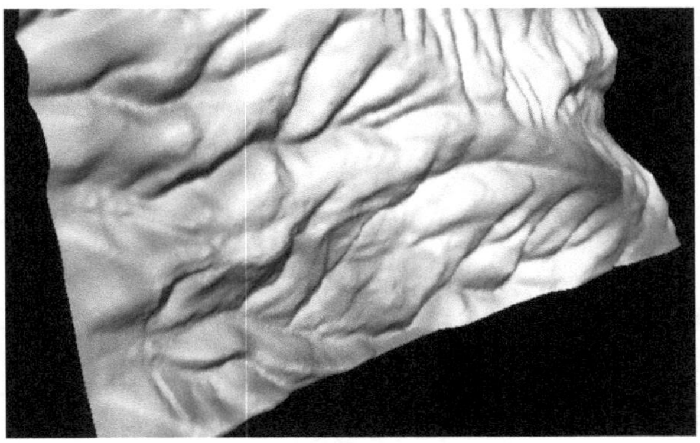

Fig. 19.8. 5m resolution hybrid DEM database: mutual coverage area on lower-left area.

19.4.3 Visualization: Intermediate Scenes and Animation

The integration product can actually be presented as an intermediate scene derived from two given scenes, i.e., source databases. We can assume that both source databases represent two epoch times: t_0=0 and t_1=1, respectively. Knowing the complete local relations between these 'times' stored in the matrix, the calculation of any intermediate position (scene) t_i is feasible (while $t_i \in [0,1]$). By interpolating along the sets (i.e., vertical interpolation), the calculation of an intermediate database at t_i is possible, for example: topographic changes or soil works. This enables complete and comprehensive 3D visualization and construction of hypothetical topography simulation that 'lies' on the path from one source database to the other - and vice versa.

Fig. 19.9 demonstrates the proposed intermediate scene visualization. Two databases are given: t_0=0.0 (produced from low resolution SPOT imagery), and t_1=1.0 (produced from digitizing contour maps). The visualization at 'times' t_i=0.333 and t_i=0.666 is presented. The intermediate scene visualization was precise and resembled accurately the morphological features existent while transforming spatially from one database to another.

Moreover, a series of intermediate scenes can be produced while visualizing from t_0=0.0 to t_1=1.0 at Δt interval scenes. Combined this scenes together will result in an animation of all transition states from one database to the other. This gives knowledge regarding the hypothetical morphological changes that are based on the correlations stored in the matrix, which occurred between the two given databases' epochs (an example carried out on the databases depicted in Fig. 19.9 can be viewed at: http://www.youtube.com/watch?v=KMjBzlQJwGE).

19.5 Related Work

Addressing the problem of modelling multiple DEM databases with various accuracies via implementing localized weighting process on their heights is presented by [22] or [23]. In both studies the databases utilized for the task were initially mutually geo-referenced, while assuming that both databases did not show any morphological local discrepancies. These can be reliable only with massive a-priori work - which is not always possible (as in the case of real-time and near real-time applications) - where in the study suggested here this is achieved by automatic geo-referencing and matching algorithms. [24, 25, 26], among others, have addressed the issue of integrating DEM databases with 2D and 2.5D vector-data focusing mainly on semantic visualization. While vector-data represent discrete entities and structures, such as networks, points, polylines and polygons, DEM geospatial databases utilized in this research represent continuous reality. [27] addressed the issue of the seam line between two adjacent DEMs by using a new conflation algorithm in order to achieve a continuous strip of DEM databases. This paper addressed only the issue of the seam line and its close surroundings, in contrast to the complete modelling suggested here.

Visualization of intermediate graphical scenes (frames) is usually restricted to 2D video texture and animation while utilizing key frames as basis. [28], for example, reuses frames to generate a seamless video while using a first-order Markov Chain model. In order to ensure motion smoothness through out the original region, [29] modified this approach while adding an automatic partial temporal order recovery algorithm to the stills in order to approximate the original scene dynamics using a second-order Markov Chain model. Another study [30] led to the statement of the path-transition paradigm, which is a general methodology for creating 2D, real-time, color animations. This paradigm relies on the use of four abstract data types: location, image, path and transition. Still, it is based solely on 2D and 3D primitives, such as cones, spheres, cubes and more, as opposed to the proposed approach were no geometric primitives' a-priori consideration is required.

19.6 Discussion

Standard on-the-shelf GIS software packages, which are designated for geospatial modelling of two or more databases, are mostly based only on their mutual coordinate reference systems. In contrast, we propose a novel modelling mechanism that is based on the geographic - as well as morphologic - information stored in those databases. In order to overcome the inaccuracies and discrepancies exist between the databases, we propose the concept of using two hierarchical modelling levels. This enabled extracting the existing mutual correlations, and hence monitoring and modelling phenomenon that are characterized only locally. These correlations - defined as geo-registration values for each modelled homogenous frame - are then stored in a matrix, later to be utilized for various accurate and reliable geo-oriented analysis tasks.

Several interpolation algorithms were introduced, which are designated for manipulating the discrete geo-registration values stored in the matrix, thus enabling continuous geo-oriented analysis tasks to be executed. It was shown that standard linear interpolation between rotation values presented by Euler angles was not adequate, so an alternative parameterization was suggested. Describing rotation as unit Quaternion and implementing SLERP and SQUAD interpolations proved to be robust. This approach enabled executing tasks where motion representations are needed, such as creating intermediate scenes.

Several experimental analysis results were presented: comparison, integration and visualization. All proved to be accurate and robust, enabling new geo-oriented capabilities to be carried out on geo-spatial databases. These capabilities proved to preserve the inner topology and morphology of the databases that were utilized. This is in contrast to a process that is based solely on their coordinates, where these correlations are not preserved leading to destroying topographic structures. The analysis capabilities are not limited only to these three examples, and other geospatial tasks that will make use of this model can be implemented. The concept and algorithms presented are straight forward, can easily be translated into 2D space, and are applicable for other fields that make use of point clouds databases. For example, time dependent differencing applications, such as MRI frames, will produce an accurate and geometrically correct analysis results.

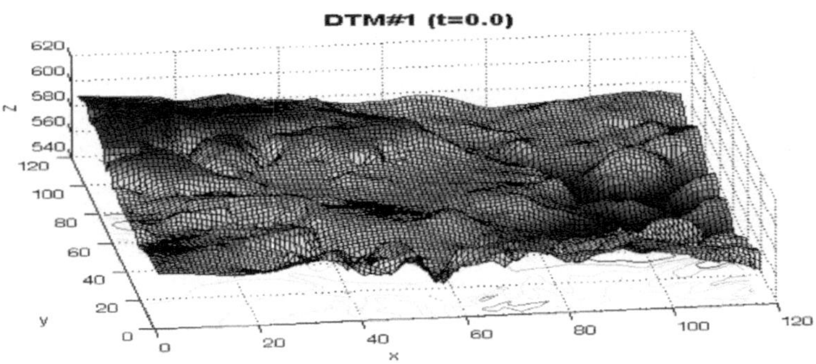

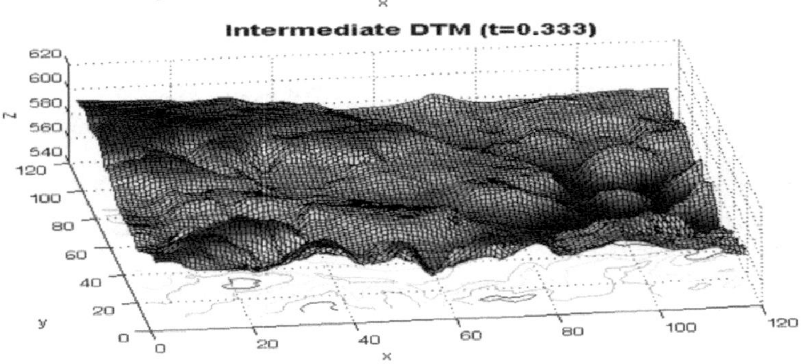

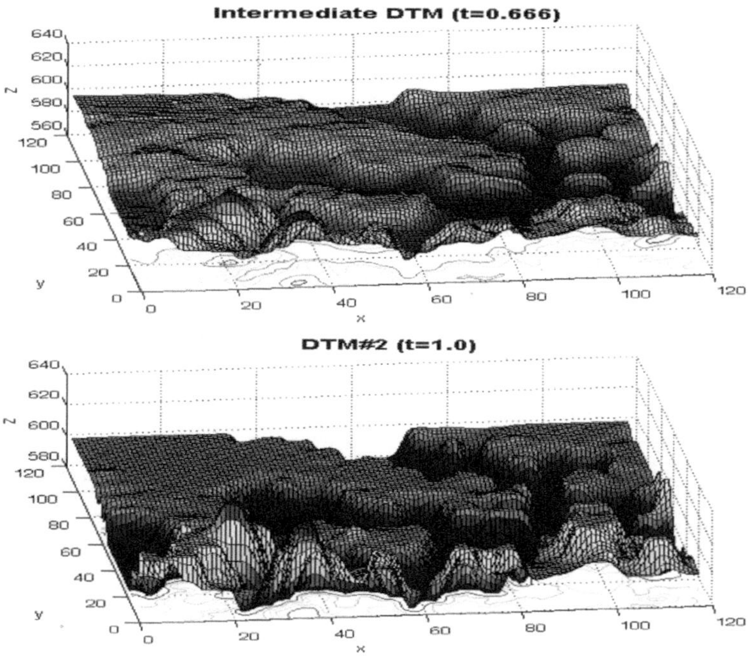

Fig. 19.9. Intermediate scenes *t*=0.333 and *t*=0.666 produced while utilizing data stored in source DEMs and geo-registration matrix (Figures are with no scale).

References

1. Hutchinson, M.F. and Gallant, J.C.: Digital Elevation Models and Representation of Terrain Shape, in Terrain Analysis: Principles and Applications, Wilson, J.P. and Gallant, J.C., Eds., John Wiley and Sons, Inc., New York, (2000), chap.2.
2. Wilson, J.P., Repetto, P.L., and Snyder, R.D.: Effect of Data Source, Grid Resolution, and Flow-Routing Method on Computed Topographic Attributes, in Terrain Analysis: Principles and Applications, Wilson, J.P. and Gallant, J.C., Eds., John Wiley and Sons, Inc., New York, (2000), chap.5.
3. Laurini R.: Spatial Multi-Database Topological Continuity and Indexing: A Step Towards Seamless GIS Data Interoperability, International Journal of Geographical Information Science, Vol. 12, No. 4, (1998) 373-402.
4. Brown, L.G.: A survey of image registration techniques, ACM Computing Surveys (CSUR), Vol. 24, Issue 4, (1992) 325-376.
5. Rusinkiewicz, S. and Levoy, M.: Efficient Variants of the ICP Algorithm, in Proc. 3D Digital Imaging and Modeling, IEEE Computer Society Press, (2001) 145-152.
6. Zhengyou, Z.: Iterative Point Matching for Registration of Free-Form Curves and Surfaces, Int. J. of Computer Vision, Vol. 13, Issue 2, (1993) 119-152.
7. Huttenlocher, D. P., G. A. Klanderman, and W. J. Rucklidge: Comparing Images Using the Hausdorff Distance, IEEE Trans. Pattern Intelligence and machine intelligence, (1993) 9:850-863.

8. Stockman, G., S. Kopstein, and S. Benett: Matching Images to Models for Registration and Object Detection Via Clustering, Pattern analysis and machine intelligence, (1982) 4(3):229-241.
9. Lamdan Y., Schwartz J.T. and Wolfson H.J.: Object Recognition by Affine Invariant Matching'. Proc. CVPR, (1988) 335-344.
10. Mount D. M., Netanyahu N. S., and Le Moigne J., Efficient Algorithms for Robust Feature Matching. Pattern Recognition 32(13) (1998) 17-28.
11. Besl, P.J. and McKay, N.D.: A method for Registration of 3-D Shapes, in IEEE Trans. on Pattern Analysis and Machine Intelligence, Vol. 14, No. 2, (1992) 239-256.
12. Gruen, A.: Least Squares Matching: a Fundamental Measurement Algorithm, in Close range photogrammetry and machine vision, Atkinson K.B., Ed., Whittles Publishing, Caithness, (1996) chap. 8.
13. Dalyot, S., and Doytsher, Y.: A Hierarchical Approach toward 3-D Geo-Spatial Terrain Merging, ISPRS Int. Archives, Commission IV Symposium, 25-30 September, Goa, India, Volume XXXVI part 4 (2006).
14. C.S. Yang, S.P. Yang, F.B. Kao and P.S Hung: Twelve Different Interpolation Methods: A Case Study of Surfer 8.0, Proc. of the XXth, ISPRS Congress, Turkey, 2004, pp. 778-785.
15. Foley, J.D., A. Van Dam, S.K. Feiner, and J.F. Hughes: Computer Graphics - Principles and Practice, Addison-Wesley, Reading Mass, 1996.
16. Doythser, Y. and Hall, J.K.: Interpolation of DTM using bi-directional third-degree parabolic equations, with FORTRAN subroutines, in Computers and Geosciences, Volume 23, Number 9, (1997) 1013-1020(8).
17. Cayley, A., An Elementary Treatise on Elliptic Functions, Dover, New-York, 1961.
18. Shoemake, K.: Animating Rotation with Quaternion curves, Computer Graphics, 19(3), 1985, pp. 245-254.
19. Bezier, P.: Numerical Control: Mathematics and Applications, Wiley, Chichester, UK, 1972.
20. B.E. Dam, M. Koch, M. Lillholm: Quaternions, Interpolation and Animation, Technical report DIKU-TR9815, Department of Computer Science, Univ. of Copenhagen, 1998.
21. Abo-Akel N., Filin S., and Doytsher, Y.: Orthogonal Polynomials Supported by Region Growing Segmentation for the Extraction of Terrain from LiDAR Data. Photogrammetric Engineering & Remote Sensing, 73(11): pp. 1253-1266, 2007.
22. Podobnikar T.: Production of Integrated Digital Terrain Model from Multiple Datasets of Different Quality, Int. Journal of Geographical Information Science, Vol. 19, No. 1, pp. 69, 2005.
23. Frederiksen P., Grum J., and Joergensen L.T.: Strategies for Updating a National 3-D Topographic Database and Related Geoinformation, Proc. ISPRS XXth Congress, Com. WG II/IV, 2004.
24. Heipke C.: Some Requirements for Geographic Information Systems: A Photogrammetric Point of View1, Photogrammetric Engineering & Remote Sensing, Vol. 70, No. 2, February (2004) 185–195.
25. Koch A., and Heipke C.: Semantically Correct 2.5D GIS Data - the Integration of a DTM and Topographic Vector Data, Fisher, P., Ed., Developments in Spatial Data Handling, Berlin, Springer, (2004) 509-526.

26. Walter, V. and Fritsch, D.: Matching spatial data sets: a statistical approach, in Int. J. of geographical information science, Vol. 13, No. 5, (1999) 445-473(29).
27. Katzil Y., and Doytsher Y.: Spatial Rubber Sheeting of DTMs, In Proc. 6th Geomatic Week Conference, Barcelona, Spain, (2005).
28. Scho D. R. Szeliski, D.H. Salesin, and I. Essa: Video Textures, Proc. ACM SIGGRAPH '00, pp. 489-498, 2000.
29. Zhouchen L., Lifeng W., Yunbo W., Kang, S.B., and Tian F.: High Resolution Animated Scenes from Stills Visualization and Computer Graphics, IEEE Transactions on Volume 13, Issue 3, May-June 2007 Page(s):562 – 568.
30. J. T. Stasko: The Path-Transition Paradigm: a Practical Methodology for Adding Animation to Program Interfaces. Journal of Visual Languages and Computing 1, 213-236, 1990.

Chapter 20
Solar Radiation over the Urban Texture: LIDAR Data and Image Processing Techniques for Environmental Analysis at City Scale

Cláudio Carneiro, Eugenio Morello, Carlo Ratti and François Golay

Abstract. A complete methodology from the extraction of Light Detection and Ranging (LIDAR) data to the environmental analysis of urban models and the visualization of results is presented. Aim of the work is to establish a process to investigate digital urban models integrating cross-disciplinary competences, like remote sensing, GIS, image processing and urban and environmental studies. Toward this goal, working on several interfaces, tools and datasets was necessary to provide a consequent structure to the introduced methodology. Case study for application was a squared area 300 metres wide in central Geneva where LIDAR data are available. The use of a hybrid approach from raw LIDAR data and vector digital maps (GIS data) of buildings footprints for the interpolation of a 2.5-D urban surface model, with a resolution grid of 0.50 by 0.50 metres, allowed to refine vertical surfaces of buildings and to process facades and roofs separately. This step reveals itself as fundamental for processing the environmental analysis of the urban texture. In particular, the implemented tool calculates solar radiation and solar accessibility on urban surfaces, in order to investigate the energy-performance of cities.

20.1 Introduction

Today's availability of 3-D information about cities offers the possibility to analyse the urban fabric in a very innovative way. In fact, even if LIDAR data permit to derive precious and precise information about the physical layout of cities, still very few

---

Geographical Information Systems Laboratory, EPFL
claudio.carneiro@epfl.ch; francois.golay@epfl.ch
SENSEable City Laboratory, MIT, Cambridge
eugenio@mit.edu; ratti@mit.edu
Human Space Laboratory, DIAP
eugenio.morello@polimi.it

applications have been implemented in order to process these data for the environmental analysis and to get a quick understanding about the performance of the urban form. For instance, the increasing interest in the quantification of energy-based indicators at the scale of the city, strongly suggests the integration of 3-D geography and urban studies in order to provide useful applications for the urban planning. In particular, we introduce a novel cross-disciplinary approach that covers the whole procedure from data acquisition from Airborne Laser Scanning (ALS) to the environmental analysis through the image processing of digital urban models.

The use of raw LIDAR data with fine density of points (at least 4 to 6 points per square metre) jointly with 2-D vector digital maps of buildings footprints, both originally on the same geodetic reference system and, later on, converted to the same projected coordinate system, allows the construction of a 2.5-D urban surface model. Due to its highest level of accuracy, the use of detailed vector buildings footprints in order to classify LIDAR points contained within buildings is crucial for the improvement of the final result of the 2.5-D urban surface model interpolated and constructed, especially along buildings facades.

Thus, using the capability of LIDAR data in order to construct an accurate 3-D urban surface model, solar accessibility analysis at different heights (for existing buildings facades) is implemented and integrated in a tool for calculating urban radiation. As presented by [9], all visibility and sunshine analysis in urban areas based on 3-D urban surface models must take into consideration the geographical relief of the terrain.

The investigation of solar radiation in architecture is not new and there are already several tools that calculate radiation performance of buildings very accurately. Nevertheless, these tools are very useful at the micro-scale of architecture (environmental performance software) or at the macro-scale of landscape and regional geography (GIS tools), but the focus on urban texture is mostly lacking. Recently, the increasing attention to environmental policies in urban studies has opened up many questions about how planners should manage those indicators in the design process. In fact, numerous authors and architects are convinced that cities play a leading role in controlling sustainability: strategies for redefining more efficient cities in terms of energy performance and environmental quality were the centre of attention in seminal work by Richard Rogers in defining policies for UK cities [15, 18] and supported the debate around the promotion of more compact cities [5, 6]. For instance, the need to overcome traditional solar laws based on trivial angular criteria, such as obstruction angles rules, would be very welcome. Anyway, a comprehensive and reliable toolkit for sustainable urban design is lacking among practicing professionals.

The tool presented here can be intended just as one tile of a larger mosaic aiming at the quantification of environmental parameters at the urban scale. In particular, the calculation of solar radiation gives us some clues about the energy-performance of the urban fabric concerning a general understanding of the different contributions of vertical and sloped surfaces in the urban context. Hence, some critical questions in urban design and planning arise: can we build denser cities without decreasing the potential for passive solar architecture? Which is the incidence of beam versus diffuse solar irradiance contributions in typical urban textures? Those questions could benefit from comparative studies between different urban design solutions using the methodology introduced here.

The area selected for the analysis is a square 300 metres wide near the Rhone River and the old town of Geneva, in Switzerland, such as presented in shadowed yellow zone of Fig. 20.1. 36 different buildings with average height of 18.5 metres were counted on site. The urban texture is quite compact and densely built, and we want to verify if buildings are sufficiently separated from each others, so that overshadowing does not limit solar accessibility of indoor spaces. The reconstruction of the model, as explained later in text is shown in Fig. 20.2.

Fig. 20.1. Pre-defined area of the case study in Geneva's city, near the Rhone River and the old town.

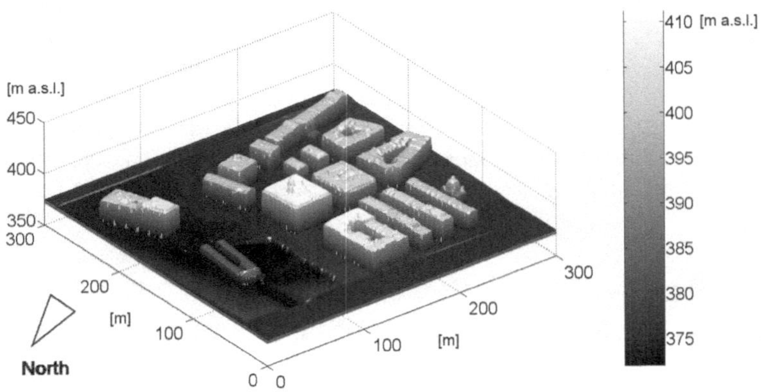

Fig. 20.2. The 2.5-D urban surface model visualized on Matlab. Heights are expressed in meters above sea level.

20.2 Research context

Reference to this work is both literature in the remote sensing, 3-D geography field and literature in the environmental and morphological analysis of digital urban models.

Previous literature on interpolation of LIDAR point clouds is vast. The advantages and disadvantages of several interpolation methods, such as triangle-based linear interpolation, nearest neighbour interpolation and kriging interpolation were presented by [19]. The most accurate surfaces are created using a grid with a sampling size that relates as close as possible to the LIDAR point density during the acquisition phase [1]. For applications where high level of accuracy is demanded, control techniques that analyse the quality of digital terrain models can be carried out [7].

A method to interpolate and construct a 2.5-D urban surface model (incorporating the geographical relief), based on LIDAR and GIS buildings data, has been proposed by [9]. As related research, there are some semi-automatic methods available to create 3-D GIS data from LIDAR data, such as the Virtual London at UCL [16] and the MapCube at Tokyo CadCenter Corporation [17].

Concerning literature about the application of raster images in urban studies, pioneers in the use of image processing techniques to analyse environmental indicators and morphology of digital urban models was a group of researchers at the Martin Centre, University of Cambridge [10, 12, 11]. Today's increasing availability of 3-D information from user generated contents and remote sensing surveys, makes this technique promising and very useful for a general understanding of the performance of our cities.

The technique is based on the use of very simple raster models of cities, called 2.5-D urban surface models. These models reproduce the geometry of the urban fabric and are produced by regularly spaced matrices of elevation values, which contain 3-D information on 2-D digital support, stored in Bitmap format. Developing software algorithms derived from image processing, it is possible to develop efficient strategic tools for analysing and planning the sustainable urban form, measuring geometric parameters and assessing radiation exchange, energy consumption, wind porosity, visibility, spatial analyses, etc. Results are extremely fast and accurate.

20.3 Methodology: The implementation of the 2.5-D urban surface model from LiDAR data to environmental analysis

The process for structuring the proposed methodology is based on four major steps as represented in the scheme of Fig. 20.3: the construction of the 2.5-D urban surface model, the environmental analysis through the image processing of urban models, the visualizations of results and their evaluation by users (this latter step will be part of future work).

For the 2.5-D urban surface model interpolation of the case study of the city of Geneva here presented three data sources are required: raw LIDAR data, 2-D vector digital maps of buildings footprints and alphanumerical data containing altimetry information about buildings heights.

First, we interpolate a digital terrain model (DTM) by classifying the LIDAR points according to the following sequential operations:

- Using a GIS software, LIDAR points contained within building polygons and in the 2 metres buffer generated from building polygons are eliminated.
- Using the classification tools provided by TerraScan software, LIDAR points whose elevation value vary significantly from surrounding points are considered to be points indicating features such as aerial points (for example, if the laser beam touches a bird), trees and vehicles, and thus are removed.

After eliminating the points indicating all these features, a DTM can be interpolated only from ground points – with good density of LIDAR points (such as in our testing areas, located in Geneva city, where LIDAR points were acquired with a density of 4 to 6 points per square metre) there is no great difference among some of the existing gridding interpolation methods that can be employed, such as the nearest neighbour binning, inverse distance weighting, triangulation with linear interpolation, minimum curvature, kriging and radial basis functions [4]. All these interpolation methods are accessible in common GIS software available on the market.

For its generalised use by the scientific community for DTM interpolation, the triangular interpolation was chosen. Secondly, we interpolate (using only the LIDAR points classified as being contained within vector building footprints) a value for each grid cell corresponding to a roof value contained within the area defined by building facades, for all the existing buildings on the areas of study. Thus, a triangulation with linear interpolation is also applied to each one of the roofs of the buildings. It is important to note that along building facades, LIDAR points whose elevation value vary significantly from surrounding points are considered to be points indicating features such as low points and are removed (using the classification tools provided by TerraScan software). For each building, and more specifically for each grid cell contained within, the building height (also defined as nDSM of buildings) is taken to be the value of subtraction of the terrain elevation (calculated in the DTM interpolated) from the building elevation. Due to the presence of obstacles, such as trees, in the few cases where a lack of LIDAR points exists in buildings roofs, the grid cells corresponding to each of these roofs have assigned an average building height value extracted from the attribute "building height" of the cadastral database.

Lastly, each building is added to the DTM as a column (whose borders are defined from the vector buildings footprints), using the building height found previously for each cell contained within, as described in last paragraph. The final result allows the construction of a 2.5-D urban surface model (2.5-DUSM), which is composed of only terrain and buildings heights information (DTM + nDSM). Data source and parameters needed for generating the 2.5-D urban surface model are shown in Fig. 20.4.

Finally, in order to complete the image enhancement of the model, we have to refine the facades of buildings that are sloped because of interpolation (Fig. 20.5c). In order to achieve this goal, we applied an image smoothing using a 3 by 3 low-pass filtering. Each building's contour pixel was deleted and then expanded again, in order to assign more constant values to facades. Among these perimeter pixels, the maximum value was later selected as the upper limit for running the computation of irradiances on facades (Fig. 20.6).

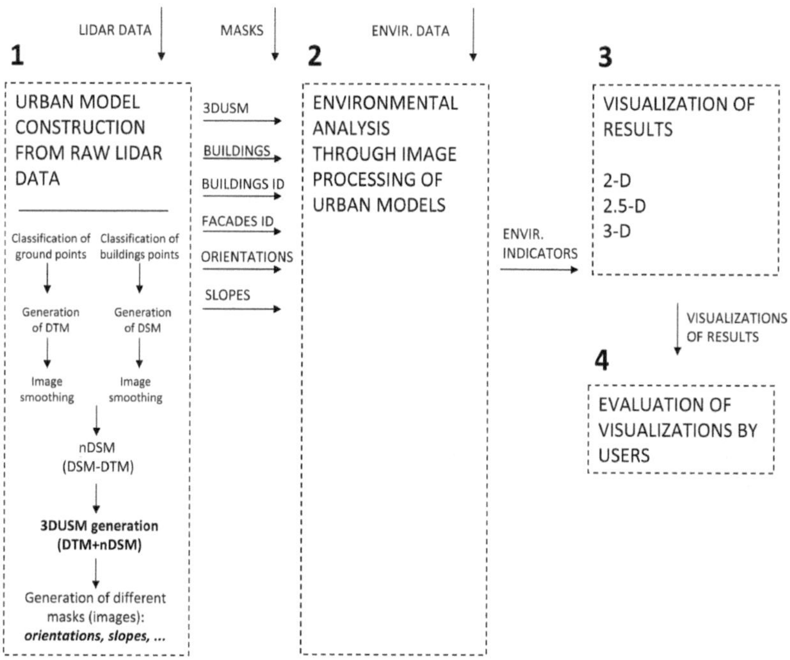

Fig. 20.3. The structure of the proposed methodology from the reconstruction of LIDAR data to the environmental analysis and visualization of results

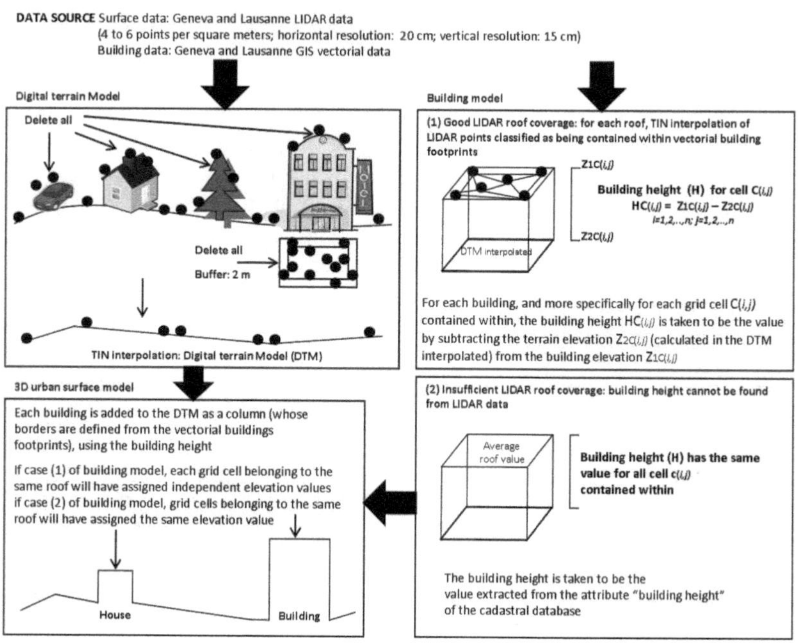

Fig. 20.4. Interpolation and construction of 3-D urban surface model

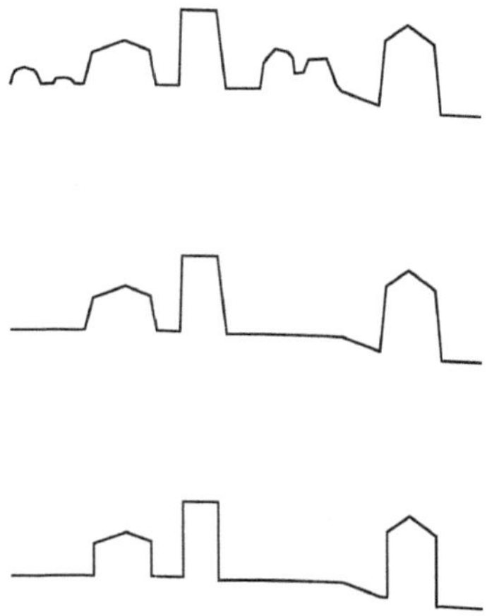

Fig. 20.5. The process of reconstruction of the urban model; from the top: (a) original LIDAR survey data; (b) removal of trees, cars and other minor details and interpolation of the terrain; (c) adjustment of building facades through image filtering, whereby the detection of facades and roofs is based on a hybrid approach using both LIDAR data and cartography

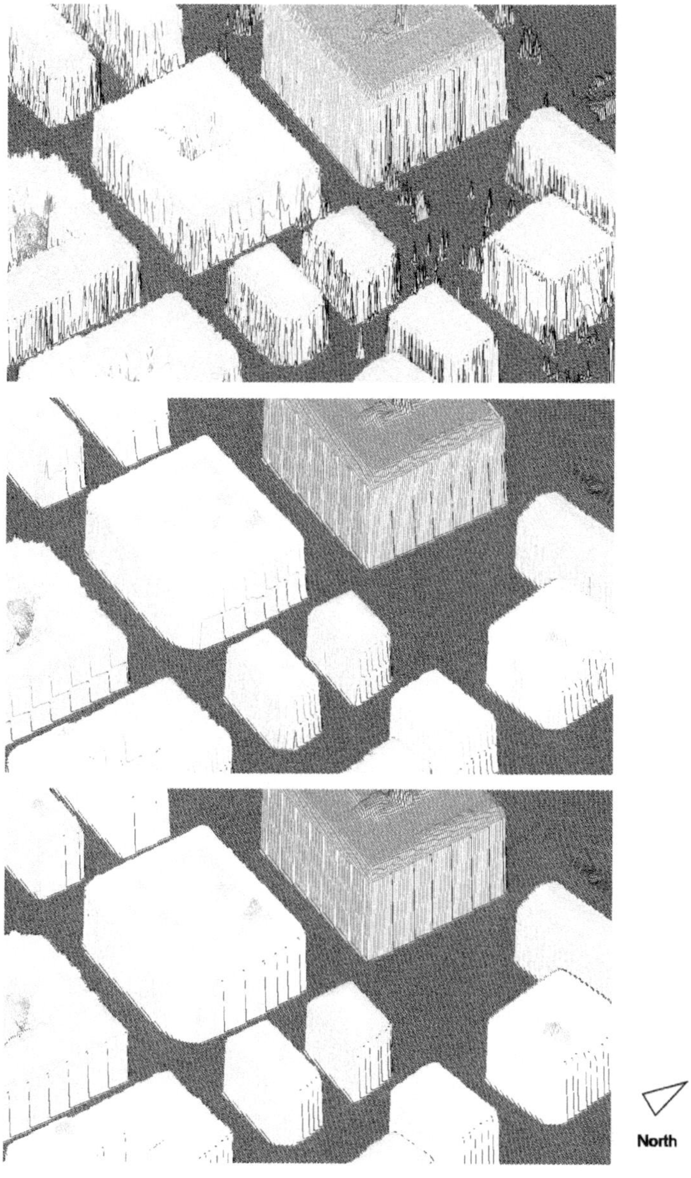

Fig. 20.6. The enhancement of the urban model; from the top: (a) the raw LIDAR data, (b) the original DEM, (c) the reconstruction of vertical facades

20.4 Image processing of raster images for the study of solar radiation

In this study the reconstruction of the 2.5-D urban model was applied for implementing a specific tool for calculating the solar irradiance intercepted by urban surfaces. Aim of the proposed application is first to determine the rate of accessibility of urban surfaces to direct solar radiation and second to exactly quantify solar irradiance (diffuse and beam) both on facades and roofs. The computation distinguishes calculations for irradiation of facades and roofs. Therefore, outputs of the tool are irradiation values stored in two different arrays: one table contains radiations computed on every storey and on every facade of each building; a second table outlines the irradiance values on roofs for each building. Finally, beside this quantitative content, visualizations of results on the physical map are provided.

The technique used in this application is based on the image processing of the provided models that are interpreted as raster images, i.e. 2-D arrays where every intensity value represents the height of the pixel in metres. The density[1] and accuracy[2] of the LIDAR data used is relatively high. Thus, a 2.5-D urban surface model (grid) of 0.50 by 0.50 metres, which has a sampling size that relates as close as possible to the LIDAR point density during the acquisition phase [1], was interpolated and constructed. The technique computes irradiance values on all target points (building facades) being visible from a viewpoint location represented by the position of the sun.

Two types of input data are required for running the tool: first, geographical and climatic data that inform the location of the case-study area and the mean daily irradiance received by horizontal surfaces for each month of the year for that location; second, a set of urban models, as result of the reconstruction process presented above.

To the first set of data we have to provide the geographical coordinates of the case-study area, the height above sea level and the physical extension of the site in metres.

Regarding climatic data, the mean daily beam and diffuse irradiances on horizontal surfaces can be obtained from statistical data collected for the city, often reported by local or national norms. In this case, irradiance values were obtained by the statistical analysis of historical data. From these letter irradiances on horizontal surfaces it is possible to derive irradiance values for every orientation and inclination of surface, according to solar geometry formulae [3]. Other minor input information can be used if provided, like for example the quote of ground-reflected radiation, which default value can be assumed as 0.2, but this can vary during the winter season in presence of snow.

A list of six masks necessary in order to run the model follows:

- The digital elevation model (DEM), here also called 2.5-D urban surface model, whereby intensity values of pixels represent the height above the sea level. This input image is the main 2.5-D information to run the script.
- The mask of the buildings alone contains the absolute height of each building and is useful in the process of slicing the model in several storeys; in this image the terrain model was removed and substituted with 0 values, so that it can also be used as mask for detecting and labelling buildings.

---

[1] 4 to 6 points per m^2

[2] Planimetric precision of 20 centimetres and altimetric precision of 15 centimetres

- The masks including labels of buildings and facades. It is possible to determine the exact identifications of buildings and facades through cadastral data (digital vector maps of buildings); a simple object recognition process in Matlab would distinguish between independent blocks, but would fail in the identification of different properties in case for example of contiguous row houses.
- The masks for describing roofs are essentially two and inform about orientations and inclinations of roof pixels. These latter masks might be also implemented as part of the Matlab code, but would require a longer calculation process.

This procedure allows to precisely distinguishing between roofs and facades of buildings and represents a noticeable innovation in digital urban analysis. For instance, all outputs in this work are stored in separated arrays in order to analyse the different contribution to solar gains of vertical and sloped surfaces in the urban context.

Solar geometry formulae allow deriving the beam and diffusing components of hourly radiations just starting from the previous mentioned inputs. In the computation of the time distribution of radiations during the day, however, we can just assess values representative for clear sky conditions, even if they tend to produce conservative estimates [3]. Therefore, even if accurate, results of the application should be used mostly in comparative studies to evaluate differences of solar caption in time.

Hence, we store a set of possible radiations in dedicated arrays that will be useful later when applied on the physical context. In particular, the typical array is a 3-D matrix (24x12x9) where the columns represent the 12 months of the year, the rows are 24 hourly intervals and the 9 z-layers are the computed values for 8 different orientations (S, SE, E, NE, N, NW, W, SW) plus the irradiance on horizontal surfaces. We computed this typical array for vertical surfaces and for sloped surfaces at intervals of 15 degrees between 0 and 90.

Once the irradiance arrays are calculated and the input images are defined, the core of the computation can run. The shadow casting routine first introduced by [13] is applied here at a specific hour of the day and is used to detect which pixels on the facades are in shadow and which collect direct sunlight. The script computes the shadow casting routine slicing the urban model at every storey (3 metres is assumed to be the average storey height) of each building. Since the technique deals with 2.5-D models, slicing each building at different heights is necessary in order to compute the real irradiance condition in the vertical direction. In other words, reciprocal overshadowing by buildings can affect only part of the facade and this method allows distinguishing at which height vertical aligned pixels on the facade start to be irradiated (see Fig. 7); otherwise, with simple shadow casting routine applied to the entire model, facades that are shaded at the basement only result totally shaded until the roof.

Moreover, we need to define the linear length of pixels on the perimeter of buildings: depending on their inclination, pixels can assume values between 1 and $\sqrt{2}$. This can be easily calculated applying sobel filters [11].

Hence, when we are able to determine for every facade at every storey, if each perimetral pixel is under shadow or light, which is its orientation and its linear extension, we can assign the incident solar irradiance calculated in Watt per square metre. For roof pixels instead, we need to add the information of slope and orientation contained in the input masks. The area of each roof pixel was calculated considering the slope of the pixel.

Finally, we have to store irradiance values in a convenient way, depending on the results we are interested in. In this work we were basically investigating about the ratio of direct solar irradiation on urban surfaces and an energy-based evaluation of collected radiation on surfaces. Therefore, results were stored listing all pixels by facade identification and by orientation; regarding roofs, results were stored following building labels.

Furthermore, it is important to provide also general morphological indicators for the case-study area, because it could be interesting to correlate urban shape indicators to environmental performance in future comparative studies. These morphological indicators include among others: built volume, built area, densities, mean and maximum height of buildings.

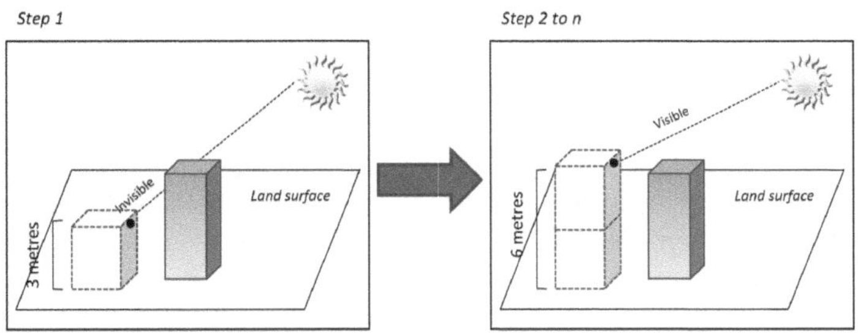

1 – White building: building under analysis
2 – Blue building: built environment
3 – The sun is considered to be the viewpoint location
4 – For each building façade under analysis (target area), every 3 metres (assumed storey height) for all grid cells (C(*i*,*j*)) on the horizontal direction of the image, evaluation of visible cells from the viewpoint (sun) location

Fig. 20.7. Method implemented (using a 2.5-D urban surface model) for the calculation of the solar radiation along each building facade at a specific date of the year. Slicing the model at different heights allows taking into consideration pixels in shadow or in light

20.5 Results of the analysis

A general overview about the energetic performance of vertical surfaces in an urban site is very useful if we want to compare different fabrics or if we want to assess the impact of new buildings inside the city. The tool presented here is accurate at the scale of the neighbourhood, if we do not consider the inevitable edge effect due to the absence of obstructions generated by buildings located right outside of the area of investigation.

Different profiles describing the solar irradiation can be traced for an urban site. We consider two main graphs, introduced later in Fig. 20.12 and Fig. 20.13. Both types of analysis were computed at different times on an hourly basis. We can observe that absolute irradiance values slightly decrease with height (Fig. 20.8a) and this is mostly caused by the parallel decrement of the amount of vertical surfaces when we move to upper storeys. Anyway, if we normalize results in respect to the areas of related vertical surfaces, the average irradiation per square metre clearly increases with height (Fig. 20.8b). This phenomenon occurs in part also with the analysis of direct solar irradiation, whereas here on the upper storeys the percentage of surfaces under direct radiation increases again and this is due to the lower reciprocal overshadowing by buildings (Fig. 20.8c).

Beside the calculation on vertical surfaces, the tool computes the irradiance values collected on roofs and their solar exposure (light or shadow) on an hourly basis (Fig. 20. 9 and Fig. 20.10). The chosen case-study area is composed by a very uniform urban fabric, whereby big gaps in the skyline do not occur. Therefore, roofs are mostly interested by proper shadows on pitch slopes opposite the sun and represent in general the location where it is easier to collect solar energy. Even if a general quantification of irradiance values on roofs is fundamental in order to have a general understanding of the passive solar potential of the city, further work should investigate limitations concerning solar energy collection occurring at pitched roofs and considering for instance the most efficient way to install solar and/or photovoltaic panels at the neighbourhood scale.

Moreover, polar plots of solar irradiances on vertical surfaces computed for Geneva show the contributions of beam and diffuse irradiance on the site (Fig. 20.11). Great part of the solar irradiation collected by vertical urban surfaces is due to diffuse irradiance, considering also the fact that the sky is not always clear. In terms of energy performance, increasing the direct solar contribution on vertical surfaces does not bring a consequent improvement of energy collection. It means that we can build more compact urban spaces and delegate roofs to collect beam radiation for the production of electric energy and/or hot water. Nevertheless, in terms of environmental quality in general, assuring a minimum amount of direct solar radiation on each building is recommendable for increasing the perceived comfort inside buildings. Anyway, if we aim at providing sufficient light for living conditions inside indoor spaces, the constant diffuse radiation is in most cases satisfactory. Concerning the case study-area in central Geneva and according to the irradiations at that latitude, the layout of buildings already guarantees good solar radiation.

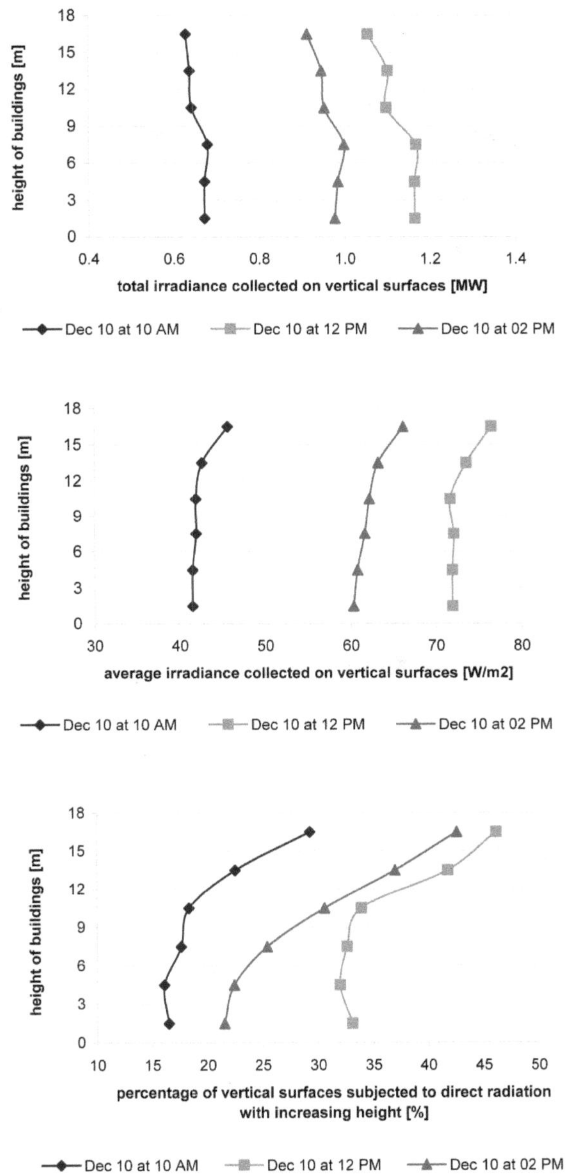

Fig. 20.8. Total irradiance falling on vertical surfaces for the case-study area on the average day of December at 10 AM, 12 PM and 2 PM with increasing heights of buildings; from the top: (a) absolute irradiances in MW; (b) mean irradiances in W/m^2; (c) percentages of vertical surfaces subjected to direct radiation

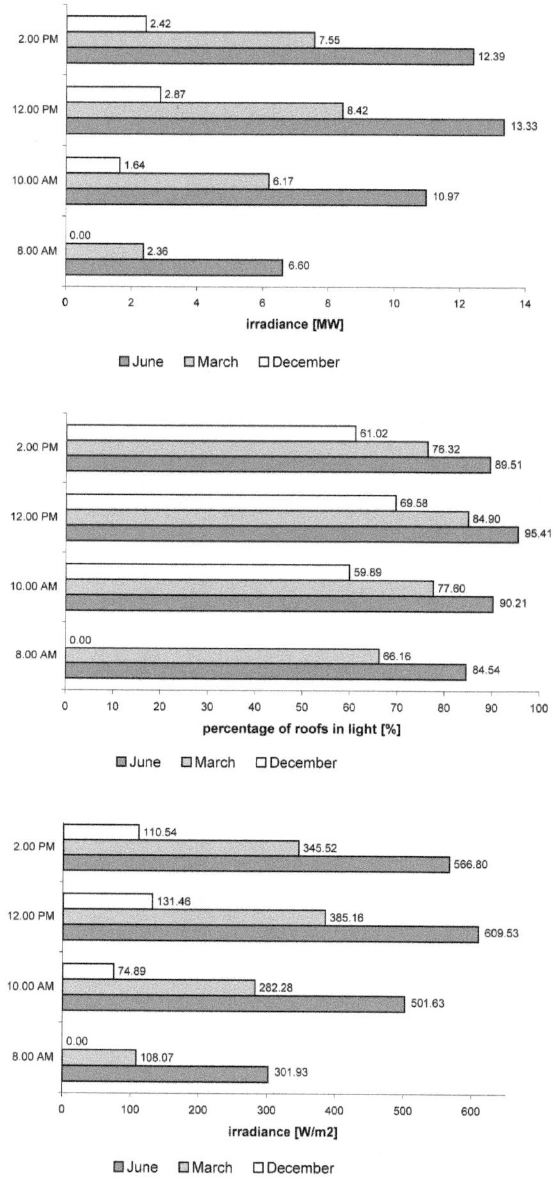

Fig. 20.9. From the top: (a) Total irradiance collected by roofs at 8 AM, 10 AM, 12 PM and 2 PM in the average day for the months of March, June and December (values are expressed in MW); (b) Percentage of roof surfaces in light at 8 AM, 10 AM, 12 PM and 2 PM in the average day for the months of March, June and December; (c) Mean irradiance collected by roofs at 8 AM, 10 AM, 12 PM and 2 PM in the average day for the months of March, June and December (values are expressed in Watt per square metre)

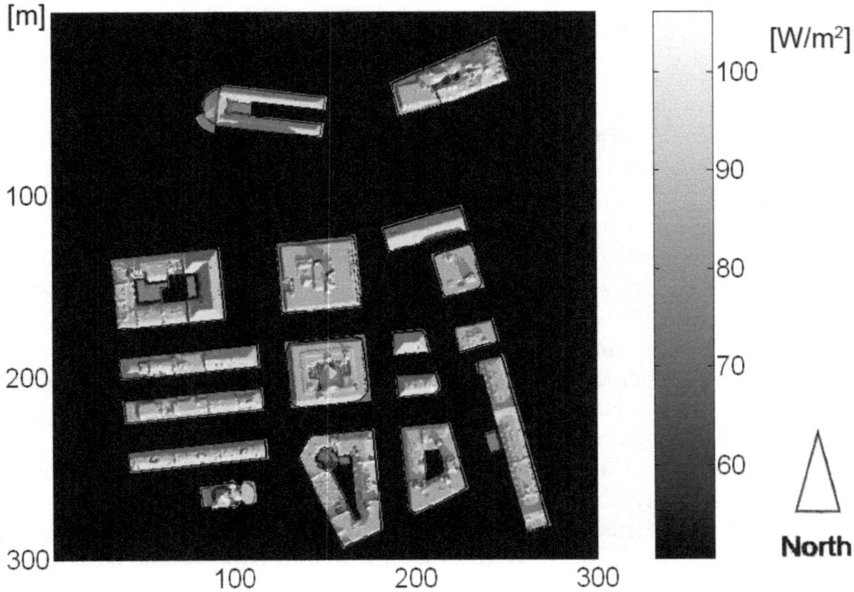

Fig. 20.10. Top view with irradiance values (in W/m^2) on roofs at 10 AM in the morning on December 10

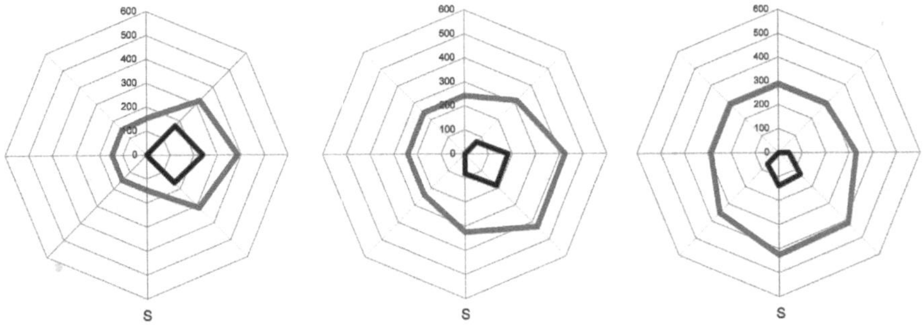

Fig. 20.11. Polar plots of solar irradiances on vertical surfaces for Geneva on the average day of June; in black beam radiation, in dark gray diffuse radiation (values are expressed in W/m^2). From left: irradiances at (a) 8 AM, (b) 10 AM, (c) 12 PM

6. Visualization of results: 2-D and 3-D displays for communicating results

According to [8], the acceptability of any visual exploratory system is strictly related to its utility (feasibility of the information to be visualized) and its usability (cognitive visual interpretation of the 3-D urban models proposed).

This combination between utility and usability determines the level of acceptability among the different users (in particular for architects and urban planners) of the proposed 3-D urban models [14].

User requirements concerning the utility and usability of 3-D urban models for communication and visualization purposes have not yet caused much attention within the world of researchers and developers. Thus, for different users and applications, it is fundamental to clearly distinguish which levels of detail (LOD) should be implemented and, based on this classification, which urban objects should be or should not be visualized. This filter of criteria is essential, in order to avoid too dense and confusing urban scenes.

The visualization of results here proposed is based on the user requirements study undertaken for the city of Geneva, such as presented by [2].

The representations of maps indicating the performance of facades on a 2-D-plot are displayed in Fig. 20.12 and Fig, 20.13: the first accounts the total irradiance (expressed in Watt per square metre) falling on vertical surfaces with increasing heights of buildings; the second plot represents the percentage of facades subjected to direct solar radiation with increasing heights of buildings. These synthetic maps reveal themselves to be very immediate because they give an omni-comprehensive understanding of irradiation conditions on the entire site.

For 3-D visualization we used the available model, developed by the DCMO[3] in Geneva. The 3-D urban model of the city of Geneva was constructed using many data sources, such as: 2-D vector data, 2.5-D raster data, alphanumerical data containing altimetric information about buildings heights, LIDAR and terrestrial laser data, aerial photos orthorectified and terrestrial photos. We used this model to present our analysis of results. Results showed in Fig. 20.14 display the average irradiance (expressed in Watt per square metre) falling on each building. Irradiance values collected on the facades of each building were divided by the total area of vertical surfaces, whereby facades that do not receive radiation at all were discarded (mostly separation walls between two buildings). Results tend to be quite uniform, due to the operation of averaging of irradiation values on all facades, lighted and shaded.

[3] DCMO, Direction Cantonale de la Mensuration Officielle, responsible for the management of the topographical, cartographical and GIS information of the city of Geneva.

Fig. 20.12. The mean solar irradiance collected by each facade on the second storey on December 10 at 12 PM, considering both beam and diffuse contributions.

Fig. 20.13. On each facade on the second storey, the percentage of surface interested by direct solar radiation on December 10 at 12 PM is displayed.

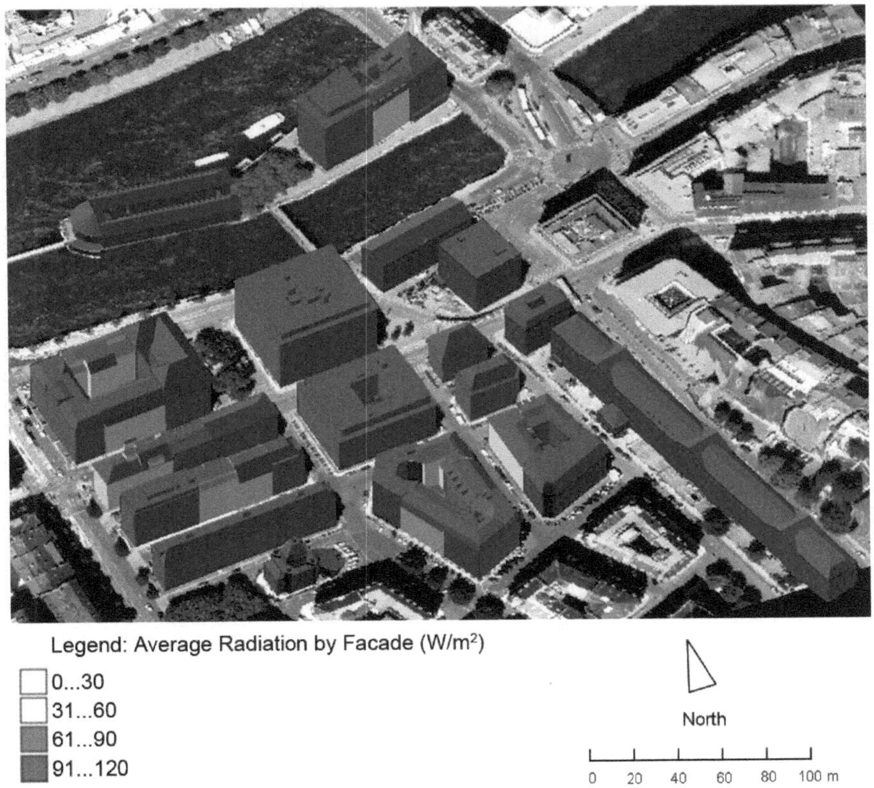

Fig. 20.14. 3-D visualization of average irradiance values per building (expressed in W/m^2) on December 10 at 12 AM

20.6 Conclusions

In this paper a complete methodology from the extraction of LIDAR data to the environmental analysis of urban models and the visualization of results was introduced. Combining different disciplines interfaces and datasets reveal constrains of today's applicability of LIDAR images for the environmental prediction of the urban form. Hence, a first result is to bring 3-D geography and urban studies together in a process that goes from data acquisition and processing to urban modelling and urban design application.

Applications of this methodology in urban design and planning are very promising. In our case study we limit the analysis to the physical built environment, but we could extend the investigation to assess the impact of new buildings in the city and use the technique for improving design schemes based on an evaluation of quantitative indicators before and after changes are introduced. The methodology has been tested for a small case-study area, but calculations can run also at a city level if LIDAR data are available.

The accuracy of results depends mainly on two issues that we tried to overcome: the quality of input LIDAR data and the constraints of the image processing technique of 2.5-D urban surface models. First, in order to enhance the precision of LIDAR data, a process of image reconstruction was implemented and a hybrid approach that integrates the use of vector building footprints data was proposed in order to reconstruct the perimeters of buildings. Second, the use of the image processing technique of raster images permits time-saving computation, but needed to be improved through multiple slicing of the model to consider different shadowing conditions on vertical aligned pixels. In general, the aim is to achieve a real understanding of the energetic performance of the urban fabric. Comparative studies conducted not only at different times, but also on different sites or on the same site introducing changes (like for example a new building or adding one storey to all buildings) could give more clues for interpreting results.

As here presented, existing vector digital maps (GIS data) can be used if available and updated, but outlines of buildings from this source of information should be always handled with special care. In fact, the 2-D outlines of buildings footprints do not have to represent the outline of the building roof. Modifications between GIS data and laser data can have numerous reasons which can not automatically be recognized. Proposals using vector digital maps as input for 2.5-D urban surface model interpolation and construction should be attentive of the fact that in some cases map information might not give the correct hints about 3-D buildings shapes.

Finally, future work should:

- improve the technique for the construction of the model through a method for roof segmentation/interpolation, in order to take into account more accurately different footprints of buildings and roofs;
- ameliorate the quality of the urban model with the integration of trees, which remains one of the most delicate aspects concerning environmental analysis using 2.5-D urban surface models;
- optimise the efficiency of the proposed code and consequently intensifying the number of slices on the model;
- provide a statistical analysis on 2.5-D urban surface models;
- Investigate the utility and usability of the proposed visualizations outputs through users' evaluation.

20.7 Acknowledgements

The authors are definitely indebted to Gilles Desthieux, from the Laboratoire Energie, Environnement & Architecture (LEEA) and to Laurent Niggeler and Adrien Vieira de Mello from the Direction Cantonale de la Mensuration Officielle (DCMO) of the city of Geneva, Switzerland, for its vast contribution to this project. Also, the authors would like to thank the Fondazione Rocca and its president, Gianfelice Rocca, for providing generous funding for this research.

References

1. Behan A (2000) On the matching accuracy of rasterised scanning laser altimetre data, in *ISPRS Congress 2000*, Vol. XXXIII, Part B2, Amsterdam, pp. 75-82.
2. Carneiro C (2008) Communication and visualization of 3-D urban spatial data according to user requirements: case study of Geneva, Proceedings of the XXI ISPRS Congress, 3-11 July, Beijing, China.
3. Duffie JA (1980) Solar engineering of thermal processes, John Wiley & Sons, New York.
4. Gonçalves G (2006) Analysis of interpolation errors in urban digital surface models created from LIDAR data, Proceedings of the 7th International Symposium on Spatial Accuracy Assessment in Resources and Environment Sciences. Edited by M. Caetano and M. Painho. Sana Hotel, 5th-th July, Lisbon, Portugal.
5. Jenks M, Burton E, Williams K (1996) The Compact City: A Sustainable Urban Form?, E & FN Spon, London, UK.
6. Jenks M, Burton E, Williams K (2000) Achieving Sustainable Urban Form, E & FN Spon, London, UK.
7. Menezes AS, Chasco FR, Garcia B, Cabrejas J, González-Audícana M (2005) Quality control in digital terrain models, Journal of Surveying Engineering, Vol. 131: pp. 118-124.
8. Nielsen J (1993) Usability Engineering, Morgan kaufmann-Academic Press, London, United Kingdom.
9. Osaragi, T, Otani I (2007) Effects of ground surface relief in 3-D spatial analysis on residential environment, The European Information Society: Lecture notes in Geoinformation and Cartography. Edited by Sara Irina Fabrikant and Monica Wachowicz. Published by Springer Berlin Heidelberg. Berlin, pp. 171-186.
10. Ratti C (2001) Urban analysis for environmental prediction, unpublished Ph.D. dissertation, University of Cambridge, Cambridge, UK.
11. Ratti C, Morello E (2005) SunScapes: extending the 'solar envelopes' concept through 'iso-solar surfaces', Proceedings of the 22nd International Conference on Passive and Low Energy Architecture, Beirut, Lebanon.
12. Ratti C, Richens P (2004) Raster analysis of urban form, Environment and Planning B: Planning and Design, 31.
13. Ratti C, Baker N, Steemers K (2005) Energy consumption and urban texture, Energy and Buildings, Vol. 37:7, pp. 762-776.
14. Reichenbacher T, Swienty O (2007) Attention-guiding geovisualization, Proceedings of the 10th AGILE International Conference on Geographic Information Science, 8th-11th May, Aalborg University, Denmark.
15. Rogers R (1997) Cities for a small planet, Faber & Faber, London, UK.

16. Steed A, Frecon E, Pemberton D, Smith G (1999) The London Travel Demonstrator, Proceedings of the ACM Symposium on Virtual Reality Software and Technology, December 20-22, pp. 50-57, ACM Press.
17. Takase Y, Sho N, Sone A, Shimiya K (2003) Automatic generation of 3D city models and related applications, International Archives of the Photogrammetry, Remote Sensing and Spatial Information Sciences (ISPRS), Vol. XXXIV-5/W10.
18. Urban Task Force (1999) Towards an urban renaissance: final report of the Urban Task Force - Chaired by Lord Rogers of Riverside, E & FN Spon, London.
19. Zinger S, Nikolova M, Roux M, Maître H (2002) 3-D resampling for airborne laser data of urban areas, In International Archives of Photogrammetry and Remote Sensing, Vol. XXXIV, 3B, pp. 55-61.

Chapter 21
Creation and Error Analysis of High Resolution DEM Based on Source Data Sets of Various Accuracy

Jari Pohjola, Jari Turunen, Tarmo Lipping and Ari Ikonen

Abstract. Digital Elevation Models (DEMs) are extensively used as basic components of geographic information systems. Our purpose in this study was to generate a high-resolution DEM based on available elevation data of varying levels of accuracy and reliability. High-resolution DEM can be used as an input for modeling landscape development of the Olkiluoto region in Finland, a future repository site for spent nuclear fuel. We also generated error models for the newly created DEM. Our approach uses thin plate approximation and Monte Carlo simulations. Our results show that our proposed framework enables validation and identification of flaws in various source data sets even if little is known a priori about their accuracy. Thin plate approximation is shown to perform well in the case of DEM generation. In the case of sparsely and irregularly located source measurements, the smoothness of the generated DEM is highly dependent on the local interpolation neighbourhood

21.1 Introduction

The study presented in this paper was motivated by the need of Posiva Oy, a Finnish company responsible for planning repositories for spent nuclear fuel, to model the landscape development in the area surrounding a potential site in Olkiluoto [1] (see Fig. 21.1). The timescale of the modeling, up to 10,000 years, takes into account both the reasonably predictable interglacial terrain development and travel times of hypothetical releases from the repository to the surface in case of a failure. In the Olkiluoto area, the current rate of postglacial land uplift is about 6 mm/year. Also, tilting of the bedrock occurs because the uplift is faster in the western part of the region. The

---

Information Technology, Pori, Tampere University of Technology,
{jari.pohjola, jari.j.turunen, tarmo.lipping}@tut.fi,
ari.ikonen@posiva.fi

changes in terrain over time can be predicted by taking the uplift into account [2] (Fig. 21.1).

The accuracy of this kind of landscape development model is normally limited by the precision of the Digital Elevation Model (DEM) that lies at the heart of the analysis. Currently, elevation data from various different sources and of variable accuracy is available for the region of interest. Our purpose was to create a uniform DEM and to evaluate its accuracy (confidence limits) based on the available data. The grid resolution of the final model was chosen to be 2.5 x 2.5 m as a reasonable scale for subsequent modelling of hydrological and ecological characteristics.

DEM is the basic dataset used in most Geographic Information Systems (GIS) and therefore its accuracy has been studied intensively by several research groups. The source data used to generate DEM usually contains various errors such as those due to measurement devices, data transfer, storage and analysis, as well as human errors. In [3] several pre-processing phases were applied to reduce the error in source data (contours, spot data, rivers, lakes and reservoirs) before estimating the interpolation parameters and generating the final DEM using an iterative process. In [4] two different datasets - lidar and satellite radar data - were combined using kriging and variogram models, which were subsequently validated with baseline validation data. Bamber et al. evaluated the accuracies of two different DEMs that were generated for the Antarctic continent using lidar data [5]. The accuracies of the models were evaluated using the surface slopes as well as the original values of the DEMs. Oksanen et al. compared a DEM model derived from contour data against lidar data and estimated the modelling error using variogram and statistical measurements [6]. Sefercik compared three different DEM models produced from geodesic measurements, photogrammetry, and Shuttle Radar Topography Mission data [7]. The author found the photogrammetric DEM to be more accurate than the others.

An important issue related to DEM accuracy is interpolation. It is often desirable that the DEM to be used in a GIS be defined on a uniform grid even though the source data might be in a very different format - contour lines, for example. Even if the source data were defined on a uniform grid (from photogrammetric data sets, for example), it might be useful to represent the final DEM at a different resolution either to save storage or analysis capacity or to match the DEM to other data sets. In [8] several aspects related to optimal sampling of a DEM were studied. Chaplot et al. compared several interpolation techniques like Inverse Distance Weighting, Ordinary Kriging, Universal Kriging, Multiquadratic Radial Basis Function and Regularized Spline with Tension [9]. The results of the different methods varied depending on the sampling densities and spatial structure variability. Also in [10] several methods were compared to determine optimal landscape visualization. Surface modelling based on the Theorem of Surfaces is presented in [11] and compared to traditional methods. The method proposed by the authors showed a slight improvement over alternatives.

Monte Carlo simulations have been used in several studies dealing with the evaluation of DEM accuracy. In [12], a Monte Carlo simulation was used to study the differences between two methods of drainage area estimation using a DEM model. Oksanen et al. and Raaflaub et al. have studied error propagation in a DEM model and its derivatives (slope, for example) using Monte Carlo simulations [13, 14]. DEMs were varied by adding a simulated random error to the grid points, or otherwise by varying the grid points according to some distribution and repeating that process 1000 or 1500 times. Based on the simulation, several statistical coefficients were calculated. In [15],

DEM reconstruction from multi-baseline InSAR data without ground control points was analysed and validated using a Monte Carlo simulation.

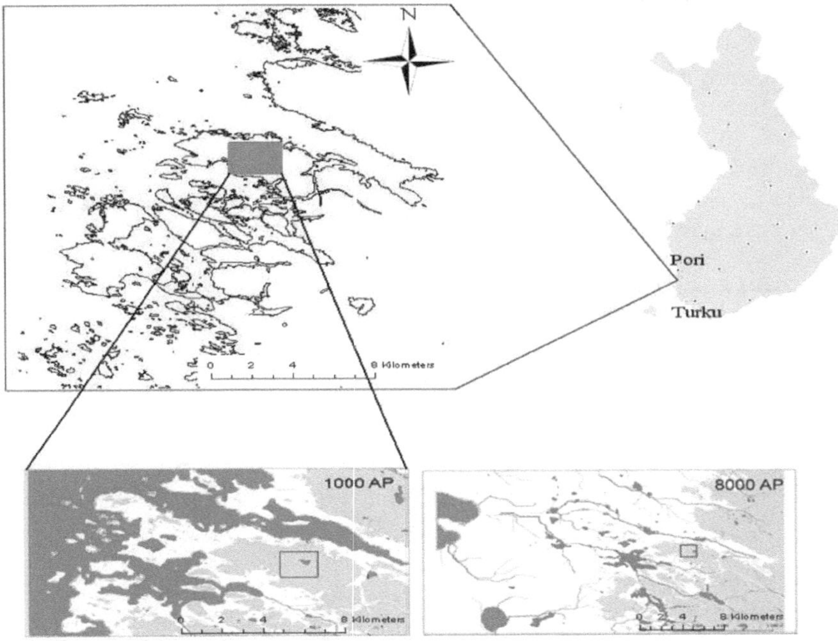

Fig. 21.1. Terrain forecasts for 1000 (left panel) and 8000 (right panel) years, respectively [2]. The current land area is shaded and the (newly formed) lakes and rivers are shown in dark grey. The square indicates the expected location of the repository containing spent nuclear fuel.

There exist studies that aim to correct the DEM model after it has been generated. Some computational methods produce pits or depressions in the model and the challenging task is to subsequently separate the artificial depressions from the real ones. In [16], an algorithm was proposed that evaluates and fixes erroneous depressions in DEM. Grimaldi et. al. proposed a method that corrects errorful pit cells in DEM by adjusting the slopes and flow direction in the associated topography [17].

In the following sections we first describe the elevation data available from the Olkiluoto research area. Subsequently, we present the framework developed to generate a uniform DEM with respective confidence limits. The methods applied include thin plate approximation and Monte Carlo simulation. Thereafter, we present and discuss the results and highlight certain problems that we encountered in combining different data sets. Finally, we draw certain conclusions from the study.

21.2 Source data sets and analysis methods

This study focuses on a test area of 1.8 x 1.5 km, while subsequently a similar analysis will be performed on a significantly larger region surrounding the test area. The different source data sets available from the test area are described in Table 21.1 (see also Fig. 21.2 for the spatial distribution of the data). An offset was added to the source data values to correct for the 6 mm/year land uplift. To perform Monte Carlo simulation, the error in the source data had to be modelled. As none of the data providers could give reliable and accurate error models for their data, we assumed the error to be normally distributed in all cases. Different error variances or confidence limits were assigned to different data sets according to the data specifications and what was known about the acquisition methods (Table 21.1).

Table 21.1. Description of the source data sets. The Finnish National Land Survey provided error measure for their data set; the 95% confidence limits for the other data sets were derived from what was known about the measurement method.

Data set	Year of assessment	Spatial distribution	Data error model
© Finnish National Land Survey data	1997	Contour lines	Standard deviation: 139 cm [18]
Combined elevation data by private consulting office	Based on data from 1991, 1995	10 x 10 m. grid	95 % conf. limit: 200 cm
Seismic measurements	2001 – 2005	Points on straight lines	95 % conf. limit: 100 cm
	2006 – 2007		95 % conf. limit: 60 cm
Acoustic-seismic measurements	2000	Individual points	95 % conf. limit: 400 cm
Sonar measurements of sea areas	2007	Cloud of points	95 % conf. limit: 100 cm
Prismarit Oy data	2001	Individual points	95 % conf. limit: 2 cm

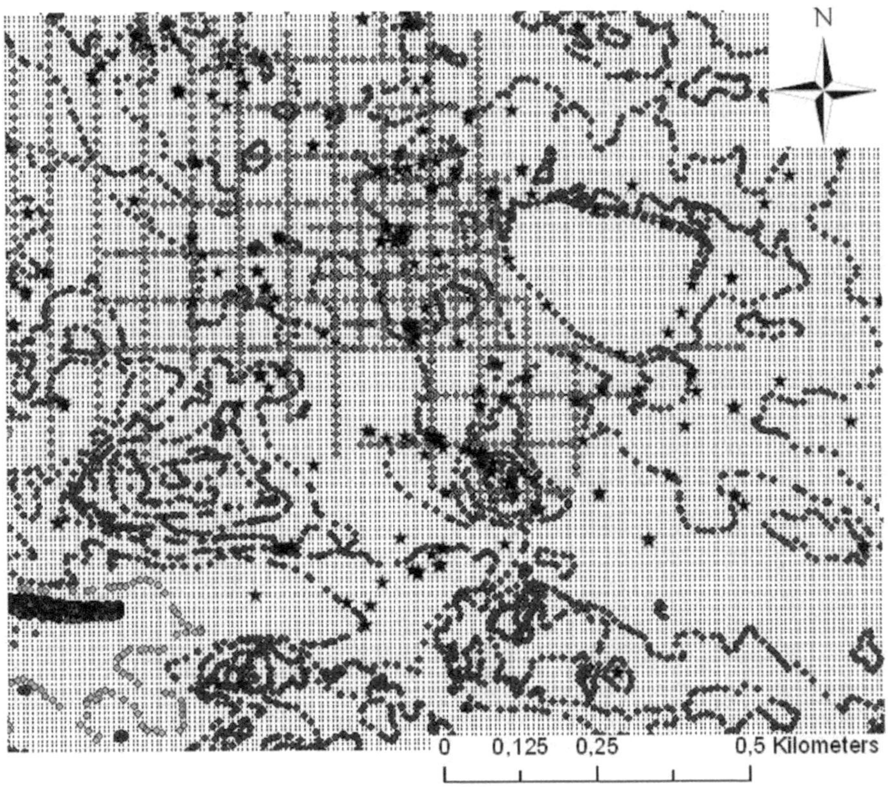

Fig. 21.2. Spatial distribution of the source data. The data provided by Prismarit Oy are shown by star-shaped signs. The 10 x 10 grid data are shown by smaller dots as otherwise it would hide the whole area.

21.3 DEM Interpolation

There are several methods for DEM surface interpolation. We considered the following methods, all of which are based on a set of data points in the surrounding area of the grid point to be evaluated: linear interpolation, parabolic fit, triangle based cubic interpolation, inverse distance weighting of orders 1, 2 and 3 (IDW1, IDW2 and IDW3, respectively), mean, median, and thin plate approximation. The accuracy and performance of the ordinary kriging method was similar to that of the IDW and therefore it was not included in the test set. As no accurate elevation data was available for reference, the behaviour of the aforementioned interpolation methods was evaluated by removing 20% of the source data points and checking the interpolated data values against the original measurements at these locations. Since there was an indication that the 10 x 10 m. grid source data was unreliable, this data set was left out while testing the interpolation methods. 1972 points in total were used to evaluate the

methods were evaluated and each point was interpolated using its 30 closest neighbours. The resulting error distributions are shown in Fig. 21.3. The 95% confidence limits as well as the proportion of points where the estimation error exceeded the ±2m limits are shown in Table 21.2 for all the methods.

Table 21.2. The results of testing the interpolation methods.

Interpolation method	95 % confidence limit (lower; upper; total) in meters	Number (percentage) of errors exceeding -2.0 … 2.0 m limits (lower; upper)
Linear	-1.41; 1.36; 2.77	16 (0.81%); 15 (0.76%)
Parabolic	-3.45; 2.15; 5.60	165 (8.37%); 56 (2.84%)
Cubic	-0.84; 1.84; 2.68	33 (1.67%); 39 (1.98%)
IDW1	-2.80; 1.96; 4.76	109 (5.53%); 41 (2.08%)
IDW2	-2.56; 1.65; 4.21	87 (4.41%); 29 (1.47%)
IDW3	-2.56; 1.57; 4.13	95 (4.82%); 28 (1.42%)
Mean	-3.36; 2.36; 5.72	154 (7.81%); 62 (3.14%)
Median	-2.76; 2.56; 5.32	258 (13.08%); 150 (7.61%)
Thin plate	-1.53; 1.24; 2.77	21 (1.06%); 18 (0.91%)

The test results show that the linear, cubic and thin plate approximations give the narrowest confidence limits while, somewhat surprisingly, the results for the IDW methods are significantly worse. The drawback of the cubic approximation method is the generation of some large outliers. It is clear that the results of this kind of comparison depend on the landscape profile as well as on the spatial distribution of the data points and can therefore be considered only indicative. However, in addition to the test results, our choice to use the thin plate approximation method was based on the intrinsic properties of the method. The surface generated by the thin plate method can have several minima and maxima and is thus capable of following the natural landscape better than most of the other methods. Unlike certain other advanced interpolation techniques, like tensor product splines, for example, the thin plate method does not place restrictions on the spatial distribution of the source data.

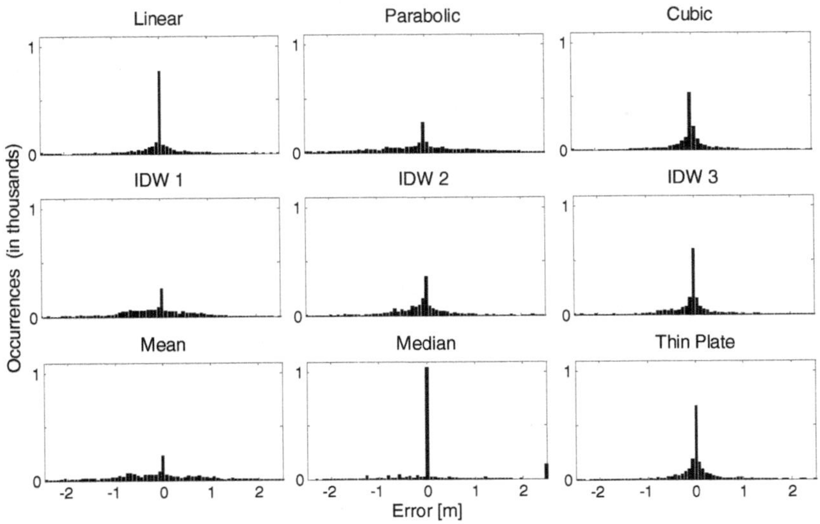

Fig. 21.3. Error distributions for the tested interpolation algorithms.

The thin plate smoothing spline approximation method employs a commonly used basis function for representing coordinate mappings. It minimizes the weighted sum:

$$pE(f)+(1-p)R(f) \tag{1}$$

where *E(f)* is an error of the form:

$$E(f)=\sum_{j}\left|z_j - f(x_j, y_j)\right|^2 \tag{2}$$

and *R(f) is a* roughness measure of the estimated surface:

$$R(f)=\int(\left|D_x D_x f\right|^2 + 2\left|D_x D_y f\right|^2 + \left|D_y D_y f\right|^2) \tag{3}$$

Here *Df* denotes the partial derivative of *f* with respect to the *i*-th component of *f*. The value of the smoothing parameter *p* lies between 0 and 1, and is chosen by the user [19, 20]. The closer the value is to 1, the more strictly the approximated surface follows the source data points. If $p = 0$, linear approximation is used, i.e., the approximating surface is flat.

21.4 Monte Carlo simulation

To evaluate the error of the newly generated high resolution DEM grid, 5000 instantiations of the thin plate surface approximation were generated using 5000 instantiations of each source data point in the neighbourhood of the particular high resolution DEM grid point. The source data instantiations were generated consistent with the assigned error distributions given in Table 21.1. The 5000 versions of the thin plate surface approximation yielded an error distribution for the new high resolution DEM. It was subsequently possible to evaluate the elevation values on the new grid as mean, median or maximum over these distributions. In our simulations, all three estimation approaches gave very similar values.

An important element of the method is the definition of the neighbourhood over which the thin plate approximation is achieved. To find the optimal number of neighbours, simulations were performed showing that the approximation error started to increase if fewer than 20 source data points were employed. However, the increase was not very sharp. Even more important was the spatial distribution of the source data points belonging to the neighbourhood. In some simulations, it occurred that at the locations where plenty of source data points were available in a certain direction moving away from a particular high resolution grid point, and none were available in other directions. In such cases, the approximation results were poor.

Another problem was evident at locations where there was a change in the set that defined the neighbourhood. Even if the change was tiny and involved only a single source data point, the approximation result changed significantly. In the generated DEM surface this phenomenon manifested itself as a scar-like artifact (see Fig. 21.5, lower left panel). Therefore, an algorithm was developed whereby the probability distribution of each grid point could be estimated based on data from four different neighbourhoods (Fig. 21.4). The neighbourhoods are based on 25 x 25 m rectangular areas, the centrepoints of which are marked by A, B, C, and D in the figure. In each neighbourhood, a search for the source data points is performed in 8 directions (the wedges corresponding to the directions are marked by numbers 1...8 in neighbourhood A in Fig. 21.4). The search area is extended to longer distances from the centre of the neighbourhood until at least two data points are found in each wedge. If the overall number of data points in this extended area exceeds 400, points located further away from the centre are eliminated from most densely populated wedges. 5000 approximations are then calculated based on the instantiations of the remaining source data points according to their a priori distributions. This procedure is repeated for all four neighbourhoods.

As a result, four distributions were obtained for each grid point in the shaded area, each of which was based on the 5000 instantiations corresponding to different neighbourhoods. The final error distribution was calculated as the weighted sum of these four distributions with the weights determined by the second order IDW with respect to the centres of the corresponding neighbourhoods.

A computer cluster was used to implement the aforementioned overlapping neighbourhoods algorithm. We estimate that using 23 computers with clock rates ranging from 1 to 3.2 GHz the calculation of the high resolution DEM for the area shown in Fig. 21.2 took approximately 30 hours. The calculation time can easily be reduced by decreasing the number of instantiations in the Monte Carlo simulation.

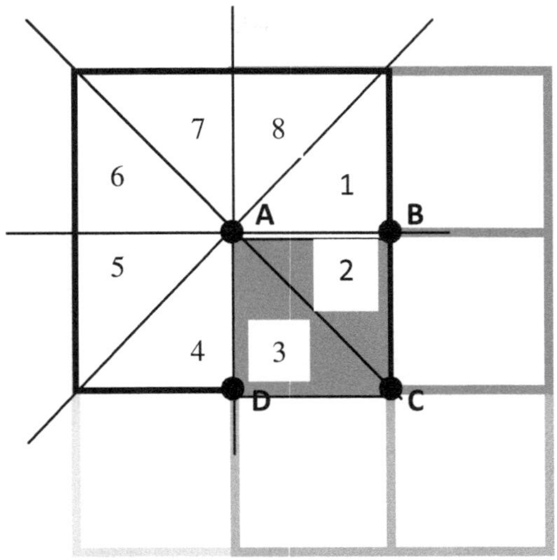

Fig. 21.4. Schematic of the neighborhood selection algorithm in the case of multiple overlapping neighborhoods.

21.5 Results and discussion

In the course of our research, we created methods to generate high resolution DEM models with various parameters and using different data sets. In this section we present our results.

The upper left panel of Fig. 21.5 shows the model that employed the 10 x 10 m grid source data and the seismic measurements data (see Table 21.1). As the 10 x 10 m source data had the highest resolution, we considered it the most reliable in the particular test area. Different from Table 21.1, a 95% confidence limit of 1 m was assigned to the a priori probability density of the data in this simulation. A neighbourhood of 15 data points was considered sufficient and the smoothing parameter was set to 0.99. These choices were justified as the data was defined on a relatively high density regular grid. It can be seen in the figure that there is a clear discrepancy between the 10 x 10 m grid dataset and the seismic measurements, indicated by small 'mountain' ranges or by series of 'valleys' corresponding to the lines along which seismic measurements were taken. A more detailed comparison was therefore performed to evaluate the accuracy of the 10x10 m grid source data.

In the upper right panel and lower left panel of Fig. 21.5 are shown the DEM models employing only the 10 x 10 m grid source data and employing all the available data sets except the 10 x 10 m grid data. In the case of using only the 10 x 10 m grid data, strong staircase-like behaviour can be seen. This indicates that although the data is defined on a fine grid, the majority of the data points are obviously derived from a small number of measurements, probably by interpolation. Relatively large flat areas also indicate that the reliability of the data might not be as good as was initially suggested. The number of neighbours used in these simulations was 20...30.

On the DEM model of the lower left panel no such staircase-like behaviour can be seen. Looking at Fig. 21.2 it is obvious that if the 10 x 10 m grid data were discarded, the sparseness and irregular locations of the data points would pose a problem. This tends to cause scar-like texturing in the DEM model (seen in the valley at the lower left of the corresponding panel, for example). Increasing the minimum number of neighbours to 70 in these simulations only slightly addressed this problem.

The lower right panel of Fig. 21.5 presents the DEM model generated using the method of overlapping neighbourhoods, as illustrated in Fig. 21.4. It can be seen that the scar-like artefact is no longer present while the steep slopes of the landscape are preserved.

Fig. 21.6 shows a cross-sectional view of the error models of the resulting high resolution DEMs. The cross-section is analysed between the locations (61° 13' 47" N, 21° 28' 27" E) and (61° 14' 37" N, 21° 28' 28" E) (see the right panel of the figure). The error models from top to bottom on the left side of the figure correspond to the DEMs of the upper right, lower left, and lower right panels of Fig. 21.5, respectively. The dashed lines indicate the 95% confidence limit of the error models.

The staircase-like behaviour as well as flat areas in the model based on the 10x10 m grid source data (uppermost window in Fig. 21.6) can be seen in the cross section as well as in the corresponding DEM surface. The confidence limit of the error models based on various data sets fluctuates greatly depending on how close to the cross-section the original data points are and their respective accuracies. At the locations where the confidence interval is extremely narrow (the resulting DEM is highly reliable), the cross-section passes through the measurement points provided by Prismarit Oy. As can be seen from Table 21.1, these data are known to be very reliable. Careful inspection of the curves indicates that there exist some sharp discontinuities seen, for example in the valley between 1100 and 1350 m in the middle window. Other erroneous data caused by the scar-like artefacts are smoothed out in the curves shown in the lower window.

In DEM generation using the thin plate approximation method an important role is played by the smoothing parameter. In the vicinity of steep slopes the method tends to generate overshoots. This effect can be seen especially well in the upper left panel of Fig. 21.5 near the (probably erroneous) high peaks in an otherwise flat area. We addressed the problem by making the smoothing parameter adaptive consistent with the standard deviation of the source data values in the neighbourhood. More specifically, if the standard deviation was high, the smoothing parameter would be relaxed (i.e., the approximation process was guided towards smoother surfaces) so that the approximation algorithm did not struggle to follow each source data value exactly. This new scheme was used in all the other models except that of the upper left panel of Fig. 21.5. It is clear that the overshooting effect was reduced significantly.

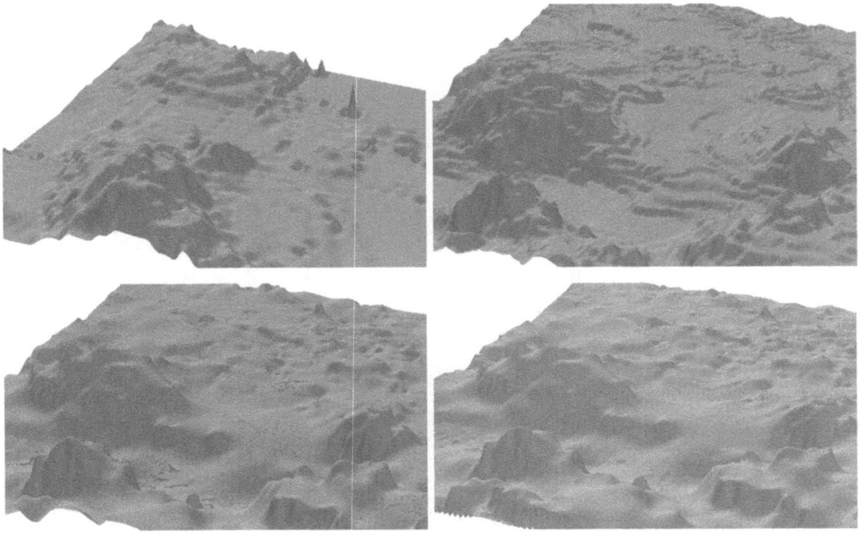

Fig. 21.5. High resolution DEM obtained using various methods and data sets. *Upper left panel*: DEM based on 10 x 10 m grid source data and seismic measurements, fixed neighbourhood of >15 data points is used; *upper right panel*: DEM based on 10 x 10 m grid source data only, fixed neighbourhood of 20...30 neighbours is used; *lower left panel*: DEM based on all source data sets described in Table 21.1 except the 10x10 grid data, fixed neighbourhood of >70 data points in used; *lower right panel*: DEM based on all source data sets described in Table 21.1 except the 10x10 grid data, the algorithm of overlapping neighbourhoods is used. Centre part of the test area (varying slightly from panel to panel) is shown in the figure. The height differences of the landscape are exaggerated for visual purposes.

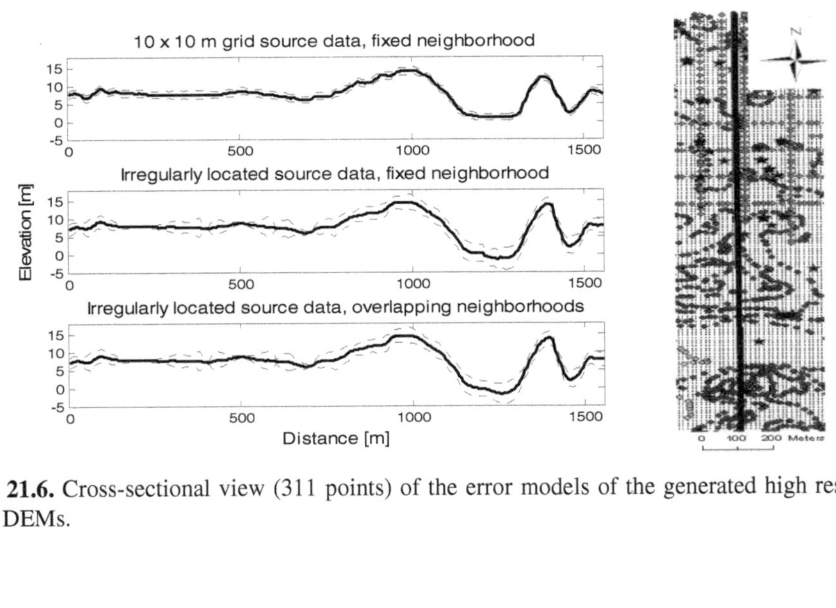

Fig. 21.6. Cross-sectional view (311 points) of the error models of the generated high resolution DEMs.

21.6 Conclusions

This paper presented a method for generating fine resolution DEM based on source data of various resolutions and accuracies. The error model of the DEM was also explored through the application of Monte Carlo simulations. The overall results indicate that the applied thin plate approximation method works well for generating DEMs even when the source data are sparse and irregular. In this case the selection of the source data points forming the neighbourhood for the interpolation procedure is of great importance. Even leaving out or adding a single data point to the neighbourhood can cause discontinuities in the model. Therefore a method of combining four overlapping neighbourhoods and taking into account the spatial distribution of the source data was developed and shown to provide smooth and reliable DEMs.

It is quite common to have various kinds of elevation data available from a specific area for which one wishes to obtain a uniform DEM. Usually the error model of the source data is not given or is specified very loosely using a statement like 'the measurements probably are within n metres.' The framework developed in this paper enables us to obtain a high resolution DEM together with a meaningful error model. Our error models should not be treated as 100% accurate, however, because the a priori error models of the source data sets were based on educated guesswork rather than on reliable specifications.

However, our results confirm that our proposed framework enables cross-validation of various source data sets even if little is known about their true reliability. Our work allows us to conduct simulations that assign different confidence limits to the available source data sets.

As of this writing, calculations are ongoing to obtain high resolution DEM of an area that measures 20 x 30 km and that surrounds the test area shown in Fig. 21.2. Our future work will apply this DEM to landscape modelling. The sensitivity of the landscape models to errors in the underlying DEM will be assessed. We also plan to compare our results with laser scanned data, which will hopefully be acquired in near future. This comparison will give us better insights into the properties and reliability of the DEM generation methods proposed in our research to date.

References

1. A.T.K. Ikonen, V. Leppänen, M. Gunia, R. Broed, 2008. Handling of climate and future human actions as scenario uncertainty within terrain ecosystems as a part of biosphere assessment for the Olkiluoto site. Radioprotection, a Supplement on International Conference on Radioecology & Environmental Radioactivity, 15-20.6.2008, Bergen, Norway (approved for printing).
2. A.T.K. Ikonen, T. Hjerpe, L. Aro and V. Leppänen, 2007. Terrain and ecosystems development model of Olkiluoto site, version 2006. Posiva Working report 2007-110. Posiva Oy, Olkiluoto, Finland.
3. Q-K.Yang, T. McVicar, T. Van Niel, M. Hutchinson, L-T. Li, X-P Zhang, 2007. Improving a digital elevation model by reducing source data errors and optimizing interpolation algorithm parameters: An example in the Loess Plateau, China.

International Journal of Applied Earth Observation and Geoinformation, 9, 235-246.
4. S. Hosford, N. Baghdadi, B. Bourgine, P. Daniels, C. King, 2003. Fusion of airborne laser altimeter and RADARSAT data for DEM generation. Proceedings on IEEE International. Geoscience and Remote Sensing Symposium, IGARSS '03, Volume 2, 806-808.
5. J. Bamber, J. Gomez-Dans, 2005. The accuracy of digital elevation models of the Antarctic continent. Earth and Planetary Science Letters, 237, 516-523.
6. J. Oksanen, T. Sarjakoski, 2006. Uncovering the statistical and spatial characteristics of fine toposcale DEM error. International Journal of Geographical Information Science, 20 (4), 345-369.
7. U. Sefercik, 2007. Comparison of DEM Accuracies Generated by Various Methods. Proceedings of 3rd International Conference on Recent Advances in Space Technologies, 2007. RAST '07, 379 - 382.
8. P. Atkinson, C. Lloyd, 2007. Non-stationary variogram models for geostatistical sampling optimisation: An empirical investigation using elevation data. Computers & Geosciences, 33, 1285-1300.
9. V. Chaplot, F. Darboux, H. Bourennane, S. Leguédois, N. Silvera, K. Phachomphon, 2006. Accuracy of interpolation techniques for the derivation of digital elevation models in relation to landform types and data density. Geomorphology, 77, 126-141.
10. J. Wood, P. Fisher, 1993. Assessing interpolation accuracy in elevation models. IEEE Computer Graphics and Applications,13, (2), 48 - 56.
11. T-X. Yue, Z-P. Du, D-J. Song, Y. Gong, 2007. A new method of surface modelling and its application to DEM construction. Geomorphology, 91, 161-172.
12. P. Pilesjö, A. Persson, L. Harrie, 2006. Digital elevation data for estimation potential wetness in ridged fields-Comparison of two different methods. Agricultural Water Management, 79, 225-247.
13. J. Oksanen, T. Sarjakoski, 2005. Error propagation of DEM-based surface derivatives. Computers & Geosciences, 31, 1015-1027.
14. L. Raaflaub, M. Collins, 2006. The effect of error in gridded digital elevation models on the estimation of topographic parameters. Environmental Modelling & Software, 21, 710-732.
15. J. Li, H. Huang, D. Liang, 2007. A multi-baseline InSAR DEM reconstruction approach without ground control points. Proceedings of IEEE International Geoscience and Remote Sensing Symposium, IGARSS 2007, 4509 – 4512.
16. A. Temme, J. Schoorl, A. Veldkamp, 2006. Algorithm for dealing with depressions in dynamic landscape evolution models. Computers & Geosciences, 32, 452-461.
17. S. Grimaldi, F. Nardi, F. Di Benedetto, E. Istanbulluoglu, R. Bras, 2007. A physically-based method for removing pits in digital elevation models. Advances in Water Resources, 30, 2151-2158.
18. URL: http://www.maanmittauslaitos.fi/tarkkuusanalyysi.htm, (in Finnish) 16.5.2008.
19. C De Boor, 2007. Matlab Spline Toolbox 3: User's guide. The Mathworks Inc., Massachusetts.
20. G. Donaldo, S. Belonge, 2002. Approximate Thin Plate Spline Mappings. Lecture Notes in Computer Science, 2352, Springer Berlin, 13-31.

Chapter 22
A Topological Analysis Method for 3D Geo-Entities Structured as Hexahedron Tessellations

GUO Jiateng and WU Lixin

Abstract. Topological spatial relationships are at the core of geographical information system, especially 3D GIS. The design of suitable models and methods for querying and analyzing topological relationships among spatial objects remains an unsolved problem and is the subject of substantial research. Unfortunately, most modern models and methods support only topological analysis for geographical objects that can be represented using simple 2D regions and lines. These approaches are not sufficient to address the complexity of real 3D geo-entities. This paper introduces a new method for 3D topological analysis. Our approach is an improvement over the nine-intersection model. We propose redefining the original interior, boundary and exterior on the basis of sixfold connectivity as represented by a hexahedron. We use SQL to compute the resulting model matrix. The approach to mapping from a quantitative computed result matrix to a qualitative topological relation is discussed in detail. Our research offers contributions in the areas of topological analysis models and methods for representing 3D geo-entities for spatial querying and analysis.

22.1 Introduction

In recent years, interest has grown in geographic information systems (GIS) that can manage huge volumes of 3D spatial data. However, almost all available systems do not support three dimensional (3D) topological querying and analysis. As a result, the capabilities of modern GIS are severely limited. With continued development of our understanding of human cognition and behavior, researchers have gradually refocused their attention from the two dimensional (2D) plane to 3D space and from the Earth's surface to underground scenarios. The variety and complexity of underground 3D

Institute for GIS/RS/GPS and Digital Mine Research, Northeastern University
guojiateng@mail.neu.edu.cn
Institute for GIS/RS/GPS and Subsidence Research, China Uni of Mining & Technology
awulixin@263.net

geo-entities cannot be adequately modeled and analyzed with simple geometric features and 2D topology. Several topological relationship frameworks [1, 4, 14] and many 3D topological data models [6, 8, 9, 10, 13] for representing topological relationships have been proposed over the past two decades. Substantial research effort has been deployed to represent 3D topological relationships [2, 4, 5, 7, 11, 12, 15]. The 9-intersection framework has proved very useful for almost all GIS data models. However, there are some limitations of this approach, such as the fact that it is often difficult to define an infinite exterior. In this paper, we attempt to make this framework applicable and computable by redefining the original exterior of a geo-entity. Our improved 9-I model performed much better in representing a gradual change of attributes between adjacent geo-entities. Our results suggest that the 9-intersection model cannot directly be used for geo-entity topological analyses without improvements. In this paper, a new model based on the 9-I model is proposed and applied to querying and analyzing topological relationships for 3D geo-entities that are structured as hexahedron tessellations. The new model definition, the computational method and the rules that define mappings between the computed result model matrix and topological relationships are described in detail.

22.2 Jordan Curve Based on Six-Connectivity Neighborhood

The Jordan curve is the fundamental basis of point-set topology that serves as the foundation for the 9-I framework. In Jordan curve theory, a closed curve partitions the plane into two parts: an infinite region and a finite one. Based on this theory, a geo-object and its topological space can be partitioned into three zones: the interior, the boundary and the exterior. In 3D raster space, the core problem is how to define the Jordan curve (surface). The neighborhood of hexahedrons must first be defined in order to formalize the connectivity between adjacent hexahedrons, after which a geo-object structured from multiple hexahedrons can be partitioned into corresponding components within the 9-Intersection framework[12-16].

In 3D space IR^3, supposing that X is a hexahedron cell of IR^3, then for arbitrary x $(i, j, k) \in IR^3$, the 6 neighborhoods (face-connected hexahedrons), 18 neighborhoods (edge-connected hexahedrons) and 26 neighborhoods (vertex-connected hexahedrons) are respectively defined in Fig. 22.1 and Fig. 22.2.

In Fig 22.1, T indicates Top, B is Bottom, W refers to West, N stands for North, and E indicates East.

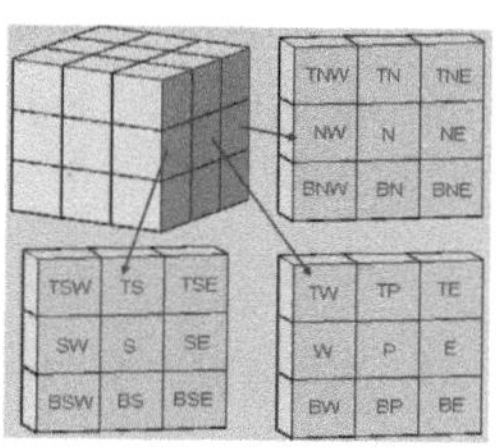

Fig. 22.1. The 3*3*3 neighborhood of a raster cell (P. K. SAHA, 1996)

$N_6(x)$ = {TP, BP, N, S, W, E};

$N_{18}(x)$ = {TN, N, BN, NW, NE, TW, TP, TE, W, E, BW, BP, BE, TS, SW, S, SE, BS};

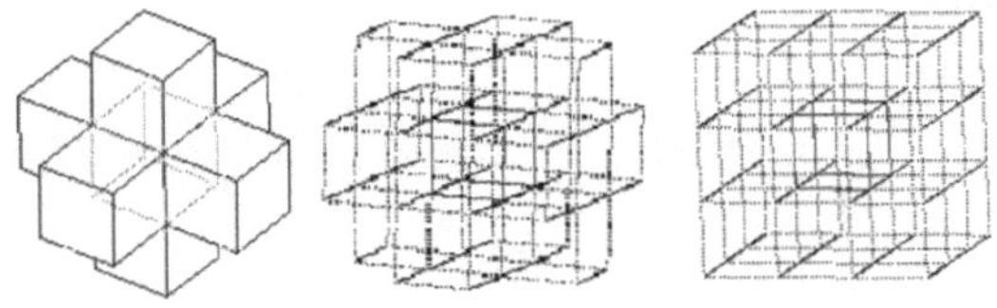

Fig. 22.2. N_6, N_{18} and N_{26} model in 3D raster space

$N_{26}(x)$ = {TNW, TN, TNE, NW, N, NE, BNW, BN, BNE, TW, TP, TE, W, E, BW, BP, BE,TSW, TS, TSE, SW, S, SE, BSW, BS, BSE}.

N_6 neighborhood is chosen in this paper for more compact relation between adjacent regular hexahedrons.

22.3 Topological Analysis Model

22.3.1 Definition of Interior and Boundary

According to the N6 neighborhood as defined above, cells from a 3D raster entity can be partitioned into two parts: the interior and the boundary. With the N_6 model, for each cell (termed e) of a 3D raster entity (E), if any cell of its N_6 is not a member cell of E, then this raster cell e is defined as a boundary cell of E. The collection of these boundary cells defines the boundary of the entity, w, which is referred to as the six neighborhood boundary (B_6):

$$B_6 = \{e \mid e \in E, \exists x \in N_6(e), x \notin E\}. \quad (1)$$

By contrast, for each cell (named as *e*) of a 3D raster entity (named as *E*), if all the cells of its N_6 are member cells of E, then this raster cell *e* is defined as an interior cell of E. The entire collection of these interior cells composes the interior of the entity, which we term the six neighborhood interior (I_6):

$$I_6 = \{e \mid e \in E, \forall x \in N_6(e), x \in E\}. \quad (2)$$

22.3.2 Definition of K-Order Six-Connectivity Neighborhood

In order to codify attribute relevancy between a certain Geo-entity and its neighboring Geo-entities, the spatial relationship between them must first be analyzed. In this paper, the k-order buffers of a Geo-entity generated using six-connectivity are termed the k-order six-neighborhood (E_6^k) and they are defined as follow:

1. E_6^1 can be defined based on I_6 and B_6, which is the collection of the N_6 cells of B6 that are not member cells of I_6 and B_6;
1. E_6^2 can be defined based on E_6^1, which is the collection of the N_6 cells of E_6^1 that are not member cells of 1-order N_6 closure (the Geo-entity itself and its E_6^1);
2. E_6^k can be defined based on E_6^{k-1}, which is the collection of the N_6 cells of E_6^{k-1} that are not member cells of (k-1)-order N_6 closure (the Geo-entity itself and its E_6^1, $E_6^2 \ldots E_6^{k-1}$).

As shown in Fig. 22.3, a vector Geo-entity which is surrounded by 114 triangles was transformed into a 3D raster Geo-entity which comprises 1652 regular hexahedrons. According to the definition of E_6^k, this Geo-entity has the following parameters: I_6(992 cells), B_6 (660 cells), E_6^1(792 cells) and 1-order closure (2444 cells).

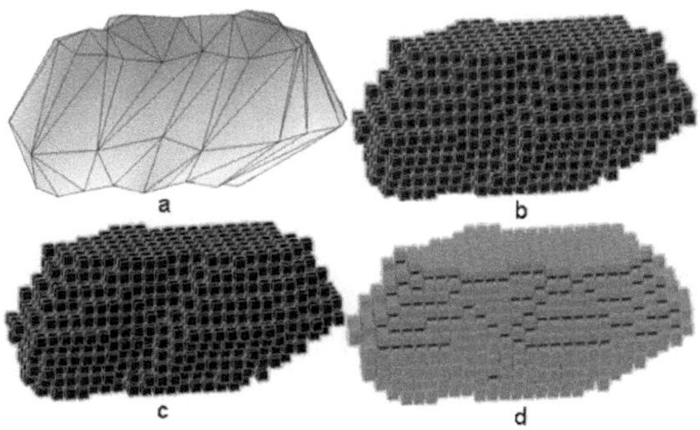

Fig. 22.3. The 1-order N_6 closure of a 3D raster entity (a. vector entity; b. raster entity; c. the I_6 and B_6 of raster entity; d. 1-order N_6 closure)

22.3.3 K-order 6 Neighborhoods-based 9-intersection model

A new topological relationship analysis model based on the 9-intersection model is presented and is termed the *k*-order six neighborhoods-based 9-intersection model (K6N9-I model). According to this model, the original interior, boundary and exterior of the 9-I approach are replaced with I_6, B_6 and E_6^k respectively. The new model is as follows:

$$\begin{bmatrix} A_6^k \cap B_6^k & A_6^k \cap \partial B_6 & A_6^k \cap B_6^\circ \\ \partial A_6 \cap B_6^k & \partial A_6 \cap \partial B_6 & \partial A_6 \cap B_6^\circ \\ A_6^\circ \cap B_6^k & A_6^\circ \cap \partial B_6 & A_6^\circ \cap B_6^\circ \end{bmatrix} \quad (3)$$

Where

A_6^k = the k-order N_6 of Geo-entity A

∂A_6 = the 6-connectivity boundary of Geo-entity A

$\mathring{A}_6$ = the 6-connectivity interior of Geo-entity A

B_6^k = the k-order N_6 of Geo-entity B

∂B_6 = the 6-connectivity boundary of Geo-entity B

$\mathring{B}_6$ = the 6-connectivity interior of Geo-entity B

22.4 Computation Method for Model Matrix

22.4.1 Data Structure for Geo-entity

Geo-entities are stored in the database using two tables.

22.4.1.1 Spatial Field Information Table

This table stores information about the spatial field that contains the geo-entities. It mainly describes the dimensions of the geo-entities, the coordinate scope of the spatial field, the starting base point of the geo-entities, and the hexahedron cell count of the geo-entities. The table structure is as follows.

Table 22.1. Spatial Field Information Table (SpatialFieldInfo)

Field Name	Data Type	Description
StartX	double	Minimum x coordinate of the spatial field where geo-entities locate
StartY	double	Minimum y coordinate of the spatial field where geo-entities locate
StartZ	double	Minimum z coordinate of the spatial field where geo-entities locate
EndX	double	Maximum x coordinate of the spatial field where geo-entities locate
EndY	double	Maximum y coordinate of the spatial field where geo-entities locate
EndZ	double	Maximum z coordinate of the spatial field where geo-entities locate
DimX	float	X dimension of hexahedron cell
DimY	float	Y dimension of hexahedron cell
DimZ	float	Z dimension of hexahedron cell
CountX	long	Hexahedron cell count in x direction
CountY	long	Hexahedron cell count in y direction
CountZ	long	Hexahedron cell count in z direction

22.4.1.2. Geo-entity Information Table

This table stores all the hexahedron cells associated with each geo-entity; every cell record is indexed using a 3D identification code. The table is structured as follows.

Table 22.2. Geo-entity Information Table (GeoentityInfo)

Field Name	Data Type	Description
ID	long	Index for each hexahedron cell in the spatial field
XID	long	Index at X direction for hexahedron cell
YID	long	Index at Y direction for hexahedron cell
ZID	long	Index at Z direction for hexahedron cell
Geo-entity	varchar	ID list of geo-entities that occupy the hexahedron cell

For example, one table record contains the follow line:

53922, 43, 25, 7, (4_B)(5_I)(3_1)(6_1)

The above expression means that a hexahedron cell identified by code 53992 is the 43rd cell in the x direction, the 25th in the y direction, the 7th in the z direction. This cell is within the boundary region defined by geo-entity 4, it is an interior cell of geo-entity 5, and it is linked to the 1-order N_6 cell of geo-entity 3 and the 1-order N_6 cell of geo-entity 6.

22.4.2 Computation Method

In Table 22.2, the data field GEO-ENTITY is used to record the ID and the component type (Interior, Boundary and Exterior) of geo-entities which occupy the hexahedron cell.

This data field must be extensible to accommodate the reality that a new geo-object can be added to the database at any time and that a hexahedron cell may be associated with several geo-entities and with multiple components. Accordingly, the data type VARCHAR was selected.

We used the LIKE query term to compute the intersection set of the model matrix.

Table 22.3. SQL for Model Computation

Model Matrix Element	SQL
$A_6^k \cap B_6^k$	select count(*) from GeoentityInfo where Geo-entity like '%(A_1)%(B_1)%'
$\partial A_6 \cap B_6^k$	select count(*) from GeoentityInfo where Geo-entity like '%(A_B)%(B_1)%'
$A_6^{\circ} \cap B_6^k$	select count(*) from GeoentityInfo where Geo-entity like '%(A_I)%(B_1)%'
$A_6^k \cap \partial B_6$	select count(*) from GeoentityInfo where Geo-entity like '%(A_1)%(B_B)%'
$\partial A_6 \cap \partial B_6$	select count(*) from GeoentityInfo where Geo-entity like '%(A_B)%(B_B)%'

$A_6^{\circ} \cap \partial B_6$	select count(*) from GeoentityInfo where Geo-entity like '%(A_I)%(B_B)%'
$A_6^{k} \cap B_6^{\circ}$	select count(*) from GeoentityInfo where Geo-entity like '%(A_1)%(B_I)%'
$\partial A_6 \cap B_6^{\circ}$	select count(*) from GeoentityInfo where Geo-entity like '%(A_B)%(B_I)%'
$A_6^{\circ} \cap B_6^{\circ}$	select count(*) from GeoentityInfo where Geo-entity like '%(A_I)%(B_I)%'

22.5 Mapping Regulation from Quantitative Result to Qualitative Topological Relations

Table 22.4. Topological relationships based on count of model element content

Model Matrix	$\begin{bmatrix} 0 & 0 & 0 \\ 0 & 0 & 0 \\ 0 & 0 & 0 \end{bmatrix}$	$\begin{bmatrix} + & 0 & 0 \\ 0 & 0 & 0 \\ 0 & 0 & 0 \end{bmatrix}$	$\begin{bmatrix} + & + & 0 \\ + & 0 & 0 \\ 0 & 0 & 0 \end{bmatrix}$	$\begin{bmatrix} + & + & 0 \\ + & + & 0 \\ 0 & 0 & 0 \end{bmatrix}$
Topological Relationship	Far Disjoint	Nearly Disjoint	Meet	Shallowly Intersect
Description	The distance is more than one cell dimension	The distance is not more than one cell dimension		Boundary into boundary
Model Matrix	$\begin{bmatrix} + & + & ? \\ + & + & + \\ ? & + & 0 \end{bmatrix}$	$\begin{bmatrix} + & + & ? \\ + & + & + \\ ? & + & + \end{bmatrix}$	$\begin{bmatrix} A_6^1 & 0 & 0 \\ 0 & \partial A_6 & 0 \\ 0 & 0 & A_6^0 \end{bmatrix}$	$\begin{bmatrix} 0 & + & + \\ 0 & 0 & \partial A_6 \\ 0 & 0 & A_6^0 \end{bmatrix}$
Topological Relationship	Intersect	Deeply Intersect	Equal	Shallowly Contained By
Description	Boundary into interior	Interior into interior		Part of A's 1-order cells are N_6 of B's boundary
Model Matrix	$\begin{bmatrix} 0 & 0 & + \\ 0 & 0 & \partial A_6 \\ 0 & 0 & A_6^0 \end{bmatrix}$	$\begin{bmatrix} 0 & 0 & 0 \\ + & 0 & 0 \\ + & \partial B_6 & B_6^0 \end{bmatrix}$	$\begin{bmatrix} 0 & 0 & 0 \\ 0 & 0 & 0 \\ + & \partial B_6 & B_6^0 \end{bmatrix}$	$\begin{bmatrix} + & 0 & 0 \\ + & + & 0 \\ + & + & B_6^0 \end{bmatrix}$
Topological Relationship	Deeply Contained By	Shallowly Contain	Deeply Contain	Cover

Where
denotation "+" means that the count of model element content is non-zero;
denotation "0" means that the count of model element content is zero;
denotation "?" means that the count of model element can be ignored.

22.6 Experimental System of Topological Queries

An experimental system for 3D topological query and analysis based on the K6N9-I Model was developed using the Visual C++ programming language and SQL (Structured Query Language). The content and count of each model element was computed, and the count and content of each model element were respectively as shown in the compute dialog and view dialog boxes (Fig. 22.5).

The set that represents the intersection of geo-entity 5 and geo-entity 3 is computed as follows:

$\begin{bmatrix} 26 & 23 & 20 \\ 22 & 19 & 7 \\ 22 & 9 & 0 \end{bmatrix}$. It is a typical instance of $\begin{bmatrix} + & + & ? \\ + & + & + \\ ? & + & 0 \end{bmatrix}$ in Table 22.4, which suggests that the topological relationship between them is that the boundary of geo-entity 5 intersects with the interior of geo-entity 3, and vice versa.

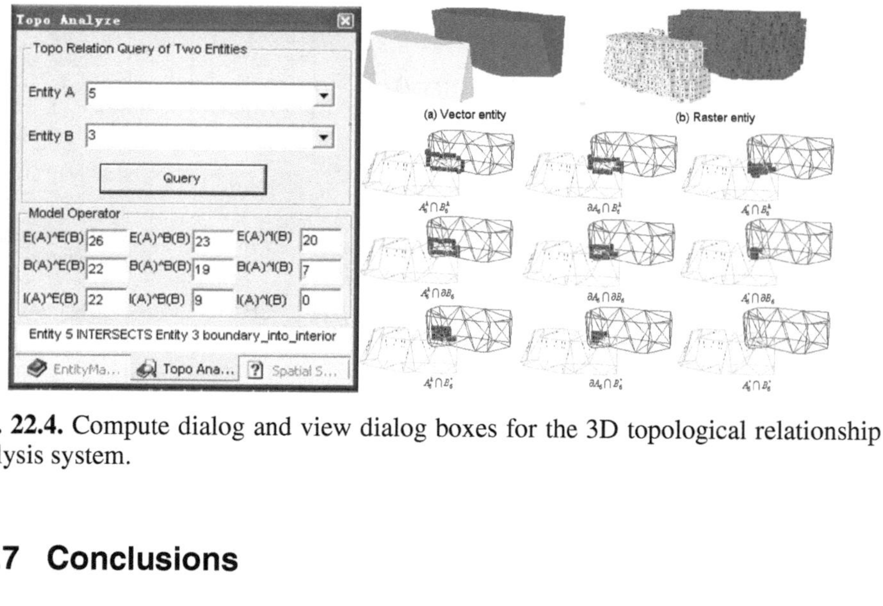

Fig. 22.4. Compute dialog and view dialog boxes for the 3D topological relationship analysis system.

22.7 Conclusions

In this paper, a new 9-intersection model (*K*6N9-I model) is presented based on the traditional 9-I model. In order to make the 9-I model applicable to geo-entities structured as regular hexahedrons, the interior, the boundary and the exterior were re-defined as I_6, B_6 and E_6^k consistent with a 6-neighborhood hexahedron-based model. Our testing proves that the *K*6N9-I model can be applied to generate topological relationship querying engines and to perform analyses of geo-entities in raster space. A computationally tractable method for topological querying and analysis based on SQL is presented. Our experimental analysis shows that the topological analysis model can be used for topological relationship queries and for the analysis of 3D raster geo-entities. Our work suggests that it can also be applied for codifying relationship metrics, direction relationships and other parameters.

The K value and hexahedron scale are two import factors associated with the K6N9-I model. However, only 1-order $N_6(E_6^1)$ of a Geo-entity was generated in this paper. In future research, K will be tuned to different values to study refined topological relationships and K-order-based neighborhood relationships between geo-entities.

22.8 Acknowledgements

The author expresses gratitude to all reviewers for their hard work and valuable advice. This research is jointly supported by the National 863 High-Tech Program of China (No. 2007AA06Z108, 2006AA12Z216), the Natural Science Fund of China for Distinguished Young Scholars (No. 50525414) and the National Natural Science Foundation (No. 40571137).

References

1. Billen, R. and S. Zlatanova, 2003, The Dimensional model: a useful concept for 3D cadastre? in: Computer, Environment and Urban Systems, Vol. 27 (2003) pp. 411-425
2. Chen Jun, Guo Wei, 1998. A Matrix for Describing Topological Relationships Between 3D Spatial Features. *Journal of Wuhan Thechnical University of Surveying and Mapping (WTUSM)*, 23(4), pp. 359-363.
3. Egenhofer M. J. and Franzosa D., 1991. Point-set topological spatial relations. International Journal of Geographical Information Systems, 5(2):161-174.
4. Egenhofer M. J. and Jayant Sharma., 1993. Topological relations between regions in R2 and Z2. In David Abel and Beng Chin Ooi, editors, Advances in Spatial Databases - Third International Symposium, SSD'93, volume 692 of Lecture Notes in Computer Science, pp. 316-336. Springer-Verlag, Singapore.
5. Kwan, M-P. and Lee, Jiyeong., 2004. Emergency response after 9/11: the potential of real-time 3D GIS for quick emergency response in micro-spatial environments, Computers, Environment and Urban Systems (in press).
6. Lee, Jiyeong ,2001a. A 3D data model for representing topological relationships betweenspatial entities in built environments. Ph.D. Dissertation, Department of Geography, The Ohio State University, Columbus, Ohio, U.S.A.
7. Lee, Jiyeong, 2001b. A spatial access oriented implementation of a topological data model for 3D urban entities. Paper presented at University Consortium for Geographic Information Science (UCGIS) Summer Assembly, Buffalo, New York, June 21-24.
8. Lee, Jiyeong, 2001c. 3D Data Model for Representing Topological Relations of Urban Features,In Proceeding of 21st Annual ESRI International User Conference, San Diego, CA.
9. Pfund, M., 2001. Topologic data structure for a 3D GIS. In: Proceedings of International Society for Photogrammetry and Remote Sensing Journal, Vol. 34. Part 2W2, May, Bangkok, Thailand, pp. 233–237.

10. Pigot, S., 1995. A topological model for a 3-dimensional Spatial Information System, Ph.D. Dissertation, University of Tasmania, Australia.
11. Pilouk, M., 1996. Integrated modelling for 3D GIS, Ph.D. Dissertation, ITC, The Netherlands.
12. Rosenfeld A, 1974. Adjacency in digital pictures. Information and Control, 26, pp. 24 –33.
Saha P. K., Chaudhuri B. B., 1996. 3D Digital Topology under Binary Transformation with Applications. COMPUTER VISION AND IMAGE UNDERSTANDING, 63(3), pp. 418-429.
13. Wu Lixin, 2004. Topological relations embodied in a Generalized Tri-Prism (GTP) model for a 3D geoscience modeling system. *Computers & Geosciences*, 30(4), pp. 405 –418.
14. Zhao Renliang, 2002. Study of Spatial Relationship Computation Based on Voronoi[D]. *Central South University,* pp. 26-28.
15. Zlatanova, S., 2000, On 3D topological relationships, in: Proceedings of the 11th International Workshop on Database and Expert System Applications (DEXA 2000), 6-8 September, Greenwich, London, UK, pp. 913-919
16. Zlatanova, S., A. A. Rahman and W. Shi, 2004, Topological models and frameworks for 3D spatial objects, Journal of Computers & Geosciences, May, Vol. 30, No. 4, pp. 419-428

Chapter 23
Constraint-based Generation and Visualization of 3D City Models

Volker Coors, Karina Hünlich and Giwon On

Abstract. In this paper, we propose a new concept for modelling 3D buildings by a set of constraints using the generative modelling language (sorry, another GML). The main advantage of GML is that the building is not fixed but can interactively be changed by the user within the given constraints. One application of this concept is the 3D visualization of a zoning map or master plan for public participation. At zoning map level, a building geometry is not precisely defined yet but some legally binding constraints are given. With the proposed approach a 3D urban model can be generated based on the zoning map. In addition, the user can interactively modify buildings and explore variations within the given constraints.

23.1 Introduction

In this paper, we propose a 3D city model where a building model is not defined by a static geometry but by a set of constraints. The building geometry itself can be interactively modified by the user with in given constraints.

State of the art 3D urban models usually represent the current status of a city. In contrast, the proposed approach supports a constraint-based interactive visualization of planned changes of the city in the (near) future. These future developments are usually documented in a Master plan or zoning map. In Germany, a zoning map defines legally binding regulation for a new build up area. Public participation is an essential part during the development of a zoning map. Within the VEPs project (http://www.veps3d.org) the possibilities of internet-based 3D-Visualization and eParticipation have been explored [5, 6, 7]. It turned out that 3D visualization improves the understanding of spatial relationships especially for the "Joe Public", a user group

University of Applied Science Stuttgar
Volker.coors@hft-stuttgart.de
Fraunhofer IGD
giwon.on@igd.fraunhofer.de

usually not involved in any planning activities. Based on this experience, we have further developed the idea of using an interactive 3D visualization for eParticipation especially for the "Joe Public" user group on zoning map level.

However, heterogeneity of the IT-systems used in the area of urban planning, the lack of a standardised data format for the exchange of these plans, actually interferes with the development of services, supporting the generation, legislation, modification, and usage of such plans. Within the XPlanning project (http://www.xplanung.de) a common XML-based data interchange format (XPlanGML) is specified to overcome this interoperability issues.

In this paper, a new concept for generating and visualizing a 3D city model based on XPlanGML, is proposed. The biggest challenge to create such a 3D city model is to show the degrees of freedom within the zoning map regulations. A simple visualization with static building geometry will be misleading and will not be very helpful for the planning and participation process.

As it is not possible to model a building by constraint within the usual data models such as CityGML [15], the generative modelling language (GML, http://www.generative-modeling.org) was chosen to describe the 3D city model. The resulting city model derived from a XPlanGML zoning map does not only show a possible building area but also gives the user the possibility to change the building geometry and location within the given constraints of the corresponding zoning map.

The paper is organized as follows. Chapter 2 gives an overview of the related work in 3D urban modelling and a short overview of the generative modelling language. The conceptual mapping of the legally binding zoning map elements to GML is explained in chapter 3. The following chapter 4 gives some details on the implementation of this concept based on the CityServer3D framework.The paper ends with some conclusions in chapter 5.

23.2 Related Works

Traditionally, a 3D city model is created using cadastral information, aerial images and airborne laser scanning. With cadastral data only, it is possible to automatically create a block model by extruding ground plans. Building height can be estimated by available attribute data such as number of storey and average height of a storey. In addition, semi-automatic analysis of aerial images and/or laser scan data leads to roof structures with exact building heights and detailed roof shape [2]. Using these techniques, the resulting 3D city model represents the real city as it is today, or more precisely as it was at the point of time when the data capturing was done. The main aim of a 3D urban model in many german cities is a more or less accurate representation of the current building geometry and related urban features like vegetation, street furniture et al. These 3D urban models will provide a 3D digital snapsot of a current situation of a city and give fundamental information for environmental simulations, planning scenarios and navigation purposes.

On the other hand, in computer graphics procedural techniques have been developed to create 3D city models of imaginary cities. These models are used for entertainment purposes as well as cultural heritage. For example, [9] used a procedural approach to create a 3D model of the ancient city of Pompeii as well as future suburbia models. In this case, the city model is generated by a formal grammar to generate streets, buildings and façade textures. [1] combine 3D city models created by extruded ground plans and procedural techniques to refine the building geometry and to generate façade textures.

While procedural techniques have the potential to create the shape of a future city based on zoning map regulations and constraints, it will result in just one city model. However, the zoning map does not precisely define the shape of a building but just gives some constraints. A city model representing a zoning map should be able to express these degrees of freedom and possible variations.

23.2.1 Generative Modelling Language

The GML is a stack-based, interpreted programming language which is very similar to Adobe's PostScript, but without any of the 2D layout operators. Its main purpose is to serve as a low-level language for the description of procedural shapes. The GML differs from other low-level shape representations in that a shape is understood as the result of a *sequence of operations* instead of just a bunch of geometric primitives like triangles [4].

An absolute advantage of GML is Gizmo. A Gizmo is a 3D object that is artificially put into a scene to represent an operation, such as a parameter change or a switch. The Gizmo can be added to 3D-objects, but it doesn't belong to the object. By using Gizmos, an object can be interactively changed. Two Gizmos which are mainly used within our work for modelling the dynamic buildings are *linear slider* and *switch ball.* These two Gizmos have the same set-up process, i.e., can be initialised by three call back functions (*click*, *drag* and *release*).

```
:dict :sliders 5.0 /std8hi
{ A A B midpoint 2pt (1.5,0,0) }        // Click
{ /A slider-point def model-update } // Drag
{ }                                     // Release
Gizmos.slider-linear
```

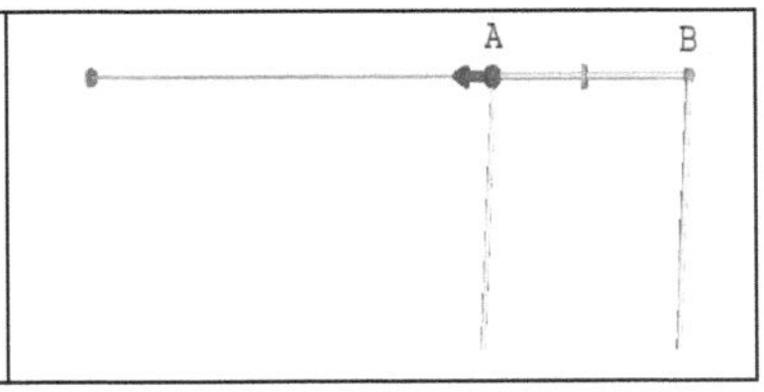

Fig. 23.1. Example of the linear slider

A linear slider is defined by three vectors: the first vector represents the point, which can be moved, the second vector is the minimal point and the last vector sets the direction and length. Within the *drag* function, the new value of the variable is defined by the variable slider-point, which represents always the actual value. Fig. 23.1 shows a linear slider which enables the point A to be moved from the midpoint of A and B on a vector with length 1.5 in the x-direction. The point A will be redefined every time in the drag-function. The command model-update initiates a new rendering of the model.

The switch ball is similar to the linear slider, but the *release* function is more important. Within the *click* function only one value must be defined, i.e., the point where the switch ball is. The *release* function contains what will be happened when the ball was clicked. Fig. 23.2 shows an example of the switch ball, in which the roof type changes when the switch-ball is clicked.

:hausdict :sliders 20 /std9hi { A } { } { /dachform dachform 1 add def model-update } Probado-Demo.Tools. Gizmos.switch-ball	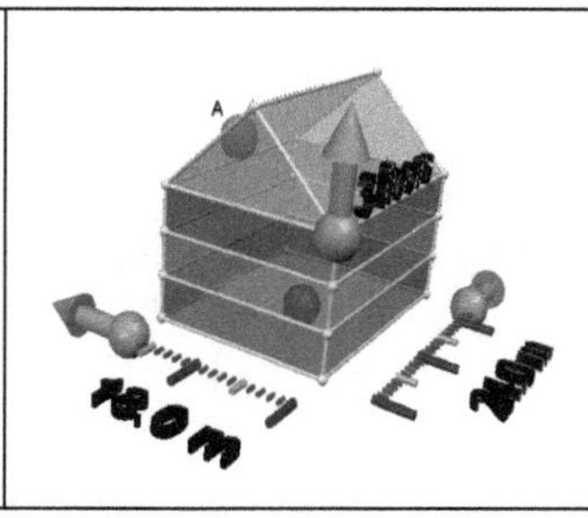	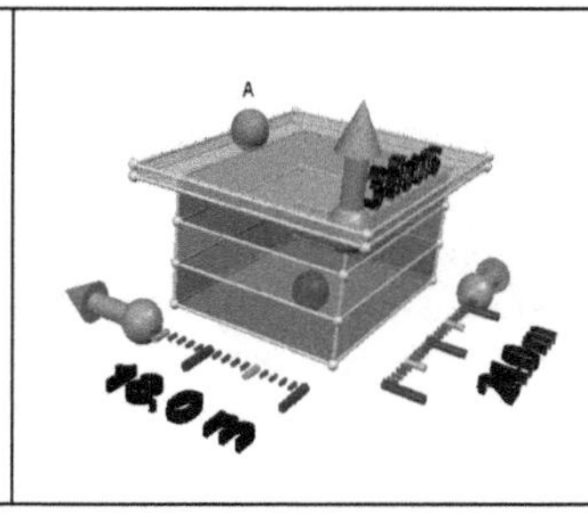

Fig. 23.2. Example of the switch ball.

23.3 Constraint-based modelling of buildings

In Germany the zoning map is legally binding and a fundamental step in a planning process within a build environment. It does not define a specific building geometry but some rules and restrictions such as the maximum height of a building, roof shape, and maximum size of the building area. Two fundamental numbers are the site occupancy index (GRZ) and the floor-space index (GFZ). GRZ defines an upper bound for the ratio of building area GR and plot area G:

$$\frac{GR}{G} \leq GRZ \quad (1)$$

The height of a building is limited by the floor-space index GFZ. It defines the ratio of plot area G and the building area GR times the number of full storeys #VG:

$$\frac{GR \cdot \#VG}{G} \leq GFZ \quad (2)$$

In addition, the zoning map defines a so called Baufenster, which is the part of the plot area where the building has to be built in. The building should not overlap this Baufenster.

Based on the zoning map, the variations of a building can not be described by a fixed geometry but an infinite set of geometries that do not validate the rules given in the zoning map.

23.3.1 Parameters and constraints for a building model

A simple building with rectangular planar ground plan is given by a 6-tupel (P_g, w_g, l_g, h_g, r) where

- P_g defines a reference point, for example the location of the lower left corner of the ground plan,
- w_g and l_g define the length and the width of the ground plan,
- h_g is the height of the building (excluding roof),
- r is the roof.

The roof itself is given by a set of parameters depending on the roof style s_r. Currently, flat, gable and lean-to roof are supported:

$$s_r \in \{FLAT, GABLE, LEAN\} \tag{3}$$

Gable and lean-to roof are specified by the roof pitch α, the height of the roof h_r, and the crown direction d_r. In case of a gable roof, the crown direction is parallel to the width or length side of the building. For a lean-to roof, the crown direction is one of the four boundary lines of the ground plan:

$$r = \begin{cases} (s_r, h_r), & s_r \in \{FLAT\} \\ (s_r, d_r, \alpha_{\min}, \alpha_{\max}), & s_r \in \{GABLE, LEAN\} \end{cases} \tag{4}$$

Under the assumption that both the Ground plan and the Baufenster are rectangles, w_g and l_g define the length and the width of the ground plan, and w_b and l_b length and width of the Baufenster.

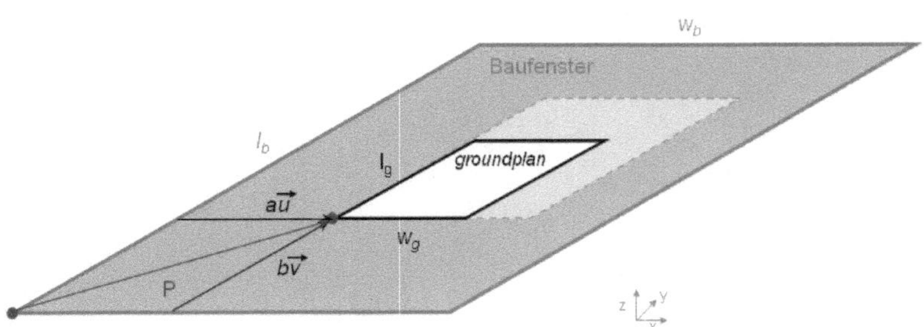

Fig. 23.3. The Ground floor has to be completely inside the Baufenster as defined in the zoning map.

The lower left corner of the ground floor geometry is given by a vector $\vec{p}$ indicating the translation of the lower of corner of the Baufenster.

$$\vec{p} = a\vec{u} + b\vec{v} \tag{5}$$

To ensure that the ground floor geometry is completely within the Baufenster, the following equations have to be fulfilled:

$$\begin{aligned} &a \geq 0 \\ &b \geq 0 \\ &|a\vec{u}| + w_g \leq w_b \\ &|b\vec{v}| + l_g \leq l_b \end{aligned} \tag{6}$$

The values w_b and l_b are given by the Baufenster, and a, b, w_g and l_g are variables. In addition, length and width of the ground plan rectangle have to be within the given limits. Obviously, the length and width of the ground plan have to be within an interval $l_g \in [l_{\min}, l_{\max}]$ and $w_g \in [w_{\min}, w_{\max}]$. In addition, the given parameter GRZ defines the maximal size of the ground floor:

$$GR = l_g \cdot w_g \leq GR_{\max} = GRZ \cdot G \tag{7}$$

It follows directly that

$$\begin{aligned} l_{\max} &= GR_{\max} / w_g \\ w_{\max} &= GR_{\max} / l_g \end{aligned} \tag{8}$$

The interval for the height of a building is usually defined by the GFZ and a given height HVG of one storey:

$$HVG \leq h_g \leq \left\lfloor \frac{GFZ}{GRZ} \right\rfloor \cdot HVG \tag{9}$$

The height of a roof depends on its style. For a flat roof, h_r is usually 0. In case of a gable and lean-to roof, h_r is limited by the specified minimal and maximal roof pitch. In case of a gable and lean-to roof the interval for the roof height is calculated by the following formula (see also Fig. 23.2). t is the length or the width of the ground plan depending on the crown orientation.

Gable roof: $$\frac{t}{2}\tan(\alpha_{\min}) \le h_r \le \frac{t}{2}\tan(\alpha_{\max}) \qquad (10)$$

Lean-to roof: $$t\tan(\alpha_{\min}) \le h_r \le t\tan(\alpha_{\max}) \qquad (11)$$

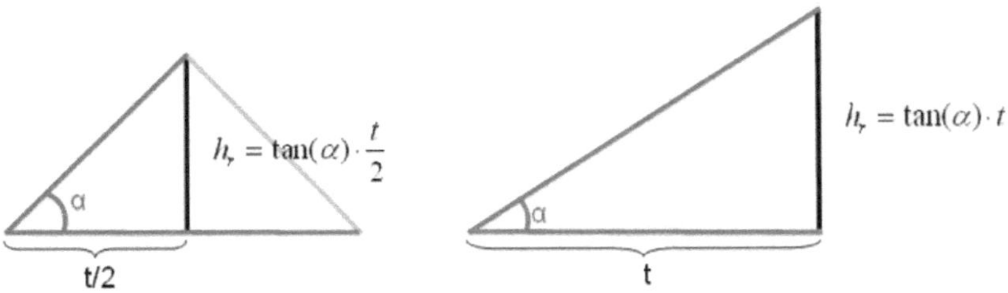

Fig. 23.4. height of the roof is calculated by the roof pitch and roof shape (left gable roof, right lean-to roof). The maximal value of roof pitch is given by the zoning map.

23.3.2 Mapping building constraints to GML

As discussed above, a building is described by a set of parameters and a set of constraints. For visualization purposes, random values for the parameters are choosen within the given constraints. However, this generates only one possible urban model out of an infinite number of possible models that fulfil the restrictions given by the defined constraints. To be able to explore the full model space, the possible variations of the given parameters are mapped to linear sliders and switch-balls.

In GML, possible roof shapes are defined in an array /roofshapes that contains the three values [/flatroof, /gableroof, /leantoroof] for each building. The variable roofshape is defined as an index of this array. A mouse click on the switch-ball selects the new roofshape by simply adding one to the array index roofshape.

{/roofshape roofshape 1 add def model-update }

As a remark, add always calculates modulo size of the array to ensure that the array index is not out of bounds.

Similar, the crown direction is mapped to a switch-ball interaction. It allows switching between 2 crown directions in case of a gable roof and four directions if the roof style is a lean-to roof. In case of the gable and lean-to roof, the interval of the height of the roof is mapped to a linear slider. In principle it is possible to map the roof pitch interval directly, but it is easier for the user to change the height directly with a linear slider. The roof pitch constraints are checked interactively by the slider, as Fig. 23.6 shows.

Fig. 23.5. Interactive variations of the roof shape by incrementing the array index roofshape.

```
% gableroof
 :rooftype /gableroof eq {
   % crown orientation
   :crownorientation /sidea eq {
     midsidea 0 0 roofheight vector3 add
     midsidea 0 0 alphamin tan width 0.5 mul mul vector3
add
     0 0 alphamax tan width 0.5 mul mul vector3
     0 0 alphamin tan width 0.5 mul mul vector3 sub
   } if
   :crownorientation /sideb eq {
     ...
   } if
 } if
```

$$midsidea + \frac{w_g}{2} \tan(\alpha_{\min})$$

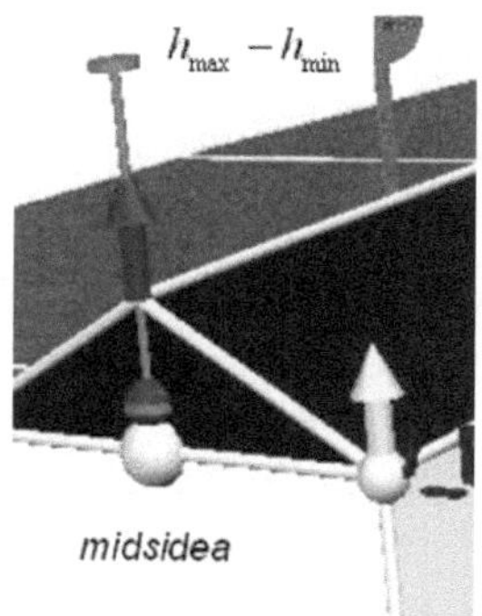

Fig. 23.6. Interactive checks of the roof pitch constraints by moving the slider.

The position, width, length and height of a building are mapped to linear sliders as well.

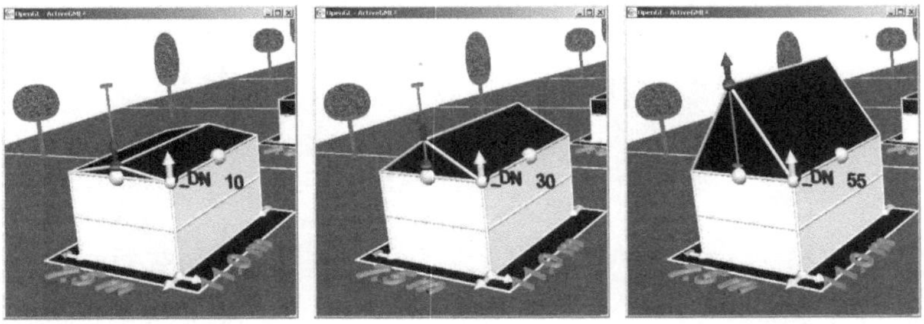

Fig. 23.7. Interactively changing of height of the roof within the given constraints of the zoning map. The roof pitch varies from 10 to max. degree in this example.

23.4 Implementation

In order to experience a 'proof-of-concept' for the concept proposed we took a technical realization. This section gives a detailed description about the architecture as well as the internals of the implementation.

As Fig. 23.8 illustrates the whole processes for the generation and visualization of the 3D-zoning map are consists basically of the four modules – import of XPlanGML, calculation of the valid boundaries based on the given constraints, export of CityGML to GML and visualization of the 3D-zoning map in the GML viewer.

To import the 2D-zoning maps and to calculate the valid ranges of the building geometry, which are the relevant parts for generating the 3D-zoning map, we use the CityServer3D [8] as implementation platform. The main reason is simplicity of implementation and of its extension, as well as possibility for an easy integration with already existing 3D models of the surrounding area as well as for planning review within the surrounding area. Additionally, the modules implemented within this work can also be used as extended modules in the CityServer3D for the generation of 3D models from the zoning maps in the XPlanGML format.

The most important requirement for importing the zoning map is to extract essential information needed for defining the (target) buildings, on the one hand, and to enable the user to change the geometry and location of the buildings within the given constraints of the corresponding zoning map. Taking these requirements into account, the import module (*importer*) reads given zoning maps of the XPlanGML format in and parses the individual building tags. It then extracts relevant elements and regulations and assigns them to the corresponding objects within the CityGML structure.

Fig. 23.9 shows a screenshot of the web-based user interface on which the user can configure the scope of the surrounding environment as well as the attributes and location of the buildings to be imported.

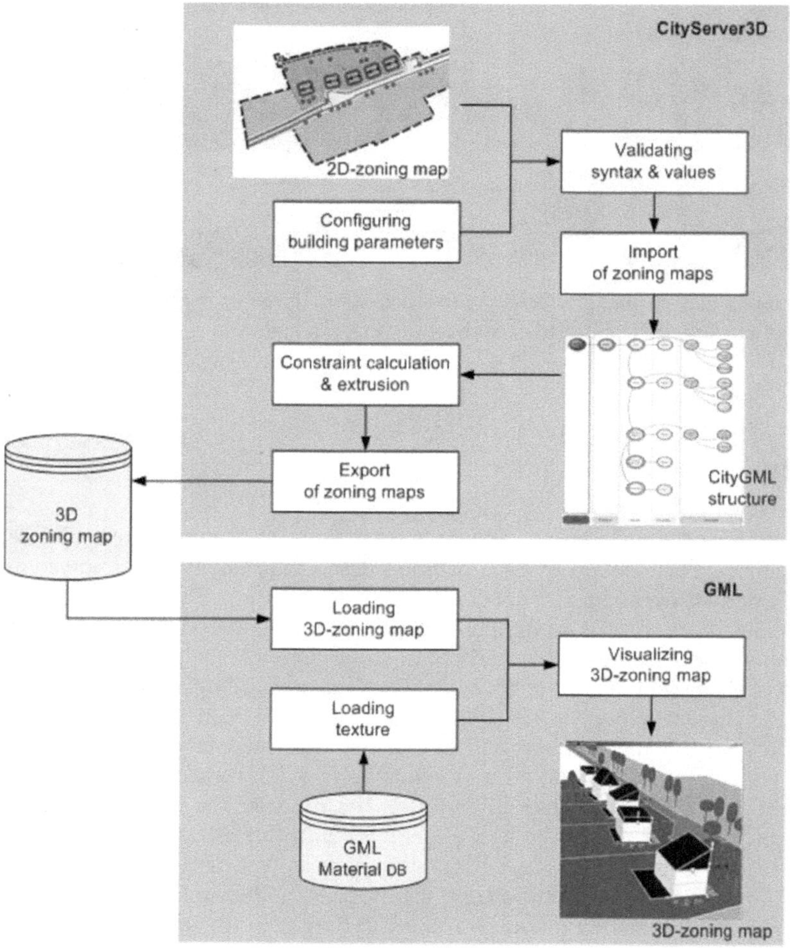

Fig. 23.8. Interactive checks of the roof pitch constraints by moving the slider.

After successful validation of the input file's syntax, the importer creates a parser (*TagParser*) which reads the single tags of each building from the XPlanGML file and creates a corresponding CityGML structure containing layer, feature, model and property as its elementary components.

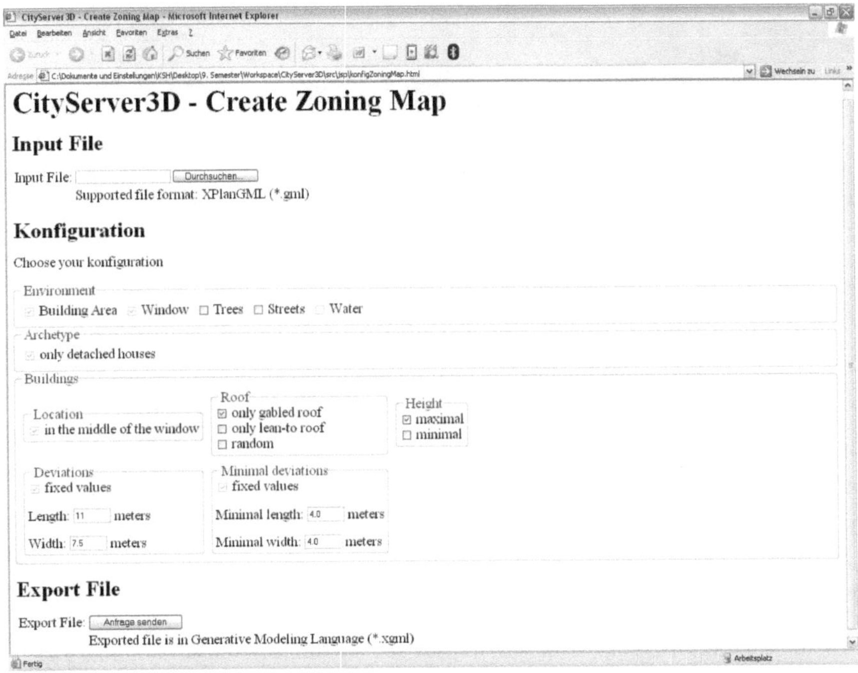

Fig. 23.9. Configuration window for setting values required by the import.

Calculation of the valid constraints is performed on the imported zoning map of the CityGML structure. To do this, the system uses the selected configuration given through the user interface, too. In the first step of the calculation, existence of a generic object as well as the fulfilment of the regulation, e.g., the validity of the estate statements is tested. In the case of an invalid statement, a correction process takes place. The system then calculates the variation of the minimal, maximal, and those for visualization, as well as the possible locations of the building. It then calculates the height of the building, which consists of the number of storey, the roof height and the given configuration values. The maximal number of storey can be validated by deploying the floor-space index, GFZ. To fix the minimal height of the building for visualization, the minimal number of the storey as well as the minimal roof pitch value is used.

The extruded 3D-zoning map is then exported into a GML (*.xgml) structure for a visualization in the GML. The in CityServer3D implemented export module contains a GML library for the 3D-zoning map, 3D-BPlan lib, which includes the required modules and operations for the visualization of the 3D-zoning map in the GML. This 3D-BPlan lib is implemented, as a user library, based on the Probado-Demo.genmod [4]. The essentials of the 3D-Bplan lib are the operations for explicit handling of the Baufenster, the ground floor and of the calculated values of the buildings. Additionally, it contains a module for handling the variations within the lean-to roof style. Fig. 23.10 illustrates how a lean-to roof can be developed.

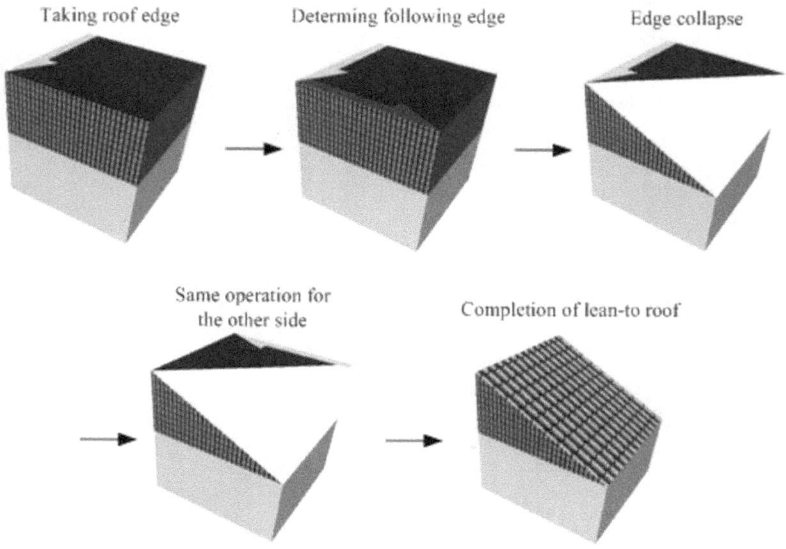

Fig. 23.10. Steps for developing the lean-to roof by using the 3D-BPlan lib.

As additional functionalities, the 3D-Bplan lib enables the user to change the building geometry such as its height or the roof pitch by using the linear slider and to make a dynamic positioning of the buildings, of course within the acceptable Baufenster. In comparison to the Probado-Demo.genmod, the 3D-Bplan lib can also handle multiple buildings in parallel, which is a mandatory function for the management and visualization of the zoning maps. Fig. 23.11 gives a screenshot of the 3D visualization of an example zoning map. The trees and streets are textured with images of the GML's material database (a folder) which is modified within this work.

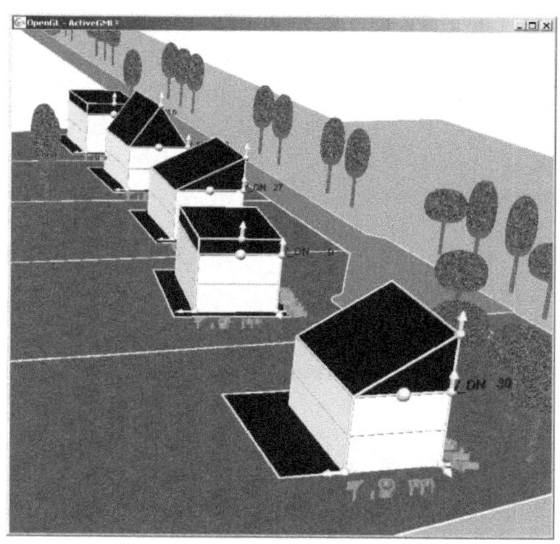

Fig. 23.11. 3D visualization of a 3D-zoning map. The geometry of each building such as the location, size, roof style and roof height can be changed interactively by moving the linear slide and clicking the switch ball.

23.5 Conclusion

In this paper, we proposed a concept for generating and visualizing 3D buildings which can interactively be changed by the user within the given constraints. The variants of those buildings are acquired from the input 2D-zoning map based on XPlanGML, which does not define a specific building geometry but instead some rules and restrictions such as the maximum height of the building, roof shape and maximum size of the building area.

To show the degree of the freedom, i.e., to be enable to explore the full model space within the zoning map regulations, we used the generative 3D modelling language, GML, especially its two Gizmos - linear slider and switch ball. Thus, for a given building, the possible variations of the building parameters are mapped to the linear sliders and switch-balls.

The usability and practicality of the proposed concept is shown through a technical realization where the system is implemented as extended modules in a well established 3D GIS, CityServer3D, as well as in the GML.

Even though the addressed build types are of rectangular planar groundplan, the resulting city models give us the possibility to change the building geometry and location within the given constraints. Furthermore, the proposed concept gives us a good possibility for visualizing urban plan data which are defined 'uncertainly', i.e., without any fixed geometry.

23.6 Acknowledgement

The authors would like to thank Daniel Holweg, Bentley Systems Germany GmbH and Thomas Eichhorn, Planning Department City of Coburg for very inspiring discussions. The XPlan data was provided by the City of Coburg.

References

1. Döllner, J., H. Buchholz, F. Brodersen, T. Glander, S. Jütterschenke, and A. Klimetschek (2005): Smart Buildings – A concept for ad-hoc creation and refinement of 3D building models. In: Gröger, Gerhard / Kolbe, Thomas H. (Hg.): Proceedings of the 1st intern. ISPRS/EuroSDR/DGPF-Workshop on Next Generation 3D City Models. Bonn 2005.
2. Förstner, W. (1999) 3D City Models: Automatic and Semiautomatic Aquisition Methods. Proceedings Photogrammetric Week, University Stuttgart, pp 291- 303.
3. Gröger, G., T. Kolbe, A. Cerwinski, and C. Nagel (Eds.) (2008) OpenGIS City Geography Markup Language (CityGML) Implementation Specification, Version 1.0, OGC project document 08-007, 20.01.2008.
4. Havemann, S. (2005) Generative Mesh Modeling, Technische Universität Braunschweig, Germany, Dissertation, 2005.
5. Knapp, S. and Y. Chen (2007) The impact of ePlanning Systems on public participation in planning processes, In: Planning for the Risk Society – dealing with

uncertainty, challenging the future, AESOP Conference 10. – 14.07.2007, Neapel July 2007.

6. Knapp, S., and V. Coors: The use of eParticipation systems in public participation: The VEPs example, In: Coors, V., M. Rumor, E.M.Fendel, and S. Zlatanova: Urban and Regional Data Management - UDMS Annual 2007, Taylor & Francis, 2008, pp 93-104
7. Knapp, S., Chen, Y., Hamilton, A., and Coors, V.: An e-Planning case study in Stuttgart using OPPA-3D, In: Handbook of Research on Strategies for Local E-Government Adoption and Implementation: Comparative Studies, Chapter 13, IGI Global, March 2009 (in print)
8. Haist, Jörg; Coors, Volker:The W3DS-Interface of Cityserver3D.In: Kolbe, Gröger (Ed.); European Spatial Data Research (EuroSDR) u.a.:Next Generation 3D City Models. Workshop Papers : Participant's Edition.2005, pp. 63-67
9. Müller, P., P. Wonka, S. Haegler, A. Ulmer and L. Van Gool (2006) *Procedural Modeling of Buildings*. In Proceedings of ACM SIGGRAPH 2006 / ACM Transactions on Graphics (TOG), ACM Press, Vol. 25, No. 3, pages 614-623.

Chapter 24
GeoVEs as Tools to Communicate in Urban Projects: Requirements for Functionality and Visualization

Mahmud Shahrear Kibria, Sisi Zlatanova, Laure Itard and Machiel van Dorst

Abstract. Urban planning is a complex decision-making process involving a large number of actors who interact intensively. Such groups often have conflicting expectations and backgrounds. Therefore, consultation and interaction is vital for the success of urban projects. A Geo-Virtual Environment (geoVE) can play an important role as a communication tool in the field of spatial planning, but such tools are still in limited use. We investigate the requirements for visualization in urban planning by analyzing user perceptions of visual materials and their needs for interaction in the different urban phases. The study is completed with the cooperation of several large municipalities in the Netherlands.

24.1 Introduction

Urban planning is the process of shaping and organizing the real world. [10] define it as follows: 'Urban planning is concerned with assembling and shaping the urban- i.e., local or municipal- environment by deciding about the composition.' Urban design acts as the interface process of design between urban planning and architecture dealing with shape and form of the geographical urban objects and the quality of the created space. Several researchers have studied how 3D visualization can be used in design processes [3, 1, 18] through the use of hand-drawn sketches, 3D CAD, GIS and VR. In urban planning, virtual 3D models can be used to visualize a designed area in the context of the existing situation to estimate the impact of planned changes. 3D models in geo-virtual environments (geoVEs) can be used as interaction tools for the

ESRI Nederland B. V.
mkibria@esri.nl
Delft University of Technology
S.Zlatanova@tudelft.nl

designers to communicate new ideas to involved actors and to minimize misunderstanding in urban renewal projects.

This paper presents our study on the use of geoVEs in the different phases of urban planning. Section 24.2 presents the taxonomy of visualization materials and functionalities of interest for urban planning. Section 24.3 presents the case study and discusses the methodology. Section 24.4 analyses the results and draws logical conclusion on visualization requirements for urban planning. Section 24.5 compares several geoVEs that might be of interest for municipalities in performing urban tasks. Section 24.6 summarises the most important findings.

24.2 Taxonomy of geoVE functionalities & visualization materials for interaction in urban planning

A Geo-Virtual Environment is broadly defined as a spatially referenced digital world that comprises visual (and non-visual) objects in an immersive and interactive 3D scene to represent and mimic reality through dynamic real-time simulation. [19] mention that 3D VR models provide flexibility in interaction and exploration. These models include active and passive interaction functionalities for users.

24.2.1 Functionalities of geoVE

There have been several attempts to define the functionalities of geoVEs. [6] defined the so-called I-factors of Immersion, Interactivity and Information Intensity (or Levels of Detail, LoD). [11] added the I-factor of Intelligence (of Objects). [16, 9] tried to extend this classification. Adopting some of these developments, we introduce a classification of geoVEs for urban planning with respect to *construction, capabilities, experience, controlling, interacting, exploring* and *components.*

24.2.1.1 Construction:

This functionality refers to the system architecture of a geoVE and the type of data used to build a 3D scene. The construction of the 3D scene is inseparable from the scale and resolution of the 3D data. These are often defined as Levels of Details (LoD). In urban planning, LoD can also be used to create different 'visual materials' (see Section 24.2.2). Different types of data can be integrated in a 3D scene through a data-fusion. The data for the 3D scene can be stored in Relational Database Management Systems (RDBMS) or other databases and visualized 'on-the-fly' in the geoVE.

24.2.1.2 Capabilities:

The capability domain consists of functionalities related to the interface of the geoVE; it enables users to interact with the interface, 3D models and attribute information. The foremost capability of the geoVE is representation and rendering capability. Visual representation deals with visual contents like color, texture, shape, rendering and geometry.

The same model can be visualized in various representations, by changing the line types, thicknesses, transparencies, colours and textures of the model. Fig. 24.1. shows the '*Poptahof* Urban Design' project in different representations keeping the same geometry. There can also be multiple representations where different 3D models (containing different geometries) are used to represent different design solutions on a single site. Fig. 24.2. illustrates the study of alternative geometry models of *Poptahof* Urban Project, Delft using same representation in geoVE to avoid bias.

Multi-dimensionality refers to the possibility that the objects in a geoVE can be visualized as text, 1D points, 2D images, graphs, maps and 3D models. Muti-layering enables various layers to be added. The capabilities of simulations trigger change in the 3D-scene by pre-defined algorithms. Animations are pre-recorded simulations usually used in presentation.

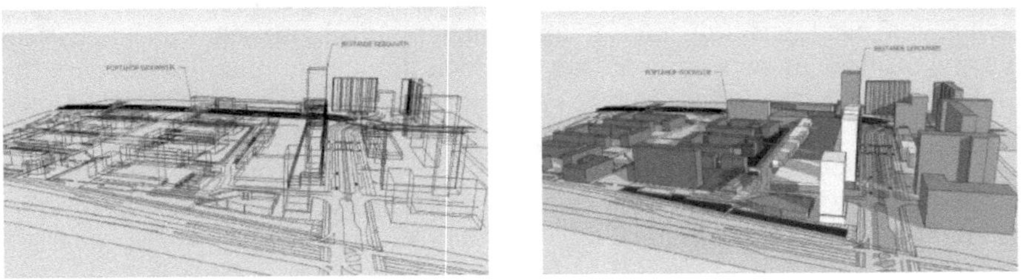

Fig. 24.1. Different representations of the *Poptahof* Urban Renewal Project from the same data (wire frame, coloured block models).

Fig. 24.2. Alternative design solutions with different geometries using the same rendering representation of the *Poptahof* Urban Renewal Project.

24.2.1.3 Experiencing:

[15, 14, 9, 5, 13] have extensively discussed the experiencing functionalities as the possibilities of a user to interact with a 3D model. We classify experiencing into passive and active. The active form relates to selection, manipulation, elaboration and exploration. The passive form of experiencing is when the user changes position in the scene but does not interact with objects.

Aerial view	Elevated oblique view	Eye level experience
City of Delft	*TU Delft Campus*	*Virtueel Apeldoorn*
Higher than 200m	200m to 2m	1.8m to 1m

Fig. 24.3. User viewpoint and experiences in an immersive environment.

Immersion in a geoVE is related to user experience and can be described as a 'psychological state characterized by perceiving and experiencing oneself to be enveloped by, included in, and interacting with an environment' [17]. Immersion can be physical and mental. Mental immersion is a state in which the human mind experiences the presence of being in the virtual world. The immersive environment of a geoVE can have three views: aerial, elevated oblique and eye level (Fig. 24.3). For urban planning, the aerial view gives an overall perception of the urban project, while an elevated oblique view only focuses on exterior forms. The eye level experience reveals architectural and interior details.

24.2.1.4 Controlling:

Controlling functionalities relate to the user's grasp on the controls of the desktop geoVE. Selection is the primary functionality of controlling. Without the ability to select data, control is not possible. [14, 15, 5] defined the controlling functionalities of a 3D scene (see Table 24.1). Fig. 24.4 shows the instruction on keyboard / mouse control on Cebra's VirtuoCity (*Virtueel Tilburg*).

Table 24.1. Controlling functionalities of the 3D scene.

Controlling geoVE	Control types	Selection of
Control of objects	Keyboard/mouse: select, click, identify.	3D objects
Control of 3D scene	Virtual controls	3D scene
	Direct-user control (in AR)	
	On-display control (menu)	

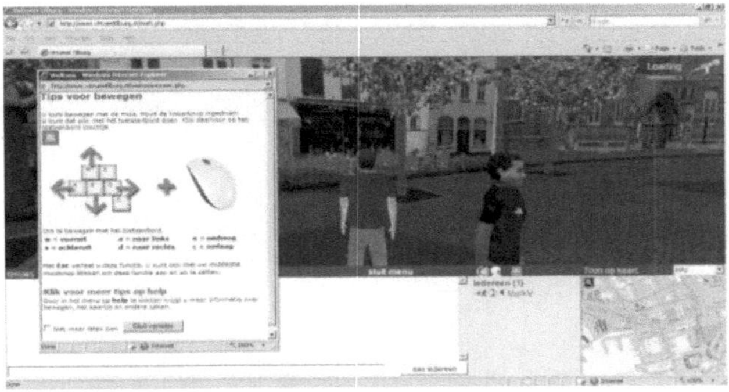

Fig. 24.4. Keyboard /mouse controls in the multi-user interface VirtuoCity.

24.2.1.5 Interacting:

[6] defines interactivity as the functionality that describes the user's ability to change the viewpoint. The most important functionality of interactivity is data manipulation. This means that an interactive environment lets the user change the object's physical characteristics (e.g., location, shape, size, colour, attributes).

[16] suggested three extra factors for the domains of geoVEs: Automated Agents, Selection and Augmented Reality. We have reclassified selection in the controlling domain. An avatar is the user's physical projection in the digital world; it gives the choice to represent oneself as one's gender, age, personality, and more to interact with other users.

[4] described five levels of interaction with users: informing, consulting, advising, co-producing and co-deciding. These last two involve collaboration. Consulting the user's email, newsgroups and weblogs are means of consulting, while tele-meeting, video-conferencing and Internet sites are means of advising. Chat-rooms help co-produce decisions and electronic voting on the planning projects lets users co-decide for or against a proposed design. These techniques were omitted from our study, as they imply different forms of communication.

24.2.1.6 Exploring:

In this research, exploration is defined as an interactive method to explore the information behind the visualized 3D data [8]. The functionalities shown in Table 24.2 can aid exploration.

Table 24.2. Exploration functionalities for a Virtual environment.

Function	Description
Data Query	Accessing spatial database, filter-out qualitative/ quantitative information (Fig. 24.5)
Dynamically linked windows/views	3D objects can be connected to external sources for additional information like website, sound, video, text, or images.

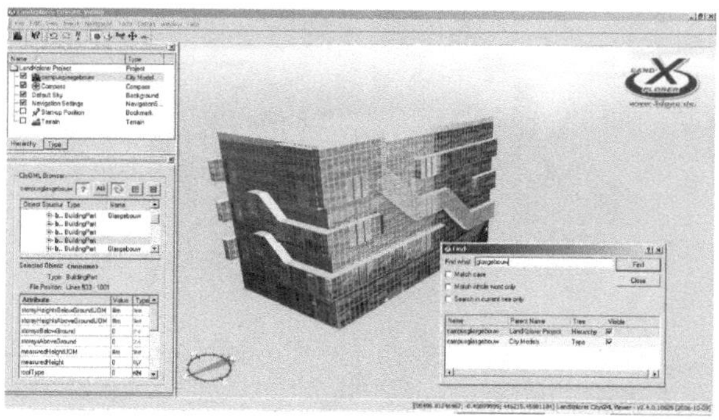

Fig. 24.5. Exploring a TU Delft building in LandXplorer, CityGML.

24.2.1.7 Components:

The components of the 3D scene are the attributes of the model, like the boundary, background, geometry (simple and complex), lighting and shading, and toolboxes. There can be also analysis components that relate directly to tools such as sun-path and shadow calculation, wind-analysis, noise and air pollution mapping, and cellular automata.

These seven functionality domains listed above are inter-related but give a background for evaluating geoVEs for urban planning [7]. Fig. 24.6 represents the hierarchical relationships. At the top of the hierarchy of the geoVE is the domain of construction of the 3D scene. The construction domain defines the capabilities of the 3D scene. These two domains are not directly related to user interaction. Experiencing is classified as the passive form of interaction that deals with simple visualization. User interaction with the 3D models begins with controlling. The interacting functionalities determine the active form of interaction in the geoVE. Exploration functionalities deal with the information behind the visualized 3D scene. Components are specific tools that enable these functionality domains to operate.

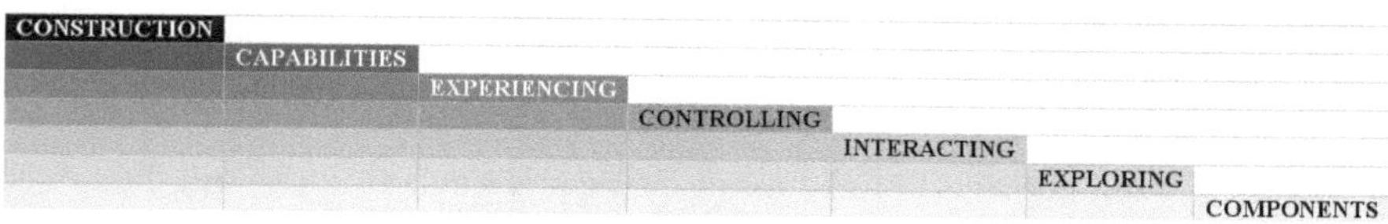

Fig. 24.6. Hierarchy of functionality domains of a geoVE.

24.2.2 Visual materials of geoVEs for urban design and planning

GeoVEs for urban planning require adequate definition visual materials to represent urban objects, including 3D models, plans, images, and textual information. A continuous line proceeding from a highly verisimilar to a highly abstract representation provides the *realism axis* as defined by [12]. The visual representation of reality can be mapped onto this axis falls somewhere between reality and abstraction. On this axis, verisimilar representations show realistic individually identifiable representations through detailed complex models. Indexed representations map the value of any concept according to classes and hierarchies. Iconic representations are naïve forms or representations that emphasize the basic true form of the object. The symbolic representation is the 2D depiction of any concept of the human mind, while finally, textual information provides sensation from the reader's memory on the essence of the object.

Visual materials classified as graphical consist of texts, graphs and images, and model representation can be 2D and 3D (Table 24.3.). The graphical representation of objects, usually multi-resolution model representations (both 2D map and 3D model), has scale and topology.

Table 24.3. Models and graphical representations.

Type	Name	
Graphical representation	Textual information (1D)	
	Images (static, dynamic) and Graphs (2D)	
Model representation	2D models	Plans & Maps (2D)
	3D models	Block models (2.5D) to textured 3D models.

24.2.2.1 Visual materials: scenic languages CityGML, KML and X3D

Internet, together with the influence of the 'game industry,' has changed the realm of visualization through Web3D's 3D modeling language (VRML/X3D). In the past, CAD models were generated on desktop computers, but VRML models came as the solution to online CAD [2]. Computer Aided Design (CAD) systems (e.g., Autodesk AutoCAD, Microstation of Bentley Systems, 3D Max, SketchUp) have been the media for planners and designers to interact with 2D/3D models for decades. However, they lack semantics and extended visualization and interaction tools. KML (Keyhole Markup Language) is a relatively new XML data-format that defines the viewing of visual objects on Google's virtual earth terrain through properties like placemarks, paths, raster images, polygons, and attributes information; it describes different LoDs and realism. While they offer a lot in graphics functionalities to create photorealistic models, they are weak in providing means for 3D geo-database object storing and representing semantics.

CityGML is a geographic information model that can be an exchange format for virtual city and regional models for municipalities based on XML (GML for geometry) based markup language. The goal of this data format is to have a rich source structure to store and exchange 3D information.

Fig. 24.7 shows 3D models in LoD1 to LoD4 in CityGML schema. In visualization terms, the LoD1 model is the volumetric representation of the building. The LoD2 represents buildings in volume with coarse details. In terms of design, LoD3 can be seen as a detailed 3D architectural model containing exterior architectural aspects. LoD4 from the outside looks like LoD3 models but contains a walkable interior space.

LoD 1	LoD 2	LoD 3	LoD 4
Block Model and no roofing	Models with roofing	Detailed textured model with roof	Detail models with interiors

Fig. 24.7. Levels of Detail (LoD) in 3D models.

Clearly, it is difficult to give a single, complete definition of visual materials for geoVEs. Through extensive study, we have defined six types of visualization materials. The visual materials are first classified as graphical (*text, graphs & images*) and model representations (2D and 3D). The model representation is further sub-divided into symbolic *'2D plan/map'* and different 3D LoD models, such as *'Volumetric block models'* for iconic LoD1, *'Volumetric envelope models'* for indexed LoD2, *'Detailed architectural models'* for indexed-verisimilar LoD3 and finally, *'Detailed models with interior'* for verisimilar LoD4 models.

24.3 Case study: *Poptahof* in Delft

The urban renewal project of *Poptahof* in the Delft municipality was taken as a case study to investigate the requirements of visualization with respect to the defined taxonomy. This urban project has provided information on the current use of tools to communicate urban projects, which helped in preparing questionnaires offering new geoVE tools.

24.3.1 Traditional approach

In the traditional method of interaction of planning, architects and planners in the *Poptahof* urban renewal project used pictures and photographs to encourage discussion. The designers found that if such reference pictures are too detailed, they seem to deter the actual purpose of the visual materials. Actors like future inhabitants inspecting agencies emphasized the characteristics that they recognize from memory, like choice of material and colour instead of relevant issues like volumetric study.

The design professionals mentioned that the external actors have difficulty translating two-dimensional maps and floor plans into three-dimensional images. Thus, they were unable to comprehend the impact of the variation of high- and low-rise structures when design was communicated in the 2D plan. The master plan for *Poptahof* was presented in 2D in a clear layout, but with abstract colours. However, these 2D maps/plans failed to transfer adequate information and were not intelligible to non-design professionals, especially when building forms were emphasized. A combination of 2D and 3D visualization was therefore necessary. 3D visualization improved the perception of the actors, but such visualization was confusing when block models were used in aerial viewpoint. Eye level experience in visualization helped the actors understand the details of the design. The case study revealed that the levels of detail of visual material are understood by actors as having a relationship with the phase/stage of the design project. Actors regularly interpreted visual materials with respect to their background knowledge. The case study of *Poptahof* helped us draw some requirements:

- Visualize information in multiple rendering, resolution, dimensionality.
- Visualize information to maximum number of actors without bias.
- Visualize comparison between present situation and designed situation.

Traditional low-tech tools cannot offer such functionalities; therefore, more high-tech computerized tools like geoVEs are needed to interact with actors.

24.3.2 3D models and Survey Population

To investigate the requirements for new geoVE tools, we generated a 3D virtual model of the city of Delft; the *Poptahof* urban project was inserted into this model. A 2D digital map and LIDAR height information helped reconstruct the city of Delft with the *Poptahof* urban area. The *Poptahof* CAD models were delivered by the Municipality of Delft and were geo-referenced inside the virtual city model. The models were created with four steps: creating landscape (DTM), creating block model from 2D footprints (LoD1), adapting CAD models (LoD2 and LoD3), and creating interior in CAD (LoD4).

Using geo-processing tools in ArcGIS 9.2 and Safe Software FME, the GIS data were prepared and 3D scenes were created in Flux Studio. AutoCAD, AutoCAD Architecture were used to detail the 3D models and FME was used to convert the geometry and geo-reference the models. Sketchup and 3DMax were used to texture surfaces. The data used in the *Poptahof* case study are a LIDAR topographic map (1:1000), a large-scale topographic map (1:1000) and CAD models of the *Poptahof* urban renewal project. [7] provides more details on the reconstruction procedure.

This model was tested in various geoVEs. Two survey questionnaires supported the tests applied in a field survey and a workshop.

The survey population for the two survey questionnaires consisted of urban planners, GIS experts, architects and planners working for municipalities and social housing companies, as seen in Table 24.4.

Table 24.4. Survey population of the research.

Urban planners from Dutch municipalities	17
GIS experts and 3D modellers	2
Communication officers and planners	5
Design professionals from social housing agencies	6
Total participants	30

24.4 Analysis

The following sections present the findings from the field survey and the workshop, organised with respect to required functionalities and visual materials. The functionalities are not classified with respect to the urban design phases since we believe that they must be available for all the actors in all the design phase. The approach to visual materials is different. The questionnaires were prepared so as to reflect the role of actor in different urban design phases.

24.4.1 Required functionalities in urban planning

Table 24.5 lists the results of the survey on required functionalities for geoVEs. The table is organised according to the taxonomy developed in Section 24.2.1. The actors have validated the listed requirements, which means that municipalities should address these functionality domains for interaction in the urban planning process when they develop their municipality interaction systems.

Table 24.5. Requirement functionalities of geoVEs for urban planning.

Domain	Functionality
Construction	Information Intensity (LoD) for 3D models
	Datasets should be stored in geo-database
	Data integration: open source solutions encouraged
	Data integration: multiple data format & interface
	Visualization through plug-in or software download
Capabilities	Multiple representations of model in various rendering
	Multiple representations of various geometry models
	Compare/statistically analyze old and new situations
	Animation and simulation for presentation & analysis
	Multi-dimensionality and multi-viewpoints
Experiencing	Minimal steps for movement and navigation
	Orientation with positioning, tracking and north arrow
	Immersion: mental semi-immersive environment
Controlling	Selection of scene/objects: click, highlight & select
	Desktop VR with user mode: multiple/single
Interacting	Ability to move & delete information
	Ability to modify, edit and change information
	Ability to add, copy and save 2D/3D data
	Ability to stop, hide, and censor information
	Real world information like interactive chat, webcams
	Automated agents (avatars) and intelligent objects
Exploring	Spatial query on geometry, semantics and attributes
	Hyper links and linked objects
	Multiple synchronized windows with multi-viewpoints
	Log-in interface for multiple users
	Measurement tool for dimensions, buffer, union, clip

Components	Adding labels & icons to highlight 3D objects
	Multi-layering by adding GIS data and map layers
	X-Ray Vision to reduce occlusion
	Geo-referencing and geo-coding components
	Lighting, camera, viewpoints change, tilt, range etc. Transparency and shading (on/off)
	Screenshots, save image and attributes, history keeping.
	Auto focus/tracking, teleportation, time component
	Basic drawing tools (mark and draw)
	Multiple layers of landmarks, nodes, zones, objects, etc.
	Reference points (home button, back and forward)
	Audio sound support, identify with attribute information
	Distance function from highlighted 3D object
	Velocity: fast, medium, slow & acceleration
	Gravity and collision (on/off) and boundary extent
	Vector/ raster data, atmosphere and visibility
	Analysis: sun-path, traffic and urban growth, etc.

24.4.2 Required visual materials for urban planning

Visual materials specified in Section 24.2.2 were investigated with respect to:

- Visual materials (in LoD) versus human perception of understanding.
- Visual materials (in LoD) versus the urban planning phases

The results are plotted in line graphs. Six visual materials form the x-axis (except for Fig. 24.8, where the x- axis is formed by perception levels) with three intervals indicating the 'degree of freedom' given to the participants. This enabled the participants to avoid crisp decisions and provided flexibility. The y-axis indicates the added maximum and minimum number of times the participants agreed with a certain decision. The graphs indicate the highest number of participants that agreed ('added agreeability') with the decision.

24.4.2.1 Relationship between visual materials in geoVEs versus human perception

The line graphs in Fig. 24.8 indicate there is a clear relationship between human cognition and understanding of design using visual materials in multiple representations, LoD and realism. This illustrates that solely textual and graphical information are inadequate to explain design. If the building is visualized in LoD1 models, the population perceives the design as a draft and fails to differentiate between various objects.

When the building is visualized in LoD2, the viewers focus on local details of the building design and think that the final design may be altered, concentrating on the outer looks like roofing and forms. When the same design is viewed in LoD3, the viewers perceive that the building will be fairly similar to the realized project. Visualizing LoD4 models with interior space implies that viewers perceive the design as final. This proves that human perception is largely enhanced when design is visualized at a higher LoD. From 2D plan/map to 3D LOD3 and LoD4 models, the human ability to perceive design increases. The result finally proves that higher levels of detail and realism in 3D models help viewers recognize and understand the area under observation with less difficulty compared to 2D maps and iconic block models (LoD1).

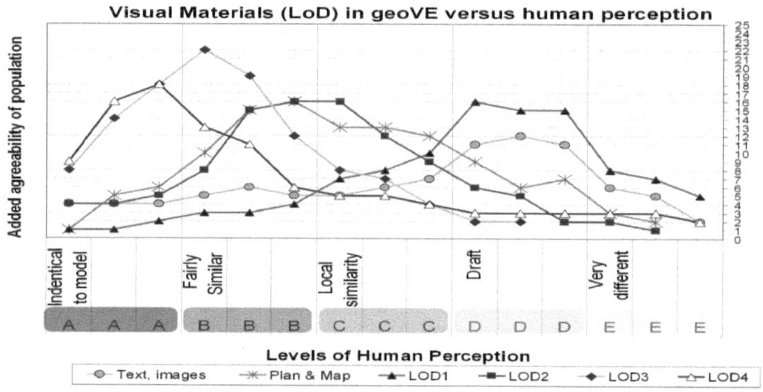

Fig. 24.8. Visual materials at various LoD and human levels of perception.

By analyzing the line graphs in Fig. 24.9, it is clear that there is a tendency for block models (LoD1) to confuse people the most, while LoD2 and LoD3 models are most suitable for design comparison. LoD2 and LoD3 models turned out to be navigation and orientation supportive.

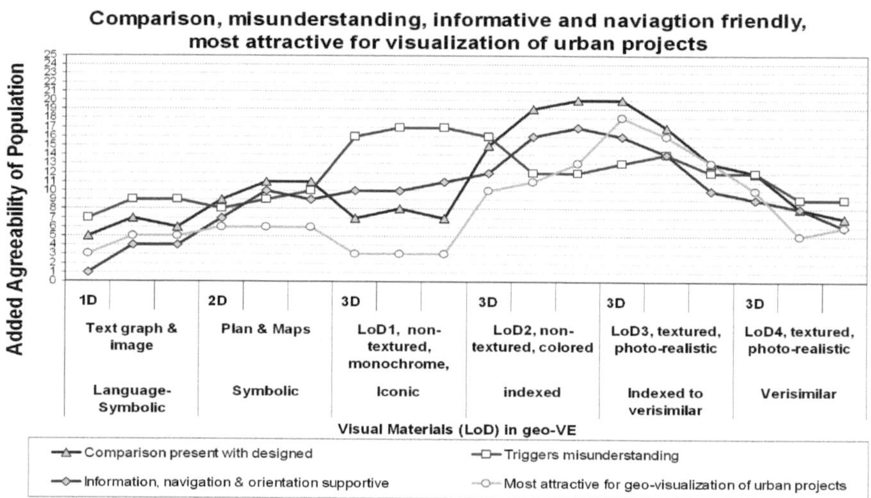

Fig. 24.9. Visual materials in various LoD and type of interaction in urban planning.

24.4.2.2 Relationship between visual materials and design phase

We studied the spatial planning phases in the Netherlands; only the major phases are mentioned here. In the field of spatial planning, a *zoning plan* is a juridical binding spatial plan in the Netherlands for permissible land use that covers parts of the municipality that vary from a large city-district to a single building block. It specifies regulations regarding what can be built and the height allowed. The *structure plan* reflects the global expectation and envisioned spatial developments for a municipality or parts of a municipality by defining the phasing of developments. A *master plan* is a large-scale comprehensive plan that provides the development concepts on built form, landscaping, space, urban texture, circulation and solutions to urban services. The *urban design plan* is the detailed worked-out plan from the master plan; it describes what will finally happen in the urban settings with building types, heights, roads, paths, street-furniture, parks, etc. When the urban plan is finalized, the urban planners transfer their goals of the building forms to the architects through the *architectonic quality plan*. The final phase is the *architectural design,* which ends up in creating a detailed building with an interior.

The graphs in Fig. 24.10 illustrate the relationships of the six major planning phases with visual materials in a geoVE. They indicate that a zoning plan should be visualized mostly as a symbolic 2D map/plan alongside textual and graphical information as attributes. Structure plans should be visualized as symbolic 2D plan/maps. A portion of the survey population argued that iconic 3D LoD1 block models should be used to visualize height restrictions of a zoning plan. The master plan tends to show real inclination to 3D, specifically the use of monochromic iconic block models (LoD1), although symbolic 2D maps/plans are also necessary. Noticeably textured higher LoD models for these three initial phases are undesirable.

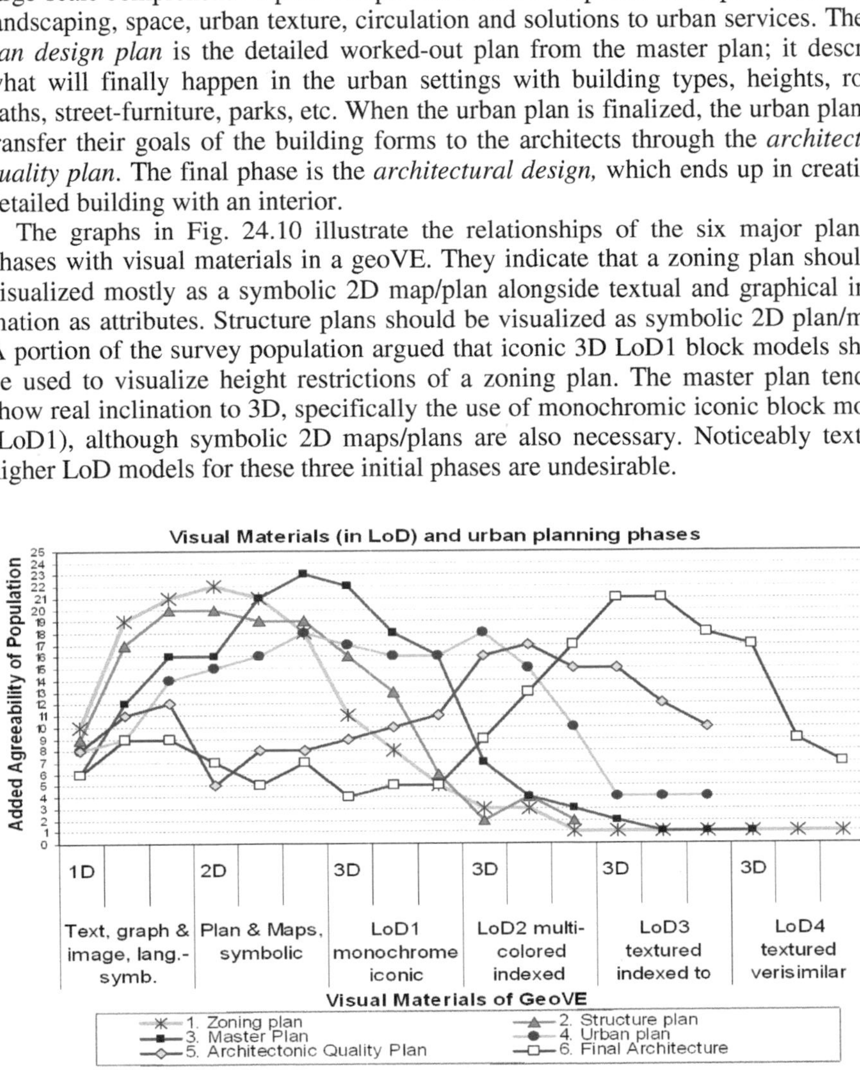

Fig. 24.10. Visual materials in various LoD and urban planning phases.

Fig. 24.10 illustrates that in lateral stages like urban plans, architectonic quality plans and final architecture designs, a limited presence of textual, graphical and image representation can be identified. However, from urban plan up to architectural plan phases, the graphs tend to show a gradual decrease in the use of text and images as media for communication and an increase in 3D. The architectonic quality plan shows a primary inclination towards indexed volumetric envelope models (LoD2) and secondarily to LoD1 models. Multi-coloured 3D models in LoD2 should visualize initial architectonic forms like roofing and exterior envelopes. In the architectural design phase, photorealistic, detailed architectural models (LoD3) and verisimilar detailed models with interior space (LoD4) are preferred over visualizing design solutions in a geoVE.

Observing these graphs, it becomes obvious that with gradual progress of spatial (urban) planning, the graphs tend to move from left to right, increasing the importance of 3D. In these initial spatial planning phases, one can identify that there is constant presence of textual, graphical and 2D plans as interaction material. In the final phases, the 3D LoD models with gradual realism dominate as interaction material for visualization.

24.5 GeoVEs suitable for municipalities for interaction in urban planning

We tested four geoVEs in this research, as seen in Table 24.6. The first three are based on widely accepted 3D formats like KML, X3D/VRML and CityGML. Cebra's VirtuoCity is chosen due to its growing popularity for designing virtual cities in the Netherlands.

Table 24.6. Four tested geoVEs of the research.

Geo-VE and data formats		Open Source
Google Earth (GE)	KML 2.1	Y
LandXplorer	CityGML	Y
Bitmanagement Internet Explorer plug-in	VRML/X3D	Y
VirtuoCity of Cebra	V3D/Python	N

These four geo-VEs are compared based on their functionality domains. The construction functionalities of the geoVEs largely depend on how the data formats are designed. In the construction domain, LandXplorer viewer's data format CityGML has a clear grammar to model in various LoD attaining thematic and geometric semantics. Google Earth's KML and X3D data formats used in Bit management Internet Explorer (IE) plug-in have a geometric hierarchy but no semantics in thematic attributes. Google Earth, LandXplorer, Bitmanagement IE plug-in and VirtuoCity connect to the Internet and import data. VirtuoCity has a multi-user interface in a three-tier client-server-database architecture, allowing streaming of 3D models in a proprietary client 'engine.' Google Earth (GE) allows streaming in Asynchronous JavaScript and XML (AJAX) for satellite images. The 3D models can be visualized 'on-the-fly' from spatial databases in GE. For LandXplorer, the CityGML-based data has to be first downloaded on a local PC. Bitmanagement IE plug-in can be used to visualize streamed 3D models from spatial databases.

These geoVEs let users passively experience 3D through movement, navigation, and orientation, and are immersive in nature. GE and LandXplorer have advanced navigation, orientation and exploration controls. LandXplorer lets the user click and select data and explore the attributes. While Bitmanagement IE plug-in and VirtuoCity use avatar-based systems, Google Earth and LandXplorer do not. Google Earth and LandXplorer allow spatial queries based on geometry and attributes. Such queries can be built on top of Bitmanagement IE plug-in.

Google Earth's KML and LandXplorer's CityGML and models in VirtuoCity are geo-referenced. The foremost difference between X3D/VRML-based Bitmanagement IE plug-in and VirtuoCity is the way in which these two geoVEs present the 3D scene. VirtuoCity is based on the concepts of social networking in real-time collaboration in an interactive environment similar to Second Life or Active World, where users can communicate with each other. On the other hand, 3D scenes made in X3D/VRML and visualized in Bitmanagement IE plug-in result in 'dead-worlds' of 'ghostly solitude'.

Implementing an X3D/VRML-based viewer for the mass public will require municipalities to build their own interactive interfaces, which may be a laborious and expensive task. However, many developments in X3D and multi-user interfaces allowing collaboration are already taking place. Media Machine has made efforts to combine X3D with Simple Wide-area Multi-user Protocol (SWMP)-3D, fulfilling the collaborative aspects by providing interactive multi-user chatting. Blaxxun Tehnologies is also involved in this research field. For urban planning, X3D fulfills the requirement to visualize the urban projects in LoD1 to LoD4. The CityGML-based LandXplorer is a professional data-mining and exploration tool rather than an interactive visualization tool. While using CityGML is positive as 3D models have semantics and can be used in multiple applications, the de-facto LandXplorer viewer falls short of fulfilling the visualization requirement for interaction in urban projects.

GE has two major advantages over the other geoVEs. First, GE is widely popular; due to its popularity, it is supported by many CAD and GIS companies to export various data formats in GE. Second, the free availability of large datasets like satellite images, web services and 3D models in GE is second-to-none. KML is accepted as Open Geospatial Consortium, Inc. (OGC) best-practice specification, which is a positive development. However, GE does not allow collaborative chatting or voting functionality for urban projects. Incorporating instant messaging functionalities in GE could overcome this limitation. The drawback of GE is that it cannot visualize interior and underground objects. For large-scale urban projects, GE is suitable to obtain an overall view of the designed area but fails in elaborating details of architectural design. VirtuoCity contains most of the relevant functionalities, but searching and exploring functionalities are limited. Eye-level experience in VirtuoCity allows the finest detail of architecture to be visualized.

We have examined the geoVEs through a positive approach to open source solutions for the municipalities. In order to use 3D models in cross application domains, semantics cannot be neglected. In a best-case scenario, 3D models should maintain semantics and LoD by following well-defined schemas like CityGML (also an OGC standard) and visualize the design in multiple Internet-based geoVEs. Given the complexity of building a custom-made collaborative multi-user X3D viewer, this makes Bitmanagement IE plug-in a lesser choice. Google Earth is a geographic visualization tool that is not specifically designed for visualizing architectural details and interior. However, due to its mass popularity to reach the maximum number of actors, GE should be used for visualizing exterior 3D models. Finally, VirtuoCity is custom-made for collaborative design process and holds solution for interactive multi-user geoVEs for urban planning in the Netherlands. However, it leads municipalities away from open source solutions. The multi-user interactivity functionalities like avatars, intelligent objects, feedback, and audio support are positive aspects of this geoVE.

24.6 Concluding remarks

In this paper, we have defined the requirements of visualization materials and the taxonomy of geoVE functionalities for urban planning. The hierarchical functionality domains are relevant for comparing geoVEs. In our case study, we found that the visualization method is important in transferring information on urban renewal and planning. We determined that increases of realism, Levels of Detail (LoD) and dimensionality increase the user perception of understanding the design linearly. This has an effect on the relationship between LoD and the planning phase. As the planning phases progress in time, the dimensionality, realism and LoD in various interaction models in geoVEs should increase. Finally, we conclude that text, images and 2D maps/plans are important in the initial planning phases. When design is at a definitive stage, 3D LoD models should be used in gradual realism.

There is a high interest to use 3D models for interaction. But appropriate solutions must be found to ensure that all actors can access the information they need in a particular urban planning phase. The tested geoVEs support only some of the required functions; it is therefore difficult to make recommendations. Some of the municipalities may need to consider in-house developments to enrich the most popular geoVE.

References

1. Al-Kodmany, K. (2001), 'Supporting imageability on the World Wide Web: Lynch's five elements of the city in community planning.' Environmental and Planning B: Planning and Design 28, pp 805-832
2. Batty M, Dodge M, Jiang B, Smith A, (2000), 'New technologies for Urban designers: the VENUE Project', ISSN: 1467-1298, CASA, UCL, Last accessed on the 15th of August 2007 at http://www.casa.ucl.ac.uk/venue.pdf.
3. Camillo S., (1965), 'City Planning According to Artistic Principles', Translated from the German by George R. Collins, and Christiane Crasemann Collins. Columbia University Studies in Art, History and Archaeology, no.2, New York: Random House.
4. Dalal, B. and Dent D. (1993), 'Sustainable Development Strategies: A Resource Book', Environmental Planning Issues No. 1 Barry Dalal-Clayton 1993, 14pp ISBN 1 84369 205 8, Barry Dalal-Clayton and David Dent 1993, 153pp ISBN 1 84369 203 1
5. Davis S. B., (1996), The Design of Virtual Environments with particular reference to VRML, Centre for Electronic Arts, Middlesex University.

6. Heim, M. (1998), 'Virtual Realism'. New York, New York: Oxford. pp. 162-167, 171, illus found at http://www.immersence.com
7. Kibria M. S., (2008), 'Functionalities of GeoVE to visualize urban projects', GIMA MSc. Thesis, conducted at OTB, TU Delft, The Netherlands, accessed in the 2008 section of the Geo-Database Management Center (GDMC) MSc. Publications at http://www.gdmc.nl/publications/
8. Kraak Menno-Jan, (2002), 'Visual exploration of virtual environments' published in 'Virtual Reality in Geography, edited by Peter Fisher & David Unwin, 2002 Published by Taylor and Francis Inc, New York
9. Lammeren R. van, Hoogerwerf T. (2003), 'GeoVirtual Reality and participatory planning', Virtual Landscape position paper, Alterra, Wageningen University, Centrum voor Geo-Informatie, CGI Report 2003-07 ISSN 1568-1874
10. Lynch, K. and G. Hack (1984), Site Planning, MIT Press, London, UK.
11. MacEachren, A.M., Edsall, R., Haug, D., Baxter, R., Otto, G., Masters, R., Fuhrman, S., and Qian, L. (1999), 'Exploring the potential of virtual environments for geographic visualization' http://www.geovista.psu.edu/publications/aag99vr/fullpaper.htm
12. McCloud, S., (1993), 'Understanding Comics', Kitchen Sink Press, 1993. Mentioned in Dykes, J A.M. MacEachren, and M.-J. Kraak, 2005, Exploring GeoVisualization by, Elsevier, Amsterdam, the Netherlands, February 2005.
13. Riedijk A., R. J van de Velde, T.C. Hoogerwerf, R.J.A. van Lameren, (2006), 'Virtual Netherlands, GeoVisualization for interactive spatial planning and decision making: From Wow to Impact', Definition study, Vrije Universiteit, Amsterdam.
14. Sherman W. R. & Craig A. B. (2003), 'Understanding Virtual Reality: Interface, Application & Design', Morgan Kaufmann Publishers, imprint of Elsevier Science, California, USA, ISBN 1-55860-353-0 pp201-280
15. Sneiderman, B. (1998), 'Designing the User Interface: Strategies for Effective Human-computer Interaction', 3rd Edition, Reading MA., Addison Wesley Longman, Inc.
16. Wachowicz M., Bulens J., Rip F., Kramer H., van Lammeren R., Ligtenberg A., Wageningen UR Centre for Geo-Information, 5th AGILE Conference on Geographic Information Science, Palma (Balearic Islands, Spain) April 25th-27th 2002
17. Witmer, B.G. and Singer, M.J. (1998), 'Measuring pressure in virtual environments: A presence questionaire. Presence' 7(3): pp 227.
18. Whyte, J. (2002), Virtual Reality and the Built Environment, Published by Oxford: Architectural Press, ISBN-10: 0750653728, September 24, 2002
19. Zlatanova S., van Dorst M., Itard L, (2007), 'The role of visual information in design tools for urban renewal', ENHR 2007 International Conference 'Sustainable Urban Areas', Rotterdam last opened on 11/10/2007 at http://www.gdmc.nl/publications/2007/

Chapter 25
Producing 3D Applications for Urban Planning by Integrating 3D Scanned Building Data with Geo-spatial Data

Yonghui Song, Hongxia Wang, Andy Hamilton and Yusuf Arayici

Abstract. Visual Information Systems for urban planning can be produced in a variety of ways. In this paper we give an account of research into integrating scanned data with urban data sets to produce 3D applications that span built environment spatial scales from building elements to the whole city. In our research, 3D laser scanner is used to capture 3D building models as a way of developing visual 3D presentations. In order to use 3D building models for urban environment, relevant standards are reviewed and the integration of 3D building models with urban scaled geo-spatial data are explored. In the recent EU-funded IntelCities(2004-2005) and Virtual Environment Planning System (VEPS, 2004-2008) research projects, the authors have worked on capturing an existing building in digital form and using 3D data in building refurbishment projects and visualisation of urban environment in planning consultation. The paper introduces the ways of producing 3D applications using integrated data from 3D scanning with geo-spatial datasets. The IntelCities application shows an application for building features survey such as windows and doors, and holistic review of the building's surroundings regarding historical building refurbishment. The application developed in the VEPS demonstrates the potential of delivering web-based integrated consultation services using 2D/3D visualisation. This can be useful to achieve holistic analysis in urban planning projects.

Research Institute for the Built and Human Environment (BuHu), University of Salford, UK

{Y.H.Song, H.Wang, A.Hamilton, Y.Arayici}@Salford.ac.uk

25.1 Introduction

25.1.1 Background research

The UK government, and governments in most of the developed world, are providing citizens with web based services that are delivered directly to the home. Governments also require Local Authorities to provide services to citizens. Many of these services are specifically concerned with urban development, such as the need to engage citizens in decisions concerning a retail park, or a new swimming pool. To deliver these services cities need to develop web applications that can show a wide range of information in a 3D environment. Furthermore these applications need to be developed in such a way that they can be incrementally updated to reflect the changing face of the city. To acheive this end application developers face a number of challenges.

Generally, 3D city models are created using CAD tools. There have been many successful projects which have produced detailed and realistic 3D city models for a diverse range of cities [8, 3, 4, 13]). These city models are created with accurate building models compiled with ortho-photographs and achieved impressive, realistic urban environment [5]. However, the creation of 3D city models using CAD tools and ortho-phtographs are facing challenges. Two of the challenges are discussed below.

The first challenge is that the creation of 3D city models with CAD tools is labour-intensive, time-consuming and expensive. Recently developed and emerging technologies have the potential to make the creation of the 3D city models more effective. GIS-based methods combine 3D techniques with GIS to improve the visualization ability of GIS. However, GIS-based 3D models are restricted to block models or block models with texture mapping, which lack of detailed information and limited quality of the visual realism due to the lack of detailed 3D spatial data for man-made objects. Aquiring 3D data is difficult, but a useful technology is photogrammetry which provides an economic mean to acquire truly 3D models based on aerial images for the coverage of a wide urban area. However this technology has limitations. For example, it seems quite difficult to capture the detailed building structure by using photogrammetry. An emerging technology for 3D data capture is the laser scanning which is a quick and efficient way of digital data capture for a building. However it should be noted that the efficient processing of scanned data to produce virtual models of buildings is still the subject of research. This will be discussed in this paper.

The second challenge is that the datasets underlying these 3D city models are in diverse formats. It is difficult to satisfy the frequent updating and extending requirement for the developing urban environments. Different tools have their own ways to describe urban environments. A standard description of the 3D city model is needed in order to support various applications. It is also necessary to develop appropriate data processing and converting mechanism, standard format of data outcome make the data model usable for urban environment based applications.

These two challenges are the focus of this paper. We will show how recent research undertaken by the authors not only addreses the challenges individually but also, by considering the way scanned data can be used in large urban models, provides for the integration of data sets across a wider spatial scale than is currently available in cities.

In this paper, two research projects in which the authors were involved, IntelCities and VEPS, have been reviewed regarding the development of 3D scanning technologies and data integration in urban models.

The Intelligent Cities (IntelCities) project (http://www.intelcitiesproject.com/) was an integrated project funded by the European Commission (EC) Information Society Technologies programme under framework 6 (6.8 million Euro, January 2004 to October 2005). The aim of the IntelCities project was the development of open, secure, interoperable and re-configurable e-Government services at the city level.

For IntelCities the authors developed the Building Data Integration System as a concept system to illustrate how digitized historic building data can be integrated with other types of city data [10]. 3D scanning techonology was used to facilitate building reverse-engineering (producing data models and plans for buildings from surveying information). Due to the 3D scanning technology having high accuracy, (e.g Reigl 390 is accurate to 5 mm with a range of 350 Metres) this allows for the development of accurate building plans. A series of laser scans were taken to capture 3D data of Jactin House in Manchester for refurbishment purpose. This work is detailed in section 25.2 and 25.4.

The second project is the Virtual Environmental Planning Systems (VEPS) project (2004-2008). VEPS project is an INTERREG IIIB funded, 4 million Euro European project (E109) was led by the Environment Agency for England and Wales (EAEW). A key activity in VEPS was the use of a variety of data sets, including high resolution three dimensional (3D) data from both aerial and ground based laser scanning, for use within Virtual Reality (VR) visualization software and the subsequent delivery of the VR environment via the Internet and World Wide Web. Through the VEPS project, issues of building data capture and building data integration were to be considered at a larger scale through investigating data standardization, modelling, interoperability and integration with other (VR) systems [6]. This will be discussed in section 25.4.

25.1.2 Overview of the paper

For the urban built environment, stakeholders rely heavily on information from various sources in the urban planning decision-making process. There is an increasing need to seamlessly integrate the relevant datasets at both building and urban scales. The capture and integration of relevant data for the urban built environment has always been a challenge due to the high cost and the heterogeneous nature of the datasets [20]. In our research, historic building data is captured using the laser scanning technologies and are processed into a CAD/IFC building model. For buildings which do not have digital format model available, it can be more effective and quicker to produce the skelecton of buildings using drawing softwares based on the 3D point clound data than drawing from scratch. Further more, the authors worked on the integration of digital building data with surrounding geospatial data which can bridge the gap between the construction and urban domain and support urban development control and construction process. This extends the use of datasets beyond their original collection purposes and facilitates the data sharing and interoperablity across domains. In this way, the construction industry and urban development can both benefit by reducing data capturing cost.

The paper is in 25.5 sections. In Section 25.2, the paper reviews 3D laser scanning technologies and how they are used; In Section 25.3, the paper discusses the needs for standards in 3D city models in order to support various applications. Specifically, the paper discussed the emerging 3D city model standard CityGML, an application scheme based on OGC's GML 3.1. In section 25.4 it reviews the integrating of 3D scanned data with 3D city models to produce a model that has a spatial scale from the smallest building elements to the whole of the city. This is through the review of two research project. In section 25.5 the paper is concluded with a review of the production of 3D applications for urban services using the integration of 3D scanned building data and 3D city data.

It should be noted that the focus of the paper is not the processing of scanning data, rather it focuses on investigation of the potential and benefit of integrating building scale 3D models derived from 3D scanning technologies to produce city scale 3D models.

25.2 Using 3D Laser Scanning Technologies for Built Environment Modelling - Jactin House 3D scanning

Laser scanning technology can be used by construction professionals to analyse 3D building data and as way of developing a highly visual 3D presentation tool [1]. In IntelCities FP6 Integrated Project (see section 25.1.1), the use of 3D Laser Scanner had illustrated how an unusual building with many curved elements can be captured in digital form with an accuracy that would not be possible without a 3D scanner.

3D laser scanning provides a way of getting accurate surface information of physical objects without contacting it. This is very useful for modelling fragile objects or unreachable objects. In some cases, normal surveying of an object can be dangerous due to physical structure etc. in all of this situation 3D scanning technologies have big advantages. That is why 3D scanning technologies have the potential to be widely used in archaeology, digital building survey, and site monitoring applications and so on.

In the IntelCities project,the 3D building data of Jatin House was captured using a 3D scanner. With the 3D scanner, it is very convenient to take internal and external measurements to provide accurate plans, sections and elevations. The potential to accurately capture information of the inaccessible and potentially hazardous areas such as pitched rooftops and workplaces makes it possible to acquire building information difficult to capture in other ways. [10]

By post-processing the captured spatial data, outputs for different purposes can be obtained such as CAD modelling, physical modelling by prototyping and visualization in different platforms. This kind of building reverse-engineering facilited by the 3D scanner can be very accurate due to the 3D scanning technology having high accuracy (e.g Reigl 390 is accurate to 5 mm with a range of 350 Metres). This allows for the development of accurate building plans.

The following Fig.25.1 shows the raw scans of the interior and exterior of the Jactin House. They are conducted with several different scan positions and then merged together.

Fig. 25.1. Exterior scans around Jactin house (left) and interior scans inside the Jactin House (right)

By having sufficient overlap scan at the entrance area of Jactin House, the exterior and interior scans were registered (merged) as well to form a complete Jactin House mesh model. A CAD model was created by a combination of using of a point cloud data software Polywork and the point cloud modeler and CAD software Microstation (Fig. 25.2)

Fig. 25.2. Jactin House CAD model is developed.

The scanning work shows that using laser scanning technologies is an effective way of producing detailed CAD building models. The authors believe that using 3D scanning technologies is one way of meeting current challenges especially quickly and efficiently capturing detailed 3D models.

It should be noted, althouth 3D scanner can obtain sufficient information of existing buildings, the post-processing can still be complex and time consuming depending upon how detailed the information needs to be. Achieving more automatisation in the post processing of the scanned data can make the process less complex and time consuming.Recent researchers in this area has made good progress it [21]. This paper however will not focus on the automatisation, instead it will put the emphasis on integrating the CAD model derived from ground based scanning with urban scale 3D model to produce 3D urban system application.

In section 25.4.1, the paper will revisit the Jactin House case study, about its integration with surrounding area 3D model to produce a 3D application.

25.3 3D city model and building model

As discussed in the section 25.1 the creation of the 3D city model using traditional CAD tools is facing challenges. The work introduced in the section 25.2 shows that using laser scanning technologies is an efficient way of producing detailed CAD building models. This section focuses on how to use CAD building models in urban environmental modelling. The emerging 3D data standards and converting between different standards are discussed.

25.3.1 3D City Model and CityGML

Section 25.1 has shown that due to the variation of 3D data formats in the market, it is difficult to satisfy the frequent updating and extension requirement for developing urban environments. Different tools have their own ways to describe urban environments. At the building scale many organisations with an interest in the construction industry have developed Industry Foundation Classes, IFCs, to provide a unified system describing all aspects of a building. At the urban scale a standard description of the 3D city model is needed in order to support various applications. The emerging 3D city model standard, CityGML, is such a standard description. These two systems are discussed in this section, with an account of recent work by the authors to integrate IFC and CityGML in order to achieve a unified model for the Built Environment across the building and urban scales.

CityGML is an application scheme based on OGC's GML 3.1. CityGML holds not only geospatial information but also supports a semantic model of city objects (http://www.citygml.org/). CityGML not only represents the graphical appearance of city models but especially takes care of the semantic representation, thematic properties, taxonomies and aggregations of digital terrain models, sites (including buildings, bridges, tunnels), vegetation, water bodies, transportation facilities and city furniture [14]. The ability of maintaining different levels of detail makes it suitable for small and large area utilizations. The underlying model differentiates five consecutive levels of detail (LoD), where objects become more detailed with increasing LoD, both in geometry and thematic differentiation [15].

25.3.2 CAD Building Model and IFC

Currently, there are many different types of CAD software in support of building design. Each CAD system from different vendors has its own method of describing geometry, both mathematically and structurally. Most CAD models use proprietary data formats and are stored as files in a file system. As one de-facto data standard for CAD applications, AutoCAD Drawing Interchange Format (DXF) is probably one of the most widely supported vector formats in the world today.

The IAI, an international cooperation of more than 650 members drawn from more than 20 countries (http://www.iai-international.org/), pioneered international technical cooperation to define a single building model as one authoritative semantic definition of building elements, their properties and interrelationships [11]. This work has largely been successful with its IFC Model now endorsed as an ISO standard (ISO/PAS 16739).

IFC is a non-proprietary set of internationally standardised object definitions for use in the construction industry The IFC model describes full 3D geometry but also relationships, process, material, cost and other behaviour data. Integrating a CAD model with IFC enables the accurate geometric representation to be integrated with structural and behaviour elements and facilitates links with external applications [7].

IFC was designed to support the whole life cycle of a facility from planning, through construction and usage, to its demolition. In contrast to common 2D and 3D geometry data formats (e.g. DXF files), IFC are capable of modelling much attribute and meta-information related to the geometry.

25.3.3 Building models in urban environment and IFC's conversion into CityGML

Buildings are one of the most important elements within urban built environment. According to Fuchs's survey, 95% of stakeholders identified three-dimensional building data to be of most interest in digital city models [9]. In a physical sense, a city is a collection of buildings. Good three-dimensional models of buildings are important for a variety of tasks including urban planning, urban management, simulation, disaster recovery etc. Digitalized 3D building models have been employed to support planning tasks by exploration, guiding, public participation and visual impact [16, 22].

Building descriptions in GIS have simple geometric descriptions, mainly, only a building's footprint. This information clearly is not sufficient to determine how a building is perceived as part of the public domain let alone the analysis and simulation for professionals. Sometimes buildings are represented by their walls and roofs. Fuchs reports high interest in roofs, as the size of the smallest element is critical for representing front details and overhanging elements [9]. More rarely windows, doors, small balconies, levelled streets and pavements are reconstructed [3]. It is nescessary to convert the detail IFC building model into the 3D city model.

In a recently completed PhD by one of the authors, an IFC building model was converted into CityGML through Building Feature Service (BFS) [21]. BFS was designed to retrieve and convert the IFC building model. The BFS can directly access building information sources which can be IFC/IFCXML files, OO building model database etc. It allows a client to query building features on the Web and convert the detailed building feature description into a CityGML document..

A prototype of BFS has been implemented [21]. The test building is Jactin House in the Ancoats area of Manchester. The Jactin House data was first captured by using a 3D laser scanner and was processed into an IFC building model. The BFS was developed as an ASP.Net web service by using C# programming language based on the Microsoft .Net framework. The process of generating a CityGML document from IFC building model included three main tasks: manipulating IFC documents, coordinating conversion and generating service output in CityGML/GML. The implmentation diagram is hown in Fig. 25. 3.

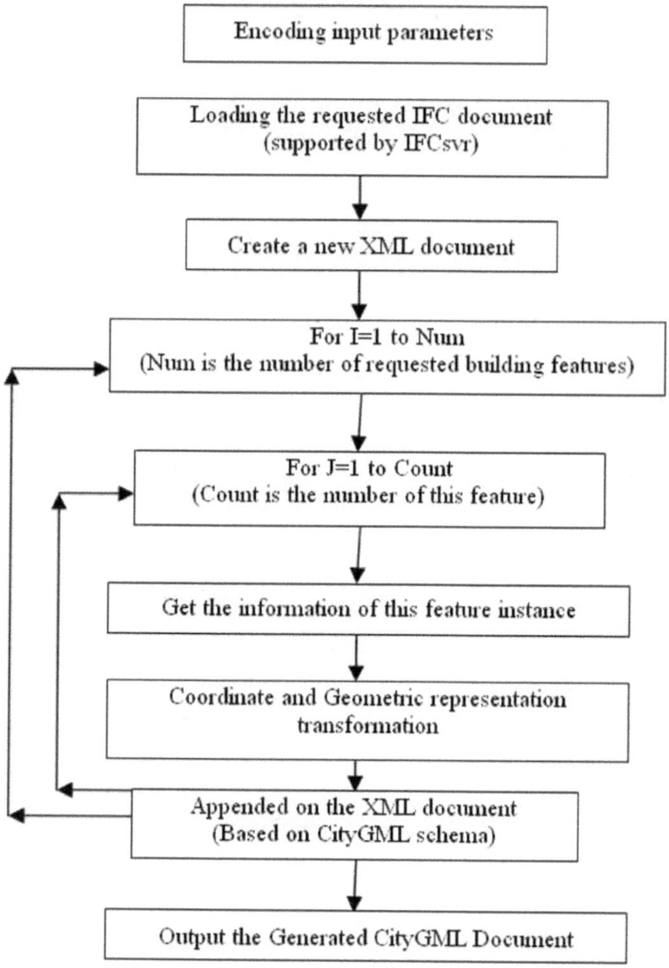

Fig. 25.3. The implementation process of IFC to CityGML

For manipulating IFC documents, there are some toolkits like IFCsvr, EDM and EuroStep etc. In this prototype, the IFC document manipulation is implemented by using a freeware IFCsvr ActiveX component [12]. The main reason of choosing it is because it is freeware and easy to embed in the development as an ActiveX components. The shortcoming is that ActiveX components can only be used in Microsoft's Window platforms.

The second task is to transform the building elements from IFC to CityGML. The first transformation is from a Local Coordination System into the World Coordination System of IFC buildings model. The transformation matrix can be calculated based on the relationship between these building elements, specifically in IfcLocalPlacement's PlacementRelTo and RelativePlacement attributes in the IFC document. The second transformation is from the geometric type in IFC into the geometric reprenstation in CityGML.

An XML document is then generated based on the CityGML schema definition. The System.Xml namespace in Microsoft .Net framework provides standards-based support for processing XML files. It is fairly easy to call the relevant functions for generation of XML documents based on CityGML schema.

The implementation details can be seen in one of the authors' PhD thesis [20]. The following Fig. 25.4 shows the integration results of the conveted 3D building model (from IFC) and surrounding building footprint (from GIS) in a CityGML viewer.

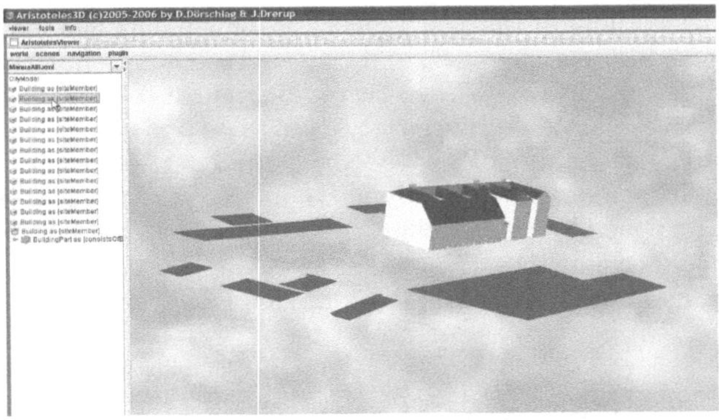

Fig. 25.4. The integrated CityGML file displayed in Aristoteles3D viewer

25.4 IntelCities and VEPS 3D application

This paper mentioned the IntelCities and Virtual Environmental Planning Systems (VEPS) project in Section 25.1. This section introduces the IntelCities and VEPS system applications from a special angle of utilizing 3D scanning data. In other words, this section shows that 3D scanning technologies are used to provide 3D data for building 3D applications for urban activities. This was done by integrating scanned 3D building data with the city scaled spatial data.

Section 25.4.1 reviewed the IntelCities application based on the integration of building CAD model from scanned data with the surrounding city 3D model, and the services the application can provide.. Section 25.4.2 is about the integration of scanning data with other VEPS data sources, and Section 25.4.3 is an introduction of VEPS application design.

25.4.1 IntelCities Application –outcome of Jactin House case study

In section 25.2, we reviewed the scanning technologies and how Jactin House was scanned and to further produce 3D CAD model. This section briefly reviews the 3D application which integrates CAD model derived from 3D scanned building data and surrounding city model.

During IntelCities, laser scanner technology was used to assist with the refurbishment process. To guide this case study in IntelCities project, a fictitious "Gaudi Bank" scenario was created as a typical multi-national development project. In this scenario, the Gaudi Bank, based in Spain, with branches in the UK, is planning to set up a training centre in Manchester for staff working in the UK. The focus building is Jactin House in East Manchester. This is the proposed location of the training centre for Gaudi Bank. The structural data of Jactin House was obtained by using a 3D scanner. The data about the surrounding area was obtained from an OS Land-line map. And other information such as public transport etc was also collected and stored in a database with reference to the 3D data objects and 2D map objects. By editing and integrating this data, a multi-dimensional model of the building and the urban area was created to be used for decision-making processes. A system called Built environment Data Integration System (BDIS) has been produced to use the combined building scale and urban scale data to give a wide scaled view of a city environment. Various spatial data from different sources have to be converted into appropriate formats necessary for the integration: 2D geo-spatial data, 3D Industry Foundation Classes (IFC) building data (captured by 3D scanning) and non-spatial data. Then the system can be used to provide functionality such as a data surveying report, and even generate a 3D representation of all windows and doors of the House [17] see Fig. 25.5.

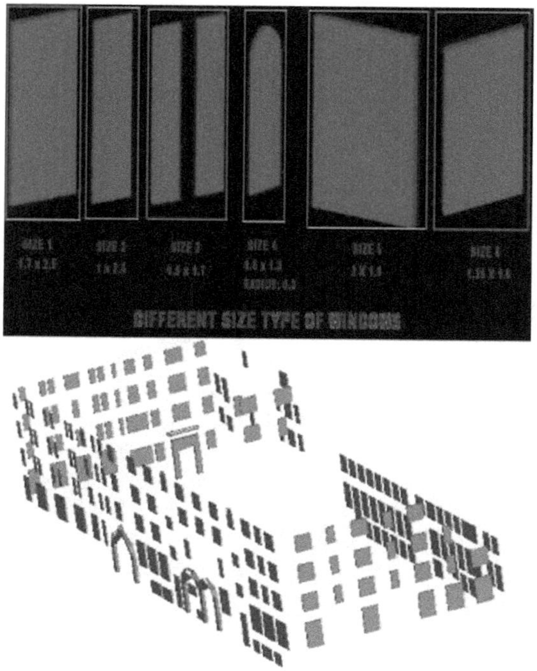

Fig. 25.5. Left, images of windows generated by BDIS; Right, 3D VRML model of just windows and doors in Jactin House

The case study shows the laser scanning greatly improved the process of data capture compared to the traditional way of surveying for refurbishment, in which data capture was a time consuming task. By means of laser scanner technology, the building data can be captured in a short time period and stored in electronic format which can be shared between the stakeholders situated in different locations, apart from providing a faster, better quality and more precise analysis and feature detection for building surveys [2]. The research pointed out that the use of the 3D laser scanner has enormous potential to benefit building surveyors and their clients in terms of accuracy, speed and productivity in plan preparation and then to extend the range of services offered through modelling applications.

The ideas behind this research is data integration [10]. The 3D scanned building models were mapped into IFC schema [1; 18; 19], and were then saved into a centralized database. The following Fig. 25.6 demonstrate the conceptual mode of the BDIS application. In this way, the 3D scanning data is integrated with other data sources. This enabled services based on the integrated data set to be provided to its users. Thus the scanned data is ready for use by a variety of users such as for building refurbishment. For instance, construction professionals are able to see the building in 3D and do a "building survey" from their office. Also for example, it is possible to generate a 3D representation of all Windows and doors of Jactin House (see Fig. 25.5) and even generate a report about windows and doors in Jactin House (See Fig. 25.2).

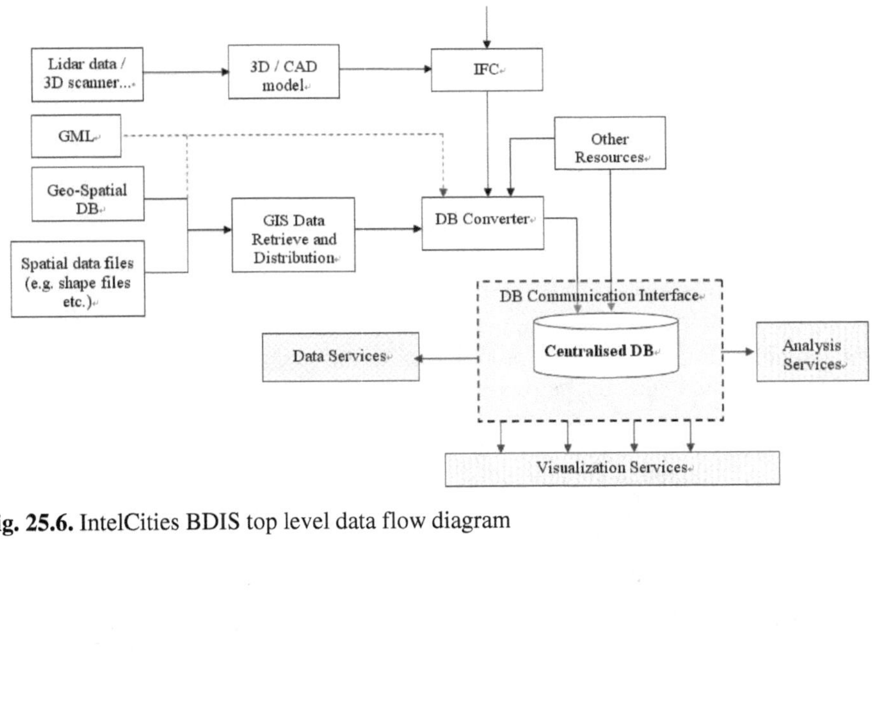

Fig. 25.6. IntelCities BDIS top level data flow diagram

Scanning-Survey Report of Jactin House Elements - Windows and doors

Generated from Centralized nD Database

There are together 199 windows, in 4 sides of Jactin House, located in 5 floors and in different size

- In each elevation of the Jactin House, the number of windows are:
  - Front elevation window-number: 82
  - Left elevation window-number: 40
  - Right elevation window-number: 22
  - Rear elevation window-number: 55
- The 199 windows are located in 5 floors (from the ground to the 4th floor)
  - The ground floor window-number: 27
  - The 1st floor window-number: 47
  - The 2nd floor window-number: 56
  - The 3rd floor window-number: 43
  - The 4th floor window-number: 26
- There are 6 different types among all these windows
  - 0) There are 34 windows of type 1: 1.7 (w) x 2.5 (h)
  - 1) There are 41 windows of type 2: 1.0 (w) x 2.5 (h)
  - 2) There are 29 windows of type 3: 0.5 (w) x 1.7 (h)
  - 3) There are 33 windows of type 4: 0.60 (w) x 1.3 (h), with a curved shape on top (radius: 0.3)
  - 4) There are 31 windows of type 5: 2.0 (w) x 1.8 (h)
  - 5) There are 31 windows of type 6: 1.35 (w) x 1.6 (h)
- Windows size illustration:
  - http://127.0.0.1/jhmodel/Win&Doors/Windows.gif

There are together 5 doors, in the front side and left side of Jactin House, in different size

- 4 of the doors are at the front of the house
- 1 of the doors is at the left of the house
- Doors size illustration:
  - http://127.0.0.1/jhmodel/Win&Doors/Doors.gif

From the survey, it is suggested that all windows and doors need to be replaced.

A VRML model is generated following this report to show just the windows and doors around Jactin house

http://127.0.0.1/jhmodel/Win&Doors/Jactin_Housewindows.wrl

Fig. 25.7. A report of the survey of windows and doors in Jactin House

From the above figure, the centralized database accepts not only 3D building information but also Geo-Spatial information and other thematic information. This information integration greatly helps people doing holistic review and analysis of a building during the decision making process; Thus it is quite possible to do a report about the features of the whole building, and its relationship with surroundings. The following Fig. 25.7 shows a report of windows and doors in Jactin House generated by the BDIS. And Fig. 25.8 shows the building model captured from the 3D scanner is integrated with urban scale 3D model and GIS 2D data.

Fig. 25.8. Merged Ground Based 3D building scanning with airborne scanning of the surrounding area (left), and converted scanning model was displayed with converted ESRI shape file (right) of Jactin House

However, the research recognized the fact that the data analysis is still problematic in terms of the automation on data processing, the post processing can still be very time consuming and depends on the required level of detail of the model.

However, there is much current research interest in fully automatic pattern matching and object recognition on both ground based laser scanning and airborne data, and some successful outcomes. Actually, the subsequent VEPS project that the authors were involved in also had a focus on object recognition [21].

25.4.2 VEPS data sources

The IntelCities 3D applitation based on the integrated building 3D scanning data and urban scale 3D data was described in Section 25.4.1. This sub-section 25.4.2 and the following sub-section 25.4.3 are going to introduce the VEPS web based 3D application design. As introduced in Section 25.1, a key activity of the VEPS project was to analyse large data sets, including high resolution 3D data from LiDAR and ground based scanning, for use within VR visualisation software and the subsequent delivery of the VR environment via the Internet and World Wide Web. In the VEPS system, post-processed and converted 3D scanning data of buildings were also integrated into urban scale 3D data. The VEPS projet was concered with the delivery of 3D data through the web. This was achieved.using Web Services.

In the VEPS project, as in most geo-spatial related development, there are three kinds of datasets used: spatial data, thematic data and administrative data. However, spatial data is the core data for the 3D City Model. In VEPS data source definition, there are two kinds of spatial data: spatial basic data and Specific spatial data. Common to all VEPS test sites is a standard set of geographic data, the so called "spatial basic data". Further specific data (like noise simulation results) might differ between test sites. This data is called "Specific spatial data". The related common standards are GML, IFC building model, CityGML (See also in Section 25.3).

In VEPS, 3D scanning data is seen as one of the available raw data collection methods. And the scanned 3D data set has to be processed and converted to be integrated with other "spatial basic data". This is actually a complex issue about data level interoperability in VEPS. XML based extended schema such as CityGML and other application schema are used as a common data model in this framework. The VEPS project saw improving the data level interoperability as a long term research ambition.

25.4.3 VEPS application

VEPS defined its user group as urban planners/architects, professionals, investors/project managers, citizens, and local authorities. The VEPS system requirements were therefore defined to include Geo-spatial planning information/commenting and communication between one to one and one to many. Requirements are: visualization and interactivity with map/3D scene, enquiry tool, secured user/data administration functions. From the users' point of view, system service is provided through the client interface. In order to provide an interface suitable for a wide range of users the VEPS system includes 3 main elements: 3D interaction, 2D interaction and communication facilities such as discussion forums and spatially located comments. These facilities are inter linked. 3D scanning data can play the role of a 3D data source to this application.

The VEPS concept architecture diagram reflects the VEPS user requirements see Fig. 25.9.

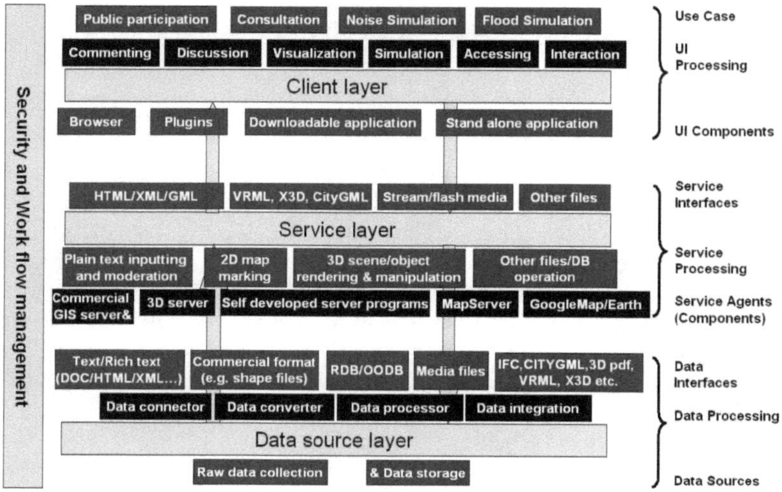

Fig. 25.9. VEPS conceptual architecture

In the Fig. 25.9, 3D scanning data is seen as means of raw data collection method. Underpinned by data sources, the system can provide various levels of functions to its users.

In VEPS project development, 2D, 3D and spatially related commenting modules were integrated to provide Internet users the ability to interact with urban scaled information. The post-processed and converted 3D scanning data of buildings were seen as a system data source to be integrated into urban scaled 3D data.

To summarize, this section reviews the IntelCities 3D application and the VEPS web 3D applications design. Both of the IntelCities and VEPS applications utilise integrated 3D scanned building data and urban scaled 3D data sets to provide service to users. The IntelCities application shows clearly services such as statistics of windows and doors of a building, and holistic review of the building and its surroundings. The design of VEPS application shows the possibilities to deliver integrated services across the Internet, and thus holistic views of whole urban environments can be made available to the public.

25.5 Conclusion

This paper reviews ground based 3D laser scanning technologies and it discusses the use and integration of 3D scanned building data with city scaled spatial data to build wide scale built environment models and to achieve holistic analysis. The authors believe this could be the solution for the two challenges that the traditional ways of creating city 3D models are facing. That brings us to the conclusion: 3D Laser scanning technology can be used to effectively produce 3D building model and as way of developing visual and 3D representations. In this review of recent research projects, the authors not only illustrated how complicated buildings can be captured in digital form with an accuracy that would not be possible without a 3D scanner, but also proved that the 3D data can be further processed and integrated with other sources of data (spatial or non-spatial) and thereby support construction professionals in making holistic analysis before coming to a decision. In the Virtual Environment Planning System (VEPS) project, it is also a proof of concept that 3D scanning data integrated with urban scaled 2D/3D data can be used to provide web based applications for participation and consultation. However, the authors still see the data level interoperability as a challenge in the integration process.

References

1. Arayici, Y., Hamilton A. (2005), Modeling 3D Scanned Data to Visualize the Built Environment, the 9th International Conference of Information Visualisation, London, 6,7,8 July 2005. pp. 509-514, ISBN: 0-7695-2397-8
2. Arayici, Y., Hamilton, A., Gamito, P., Albergaria, G., (2004) The Scope in the INTELCITIES Project for the Use of the 3D Laser Scanner, Proceeding of ECT2004: The Fourth International Conference on Engineering Computational Technology, 7-9 September 2004, Lisbon, Portugal. ISBN 0948749962 pp 111-112.
3. Arena 2000 project, (1999) Virtual Helsinki, http://www.virtualhelsinki.net/english/, Accessed in Nov 2004.
4. Bulmer, D. (2001). How Can Computer Simulated Visualizations of the Built Environment Facilitate Better Public Participation in the Planning Process? OnLine Planning Journal.
5. Chan, R., Jepson, William., Friedman, Scott. (1998). Urban Simulation: An Innovative Tool for Interactive Planning and Consensus Building. Proceedings of the 1998 American Planning Association National Conference, Boston, MA, USA.
6. Counsell, J., Littlewood, J., Arayici, A., Hamilton, A., et. (2008). Future Proofing, Recording and Tagging Historical Buildings: a Pilot Study in Wales, UK. Accepted for 2008 RICS COBRA Conference.
7. Ding, L., Liew, P, Maher, ML, Gero, JS and Drogemuller, R (2003). Integrating CAD and 3D Virtual Worlds Using Agents and EDM. CAAD Futures.
8. Dodge, M., Smith, A., Fleetwood, S. (1998). Towards the Virtual City: VR & Internet GIS for Urban Planning. Virtual Reality and Geographical Information Systems, Birkbeck College.
9. Fuchs, C., E. Gülch and W. Förstner (1998). OEEPE Survey on 3D-City Models. Frankfurt, Bundesamt für Kartographie und Geodäsie
10. Hamilton A., Burns, M., Arayici, Y., Gamito, P., Marambio, A. E., Abajo, B., Pérez, J., Rodríguez-Maribona, I.A., (2005), Building Data Integration System, Final Project Deliverable in the Intelligent Cities (IntelCities) Integrated Project, IST – 2002-507860, Deliverable 5.4c, Final Report, September 2005, University of Salford, UK.

11. Howell, I., Batcheler, B. (2003). Building Information Modeling Two Years Later - Huge Potential, Some Success and Several Limitations. http://www.laiserin.com/features/bim/newforma_bim.pdf.
12. IMS (2002) IFCsvr ActiveX Component Object Reference, VTT (Finland) and SECOM, Inc (Japan) http://cic.vtt.fi/projects/ifcsvr/ifcsvrr200/default.html
13. Jepson, W. H., Liggett, Robin. S., Friedman,Scott. (2001). An Integrated Environment for Urban Simulation. Planning Support Systems: Integrating Geographic Information Systems, Models, and Visualization Tools. R. K. Brail and R. E. Klosterman. Redlands, California, USA, ESRI Press: 387-404.
14. Kolbe, T. H., Gröger, Gerhard., Plümer, Lutz. (2005). CityGML – Interoperable Access to 3D City Models. International Symposium on Geoinformation for Disaster Management GI4DM 2005, Delft, Netherlands,, Lecture Notes in Computer Science, Springer.
15. Kolbe, T., Bacharach, Sam. (2006). CityGML: An Open Standard for 3D City Models. Directions Magazine.
16. Shiode, N. (2001). 3D Urban Models: Recent Developments in the Digital Modelling of Urban Environments in Three-Dimensions. GeoJournal 52(3): 263-269.
17. Song, Y., Hamilton, A. and Wang, H. (2007) Built environment data integration using nD modelling, ITcon Vol. 12, pg. 429-442, http://www.itcon.org/2007/28
18. Tanyer, A. M., Tah, J. H. M., Aouad, G. (2005). Towards n-Dimensional Modelling in Urban Planning. Innovation in Architecture, Engineering & Construction (AEC), Rotterdam.
19. Wang, H., Hamilton, A. (2005). Data Integration Issues within nD Information Model for Urban Planning. 5th International Postgraduate Research Conference, Salford, UK, Blackwell Publishing.
20. Wang, H., Hamilton, A., Counsell, J. and Tah, J. A web-based framework for urban data sharing and dynamic integration, in ACE: Architecture, City and Environment ISSN: 1886-4805 Publisher: Universitat Politecnia de Catalunya, Barcelona, Spain.
21. Wang, Y. (2008) a further *discussion of 3D building reconstruction and roof reconstruction based on airborne LiDAR data* by VEPS' partner, the Department of Remote Sensing and Land Information Systems, Freiburg University. http://portal.uni-freiburg.de/felis/Members/yw78/Topic2?set_language=en&cl=en
22. Yao, J., Tawfik, H., Fernando, T. (2003). Supporting Collaborative Urban Planning with GIS and Virtual Reality. GISRUK, City University, London.

Chapter 26
3D Dynamic Simulation and Visualization for GIS-based Infiltration Excess Overland Flow Modelling

Izham Mohamad Yusoff, Muhamad Uznir Ujang and Alias Abdul Rahman

Abstract. Effective GIS-based Infiltration Excess Overland Flow (IEOF) simulation and visualization requires good knowledge of GIS core concepts and prediction of soil infiltration rates due to impervious area coverage. The success or failure of GIS-based IEOF simulation and visualization resides initially with the georeference system used. Cartographers have long complained about the poor quality of the output from GIS, which today is generally due not to limitations of the GIS itself but instead to a lack of understanding of cartographic principles among hydrologists and environmentalists. Implementation of soft geo-objects representing flow elements such as streams, mudflows, and runoff provides better dynamic visualization in terms of velocity and direction. Inclusion of volumetric overland flow would help in determining the volume of runoff that hits the flood-plain areas, estimating channel flow capacity, and routing and diversions to reduce effects from flooding. With rapid urbanization, industrialization, and climate change, historical runoff and infiltration rates would provide an improper guide for future enhanced visualization of the current 2D land use surface. This study aims to visualize the influence of georeferencing on IEOF simulation when represented by volumetric soft geo-objects within a 3D environment, which is driven by the physically based Green-Ampt method. Visualization is analyzed by focusing on infiltration and overland flow processes using the conformal-based Malaysian Rectified Skew Orthomorphic (MRSO) and the equidistant-based Cassini-Soldner projection. Appropriate usage of a georeferencing system to visualize 3D dynamic IEOF simulation may see high demand from civil engineers, environmentalists, town planners, geologists, and meteorologists as a basis for producing scientific results in flood management control, sustainability for long-term development purposes, stream restoration, rehabilitation, and hydrologic impact assessment.

Department of Geoinformatics, Faculty of Geoinformation Science and Engineering, University Technology Malaysia
{izham1,uznir,alias}@utm.my

26.1 Introduction

GIS is very well adapted for spatial data organization, visualization, querying, and analysis, and it is also helpful in the context of simulation and modelling of spatial phenomena (e.g., floods, subsurface flow, evapotranspiration and groundwater flow) [20, 5]. The use of GIS in infiltration modelling has given benefits in expanding various kinds of simulation basis, spatial representation, and temporal representation models to display results based on site-specific measurements and experiments [10, 4, 12]. Such models have been used by many civil engineers, environmental scientists, town planners, geologists, and meteorologists to handle, analyze, and manage spatial information such as the rate of stormwater infiltrated into a soil profile, total surface runoff generated by urban growth, floodplain analysis, total Non-Point Sources (NPS) pollutant load, and channelizing [14]. Many catchments in Malaysia and other countries are now under intense pressure from urban, industrial, and infrastructural development. Downstream receiving water bodies such as rivers, lakes, ponds, reservoirs, and estuary and coastal waters are facing increased rates and volumes of runoff and pollutant discharge [18]. Urbanization increases the percentage of impervious area in a watershed, causing the infiltration rate in a post-development area to be less than in a pre-development area, particularly in the western states of Peninsular Malaysia.

Existing GIS-based overland flow models such as Agriculture Nonpoint Source (AGNPS), Source Water Assessment Tools (SWAT), QUALHYMO, Long-Term Hydrologic Impact Assessment model (L-THIA), LISFLOOD, and Storm Water Management Model (SWMM) focus on modifying a hydrologic algorithm to delineate watershed, flow accumulation, flow direction, runoff flow path, runoff volume, peak flow, NPS pollutant load, and sediment loading based on empirical, physical, kinematic, and dynamic equations. These models have not focused on the influence of georeferencing and transformations of map projections towards spatial properties of hydrologic parameters [10]. Implementing GIS-based applications requires careful understanding in terms of selecting appropriate georeferencing systems and using appropriate transformations, scales, and grid resolutions when entering spatial data [23, 9]. In planning the mapping of a limited area, one of the first decisions to be made is choice of a map projection [16].

Recently, GIS have started to move from 2D basis spatial hydrologic information systems towards 3D applications, and most recently from static (3D) systems towards dynamic systems that incorporate a temporal element, i.e., 4D [5]. Modelling dynamic overland flow simulation helps specialists and decision makers to understand, analyze, and predict natural disasters (e.g., flash floods, landslides, water pollution) to reduce related damages. Due to limitations in analyzing and modelling multidimensional data sets within GIS software, hydrologic modellers such as [17] linked HEC-HMS (Hydrologic Engineering Centre-Hydrologic Modelling System) and HEC-RAS (Hydrologic Engineering Centre-River Analysis System) with ArcGIS in a case study of flood simulation for Rosillo Creek in San Antonio, Texas. Although the commercial GIS software package was good at representing 2D spatial features, it did not support dynamic, probabilistic modelling.

Existing approaches for 3D GIS modelling can be classified into three geometry types as mentioned by [11]: surface-based (e.g., grid modelling), volume-based (e.g., tetrahedron network (TEN) modelling), and hybrids (e.g., TIN-Octree modelling).

Previous work has shown that these approaches can appropriately represent rigid geo-objects such as mountains, roads, and buildings [19], but soft geo-objects are less represented. [21] developed a method for representing 3D soft geo-objects representing overland flow using the GIS flow element (FE) concept, which can be realized using particle systems and the metaball approach. However, volumetric dynamic flow is not visualized, and the influence of different map projections towards simulation results is not clear. The visualization techniques in this study improve the IEOF process determination and allow hydrologists, environmentalists, and other professionals to be included in the decision-making process. Moving from a 2D map to a 3D landscape image can help end-users envision complex infiltration and overland flow information.

This paper describes the influence of conformal-based MRSO and equidistant-based Cassini-Soldner projection for GIS based IEOF modelling visualized in 3D dynamic simulation using volumetric soft geo-objects. Simulation is performed within IEOF boundaries driven by the physically based Green-Ampt method using the metaball approach. Section 26.2 introduces the concepts of georeferencing for 3D dynamic IEOF simulation and volumetric soft geo-objects. Section 26.3 presents experiments in determining infiltration and volumetric soft geo-objects overland flow simulation within IEOF boundaries. Determination of IEOF area is explained in section 26.4, 3D dynamic simulation results are visualized in section 26.5, and conclusions are stated in section 26.6.

26.2 GIS for Infiltration Excess Overland Flow

Soils and soil properties are fundamental to the partitioning of water inputs at the earth's surface. There is a maximum limiting rate at which a soil in a given condition can absorb surface water input. Important infiltration factors include soil surface conditions, subsurface conditions, hydrophobic, flow characteristics of the fluid, and factors that influence soil surface and subsurface conditions [26].

The land use surface is a dynamic zone, representing at any time a net balance between changing processes and landforms, with complex scale-dependent interactions. Current policy initiatives in Malaysia, such as Urban Storm water Management (USM), recognize this dynamism and are encouraging longer planning horizons and improvement of the storm water runoff process [18]. Flood management initiatives include the general public in the decision-making process through consultations, but flood management remains a difficult task because of the dynamic complexities of the land surface. Identifying present or historic patterns of surface runoff processes may be less appropriate due to changes in driving forces and climate. Current GIS commercial software packages are adapted for handling 2D and 2.5D aspects of TIN and raster surfaces with single z values for each data point. Advanced GIS software could represent data well as 3D objects on surfaces, but to access true volumetric analysis, specialist domain-specific packages are required [5].

Existing GIS products use a very static map-based analysis and have been successfully implemented for managing natural and physical resources as assets. However, the IEOF process is fuzzy, uncertain and dynamic. Successful simulation of the GIS-based IEOF process may require multi-dimensional space-time modelling. There are some encouraging trends in ArcGIS software to support some dynamic simulation capabilities through scripting [5].

26.2.1 Dynamic GIS based IEOF Visualization

Dynamic representation of geographic features in GIS began in the late 1980s and early 1990s. Several researchers proposed event-based models [27] that move from geographic feature identification and location characterization to an explicit focus on changes. Appropriate temporal data streams from monitoring and sensor networks provide tremendous scientific value but are not fully exploitable due to difficulties in integrating across the heterogeneous spatial and temporal sampling regimes and assimilating across a large multi-variant space. GIS have been useful tools for investigating spatial patterns but have suffered from a lack of abilities to explore the dynamic aspects of geographic phenomena. The event-based Green-Ampt method provides the foundation of dynamic infiltration and overland flow phenomena. The method requires new visualization methods to fully address the spatial detail of shape movement and changes.

Furthermore, the method allows interpretation of future uncertainty through the use of different IEOF management strategies in combination with climate change scenarios. The link between the visualization system and the simulation model is provided by a GIS, which allows querying of the model data, integration with other datasets using a common format, and then transfer between the modules used for visualization. By developing IEOF model-based simulations, scientific analysis can potentially provide a replicable, rational, and transparent method to explore the complex processes of the infiltration and overland flow process within a structured framework.

26.2.2 Georeferencing for Dynamic IEOF Visualization

[9] stated that the main component of any GIS usage is the adaption of georeferencing systems to retrieve the actual positions of features in the real world. Information regarding georeferencing and transformation are often lacked by civil engineers and hydrologists when using a GIS approach [10]. [16] stated that representations of spatial features larger than 10 km^2 are often distorted on a projected map. Cartographers have long complained of low-quality output from GIS, which today is generally not due not to limitations in the GIS itself but to a lack of understanding of cartographic principles by users [7].

There are two types of ellipsoids used for georeferencing purposes in Peninsular Malaysia: the Modified Everest ellipsoid for the existing Malayan Revised Triangulation (MRT) system, and the Geodetic Reference System 1980 (GRS80) ellipsoid for the new Geocentric Datum of Malaysia (GDM2000) system [15]. Both systems derive the Malaysian Rectified Skew Orthomorphic (MRSO) and Cassini-Soldner coordinate system for mapping in Peninsular Malaysia. The MRSO coordinate system preserves the shape of spatial features for mapping topographic layers, while Cassini-Soldner preserves distances between spatial features for mapping cadastral lots based on origin within each state in Malaysia [25]. Considering georeferencing influences for hydrologic modelling would benefit the actual GIS core concepts and their applications.

The IEOF process depends on soil properties and land use. Transforming between MRSO and Cassini-Soldner causes distortion on the shape, area, distance, and direction of the original position [25]. The MRSO projection is characterised as a conformal based system where the shape of each feature in the map are preserved, but other physical characteristics of those features, such as area, distance, and direction, are distorted. The Cassini-Soldner projection is an equidistant-based system where the distance and area between features on the map are maintained, but the angle of those features are distorted. Thus would result in influences on terrain, flow direction, and volume of infiltrated stormwater and overland flow being represented by volumetric soft geo-objects. Such situations may lead to inappropriate IEOF simulation results. Water management and environmental practitioners may need to verify the physical properties of spatial features that need to be preserved for modelling the IEOF process within a GIS.

26.2.3 Mathematical Transformation between MRSO and Cassini-Soldner Map Projection.

The MRSO provides an optimum solution in the sense of minimizing distortion of spatial objects whilst remaining conformal for Malaysia [15]. The Cassini-Soldner projection system for the Peninsula is based on several local data and realized by their published equations and coordinates of their respective State origin. The existing Cassini-Soldner projection for cadastral mapping is based on the MRT system referenced to the Modified Everest ellipsoid. It is useful for mapping areas with limited longitudinal extent. The projection has a straight central meridian along which the scale is true, all other meridians and parallels are curved, and the scale distortion increases rapidly with increasing distance from the central meridian.

Transformation of coordinate system between MRSO and Cassini-Soldner is done via two methods: the general method or a polynomial equation [25]. General transformation is done by changing a coordinate in its existing projection to geographical coordinates as in (1) and then re-computing them to the coordinate grid in the targeted map projection.

$$(X,Y) \qquad (Q,L)P \qquad (x,y)p \tag{1}$$

The polynomial solution is used when the numbers of coordinate points are high. In this method, a relationship is established as in Equations (2) and (3).

$$X = C_1 + x.C_2 + y.C_3 + xy.C_4 + x^2.C_5 + y^2.C_6 + \ldots \tag{2}$$

$$Y = D_1 + x.D_2 + y.D_3 + xy.D_4 + x^2.D_5 + y^2.D_6 + \ldots$$

where x,y are the coordinates in the existing map projection, X,Y are the coordinates in the targeted map projection, and C_i,D_i are the parameters of the transformation of the projections. Transformation of MRSO into the Cassini-Soldner coordinate system is done by using the equations in (4) and (5). The reverse process of coordinate system transformation from Cassini-Soldner to MRSO is performed using the equations in (6) and (7).

$$N_{cs} = N_0cs + X - (R_1 + xA_1 + yA_2 + xyA_3 + x^2A_4 + y^2A_5) \tag{3}$$

$$E_{cs} = E_0cs + Y - (R_2 + xB_1 + yB_2 + xyB_3 + x^2B_4 + y^2B_5)$$

where $X = N_{rso} - N_{0rso}$; $Y = E_{rso} - E_{0rso}$; $x = X/10000$, $y = Y/10000$; N_{rso}, E_{rso} is the state coordinate in MRSO; N_{0rso}, E_{0rso} is the state origin coordinate in MRSO; N_0cs, E_0cs is the state origin coordinate in Cassini-Soldner; and R_i, A_i, B_i where i = 1,2,..5 are the transformation parameters.

$$N_{RSO} = N_0RSO + X + R_1 + xA_1 + yA_2 + xyA_3 + x^2A_4 + y^2A_5 \tag{4}$$

$$E_{RSO} = E_0RSO + Y + R_2 + xB_1 + yB_2 + xyB_3 + x^2B_4 + y^2B_5$$

where $X = N_{cs} - N_{0cs}$; $Y = E_{cs} - E_{0cs}$; $x = X/10000$, $y = Y/10000$; N_{cs}, E_{cs} is the state coordinate in Cassini-Soldner; N_{0cs}, E_{0cs} is the state origin coordinate in Cassini-Soldner; N_0RSO, E_0RSO is the state origin coordinate in MRSO; and R_i, A_i, B_i where i = 1,2,..5 are the transformation parameters.

26.2.4 Volumetric soft geo-objects in GIS

Geo-objects can be represented by two approaches: soft geo-objects, which display streams, fire, and mudflows, and rigid geo-objects, which display buildings, bridges, and mountains [21]. Simulating soft geo-objects can be performed using a particle system, which uses small particles as basic elements representing soft geo-objects; and the metaball approach, which displays different formations where more metaballs collide with each other. [21] introduced the soft geo-objects concept by performing GIS FE based on pixel imagery and controlled by geoscientific models. The GIS FE concept has position, velocity, and direction but neglects volume. Hence, this paper intends to model IEOF processes dealing with the calculation of total infiltrated stormwater and formation of overland flow volume towards streams. Inclusion of volumetric soft geo-objects, controlled by the physically Green-Ampt method under conformal-based MRSO and equidistant-based Cassini-Soldner projection, would provide proper usage of the georeferencing concept, guidelines for enhanced visualization of the current land use and soil surface, designation of channel capacity, and diversion to improve future hydrologic impact assessment. Volumetric soft geo-objects are simulated using the metaball approach, which visualizes the continuous surface that is formed when various overland flow sources meet. The contribution from all volumetric soft geo-objects are collected and merged into ordinary rendered settings and represented as volume, flow direction, and flow discharge.

26.2.5 IEOF Modelling

[8] stated that two conditions must be fulfilled for distribution of IEOF: the delivery of surface water input must be in excess of the hydraulic conductivity on the soil surface, and the duration of precipitation must be longer than the time required saturating the soil surface as in Fig. 26.1. Due to spatial variability of the soil properties affecting infiltration capacity and surface water inputs, IEOF does not necessarily occur over a whole drainage basin during rainfall events [24]. Moreover, the exception to localized Horton flow in temperate areas occurs on exposed bedrock [1], anthropogenic effects such as urban development [6], agriculture, and removal of vegetation due to air pollution [3]. IEOF produced on catchment ridges and extended down slope until the entire catchment generated runoff, low slope angle and high saturated hydraulic conductivity [2].

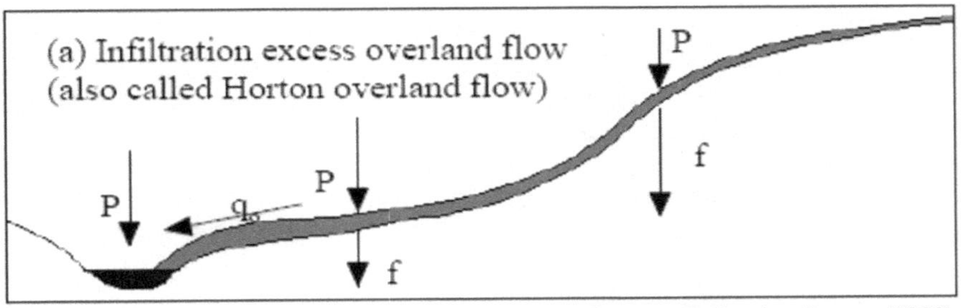

Source : Following Beven, (2000)

Fig. 26.1. Generation of Infiltration Excess Overland Flow mechanism.

26.2.6 Mathematical Green-Ampt Infiltration Equation

[13] developed the Green-Ampt method for determining the amount of precipitation infiltrating into the soil during a precipitation event. The Green-Ampt infiltration model is a physical model which relates the rate of infiltration to measurable soil properties such as the porosity, hydraulic conductivity, and moisture content of a particular soil based on simplified solutions to the Richards equation. This approach was developed for three reasons: (a) the solution of the Richards equation is difficult and not justified given that this equation is, at best, only a rough approximation of the actual field infiltration; (b) a simplified solution still produces the exponentially decreasing relationship between infiltration capacity and cumulative infiltration; and (c) the parameters of the methods can be related to soil properties that can be measured in the laboratory, such as porosity and hydraulic conductivity [24].

[13] developed a flow equation for infiltration under constant rainfall based on Darcy's law, assuming a capillary tube analogy for flow in porous soil:

$$f = K(H_o + S_w + L)/L \quad \textbf{(5)}$$

where K is the hydraulic conductivity of the transmission zone, H_o is the depth of flow ponded at the surface, S_w is the effective suction at the wetting front, and L is the depth from the surface to the wetting front. The method assumes piston flow (water moving down as a front with no mixing) and a distinct wetting front between the infiltration zone and soil at the initial water content. [22] stated the basic Green and Ampt equation for calculating soil infiltration rate as:

$$f = K_s (1 + [\Psi\theta / F]) \quad \textbf{(6)}$$

where K_s is the saturated hydraulic conductivity, Ψ is the average capillary suction in the wetted zone, θ is the soil moisture deficit (dimensionless), equal to the effective soil porosity times the difference in final and initial volumetric soil saturations, and F is the depth of rainfall that has infiltrated into the soil since the beginning of rainfall.

Proper information delivery and prediction of Green-Ampt equation parameters are vital for integrating GIS techniques to compute the infiltration rate and overland flow. The capability of GIS techniques to analyze IEOF, as mentioned by [1], [6], [3] and [2], may produce certain features on each layer with which to perform overlay, buffering, intersection, union, and merge analysis to produce a new layer as the IEOF area.

26.3 Modelling 3D Dynamic IEOF Visualization

26.3.1 Study Area

The Pinang River basin is located between Latitudes 5° 21' 32" and 5° 26' 48" and Longitudes 100° 14' 26" and 100° 19' 42". Pinang River is the main river system in Penang Island, with a catchment size of approximately 52 km^2, as illustrated in Fig. 26.2. The Pinang River basin has been selected for the IEOF process due to the continuous development that has affected land use and soils, degraded water quality, and increased water quantity in the entire basin. Flash floods and water pollution are the major problems faced by highly urbanized areas such as Georgetown, Jelutong, and Air Hitam.

In this study, the procedure for linking GIS and infiltration model parameter components involves the following steps: (1) acquisition and development of GIS map data layers of the Pinang River basin in MRSO and Cassini-Soldner projection; (2) pre-processing of Green-Ampt model input data, parameter and computation results; and (3) post-processing of all infiltration components results for the 3D simulation, with volumetric soft geo-objects displaying IEOF area, volume of infiltrated stormwater, and overland flow. The Green-Ampt model parameters are linked into the PC-based GIS package called ArcGIS and commercial 3D modelling software packages, both of which are used to store and visualize dynamic GIS-based IEOF volumetric soft geo-objects simulation results.

(a)

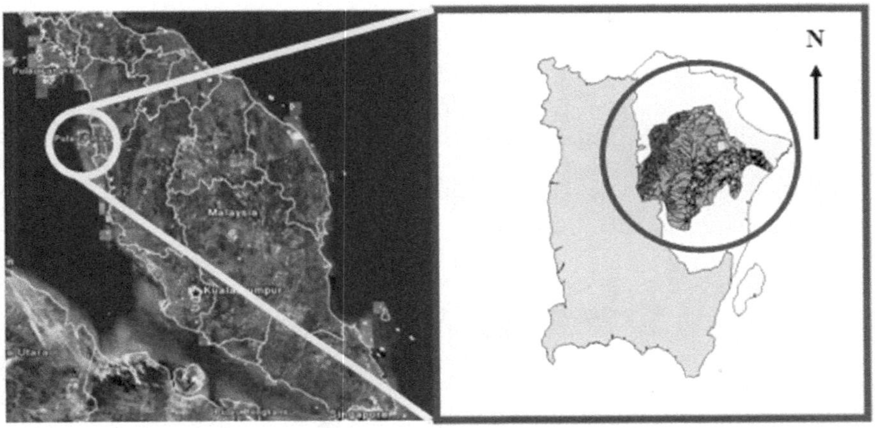

(b)

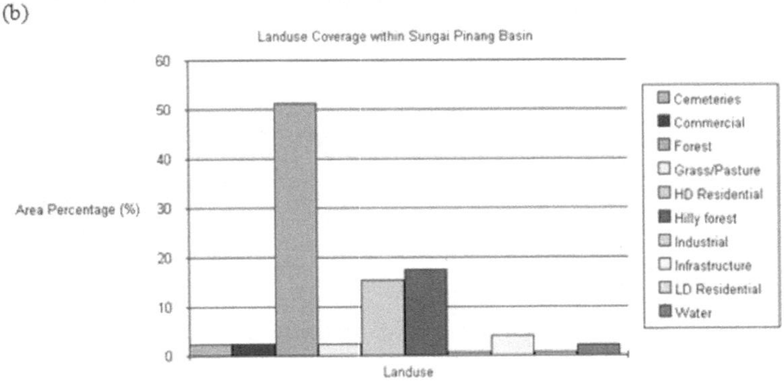

Landuse	Cemeteries	Commercial	Forest	Grass / Pasture	HD Residential	Hilly Forest	Industrial	Infrastructure	LD Residential	Water
Area (%)	2.60	2.50	51.25	2.54	15.47	17.54	0.87	4.12	0.89	2.22

Fig. 26.2. Location of Pinang River basin (a) and its land use for 2006 (b).

A digital topography map with 1 : 25 000 scale is used to extract layers consisting of Buildings, Contours, DEMs, Road networks, and River networks. The land use and soil map is obtained from the Department of Agriculture for evaluating the soil condition at the locations of interest. In this study, the rainfall data on the 18th of June, 2006, with a duration of 60 minutes, is used to determine infiltration rate and overland flow generated from IEOF areas using grid data layers with 20 meter and 5 meter resolution. Topographic information such as slope, aspect, flow length, contributing area, drainage divides, and channel network can be reliably extracted from DEM.

Square-grid DEMs are used in the IEOF modelling because of its simplicity, ease of processing, and efficiency for computational purposes.

26.3.2 Determining Potential Area of IEOF using MRSO and Cassini-Soldner Projection

Raster-based layers with 20 meter and 5 meter grid cells are used within MRSO and Cassini-Soldner projection, respectively. Analysis is performed in two phases. The first phase is to model spatial data layers by overlaying raster layers of Precipitation, Land use, Slope, Soils, Buildings, and Road network based on criteria mentioned by [8, 6, 3, 2] to map the potential IEOF area. The second phase is to intersect these raster layers to map the potential location of IEOF and its area. The schematic diagram for determining IEOF area is presented in Fig. 26.3.

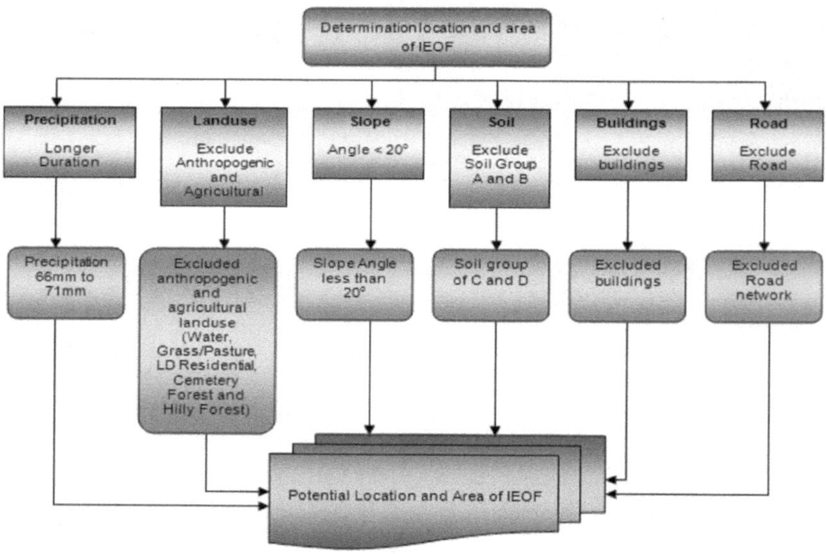

Fig. 26.3. Schematic diagrams for determining IEOF area.

26.3.3 Computation and 3D Dynamic Visualization of Infiltration and Overland Flow Volume Simulated within IEOF Area

The parameters required for the Green-Ampt method — soil hydraulic conductivity (K_s), soil percent impervious (R_s), percent effective soil area (*Eff*), the initial abstraction (I_a), land percent impervious (R_l), percent vegetation (*Veg*) and the degree of saturation (dry, normal or saturated) — are obtained through ground observation and assigned to each pixel [22]. Simulations of volumetric soft geo-objects are performed based on the soil infiltration rate, referring to equations (8) and (9). The total overland flows within IEOF areas are computed by subtracting rainfall volume with the infiltrated rainfall volume. Fig. 26.4 shows a flow diagram of determining 3D dynamic volumetric soft geo-objects IEOF simulation, which is represented by a fine cylinder.

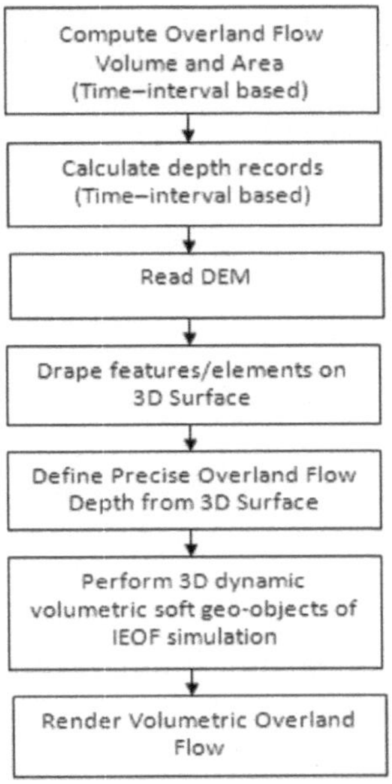

Fig. 26.4. Flow diagrams for rendering 3D dynamic IEOF volumetric soft geo-objects simulation.

26.4 Potential IEOF Areas

Fig. 26.5 illustrates our experiment in determining the IEOF area. The IEOF area is shaded in white.

Approximately 5.2 km^2 of IEOF area is identified. Most of the IEOF coverage lies in areas of Paya Terubong, Air Hitam, Air Terjun River, Kebun Bunga, Green Lane, and partly in Gelugur and Jelutong. Table 26.1 summarises the differential amount of calculated IEOF area using 20 meter and 5 meter grid size under MRSO and Cassini-Soldner map projection. The location of IEOF lies in the sub-humid to humid regions, which are the major controls on the various runoff processes, based on climate data, land use, soil topography, and rainfall characteristics as stated by [24, 26].

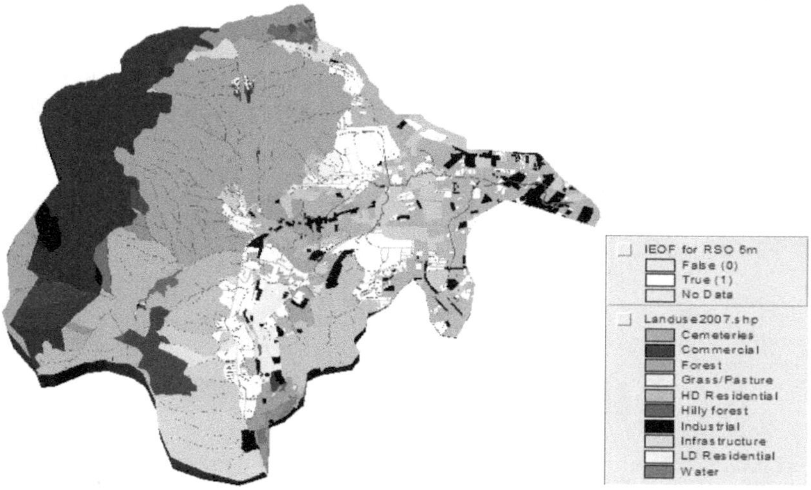

Fig. 26.5. Potential area of IEOF within Pinang River basin.

26.5 3D Dynamic Simulation of Infiltration and Overland Flow Volume

Approximately 355,000 m^3 of rainfall volume were recorded within the IEOF boundaries with 60 minutes of precipitation time. The estimated volume of rainfall infiltrated into soil is 129,700 m^3. The total overland flow simulated by volumetric soft geo-objects within the IEOF area is estimated to be approximately 223,700 m^3. The results obtained are illustrated in Fig. 26.6 and Fig. 26.7.

Fig. 26.7 illustrates the large overland flow recorded in the areas of Georgetown, Jelutong, and Air Hitam. Construction of apartments, flats, the Jelutong Coastal Expressway, and shop lots increase the proportion of land cover with impervious areas, the main factor contributing to flood risk. Existing river networks and drainage systems of the Pinang River basin dealt with degrading of water quality and NPS pollutant load. This has caused the existing rivers and drainage systems to lack the capability to shift runoff volumes from highly urbanized areas.

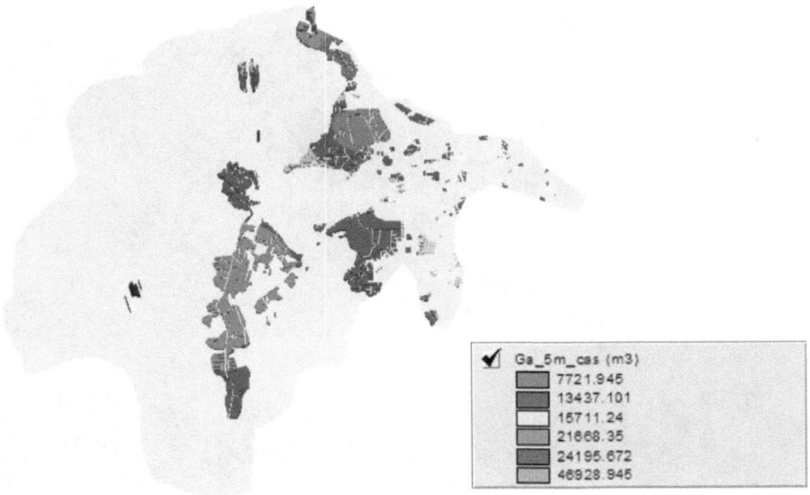

Fig. 26.6. Infiltration Volume based on Green-Ampt equation within IEOF area.

Fig. 26.7. Amount of Overland Flow generated from IEOF area using Green-Ampt equation based on 18th of June 2006 precipitation data.

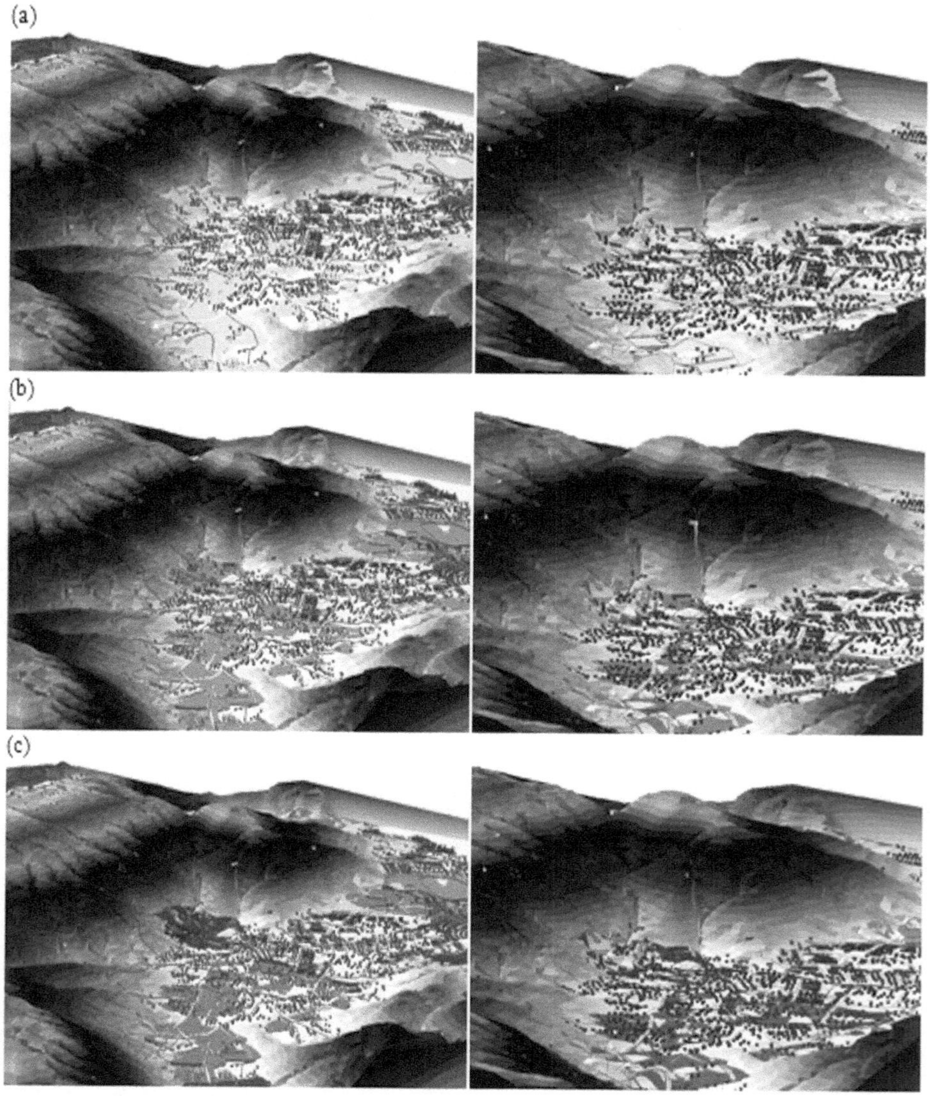

Fig. 26.8. 3D dynamic volumetric soft geo-objects simulation visualized at (a) 20 minutes, (b) 40 minutes and (c) 60 minutes precipitation time using conformal MRSO projection (left) and equidistant Cassini-Soldner (right) for Pinang River basin, Penang.

Fig. 26.8 shows the effects of georeferencing system on infiltration and overland flow volume visualization. Continuous input from precipitation increases the height and coverage of overland flow volume, mainly on down slope and flat areas. The volume of infiltrated storm water and overland flow is proportional to the Green-Ampt method and physical characteristics of the conformal MRSO and equidistant Cassini-Soldner projections, which result in differentials in area, shape, flow path, slope, and deformation of volumetric soft geo-objects within the basin boundaries. These changes cause dissimilarity of high and low infiltration and overland flow volume for each sub-basin.

Table 26.1. Summary of identified IEOF area, precipitation, infiltration and overland flow volume under MRSO and Cassini-Soldner projection with 20 meter and 5 meter grid cell resolution.

Analysis and Results	Raster based	
	20 meter	**5 meter**
1. Total area of IEOF under RSO Projection (m^2)	**5238000.0000**	**5253950.0000**
Total area of IEOF under Cassini Projection (m^2)	**5265600.0000**	**5255650.0000**
Different (20 meter and 5 meter RSO Projection) (m^2)	± 15950.0000	
Different (20 meter and 5 meter Cassini Projection) (m^2)	± 9950.0000	
Different (20 meter RSO and Cassini Projection) (m^2)	± 27600.0000	
Different (5 meter RSO and Cassini Projection) (m^2)	± 1700.0000	
Different (20 meter RSO and 5 meter Cassini Projection) (m^2)	± 17650.0000	
Different (5 meter RSO and 20 meter Cassini Projection) (m^2)	± 11650.0000	
2. Total Precipitation Volume for RSO projection within IEOF area (m^3)	**352849.60000**	**353944.17500**
Total Precipitation Volume for Cassini projection within IEOF area (m^3)	**355180.80000**	**354057.92500**
Different (20 meter and 5 meter RSO Projection) (m^3)	± 10945.5750	
Different (20 meter and 5 meter Cassini Projection) (m^3)	± 1122.8750	
Different (20 meter RSO and Cassini Projection) (m^3)	± 2331.2000	
Different (5 meter RSO and Cassini Projection) (m^3)	± 113.7500	
Different (20 meter RSO and 5 meter Cassini Projection) (m^3)	± 1208.3250	
Different (5 meter RSO and 20 meter Cassini Projection)	± 1236.6250	
3. Total Infiltration Volume (Green-Ampt Equation under RSO Projection), (m^3)	**129830.9597**	**129679.1689**
Total Infiltration Volume (Green-Ampt Equation under Cassini Projection), (m^3)	**129923.2375**	**129663.2530**
Different (20 meter and 5 meter RSO Projection) (m^3)	± 151.7908	
Different (20 meter and 5 meter Cassini Projection) (m^3)	± 259.9845	
Different (20 meter RSO and Cassini Projection) (m^3)	± 92.2778	
Different (5 meter RSO and Cassini Projection) (m^3)	± 15.9159	
Different (20 meter RSO and 5 meter Cassini Projection) (m^3)	± 167.7067	
Different (5 meter RSO and 20 meter Cassini Projection) (m^3)	± 244.0686	
4. Total Overland Flow (Green-Ampt Equation under RSO Projection), (m^3)	**223981.0403**	**223436.8311**
Total Overland Flow (Green-Ampt Equation under Cassini Projection), (m^3)	**224414.3625**	**223391.4970**
Different (20 meter and 5 meter RSO Projection) (m^3)	± 544.2092	
Different (20 meter and 5 meter Cassini Projection) (m^3)	± 1022.8665	
Different (20 meter RSO and Cassini Projection) (m^3)	± 433.3222	
Different (5 meter RSO and Cassini Projection) (m^3)	± 45.3341	
Different (20 meter RSO and 5 meter Cassini Projection) (m^3)	± 589.5433	
Different (5 meter RSO and 20 meter Cassini Projection) (m^3)	± 977.5314	

Simulated results visualize the changes on the calculations of potential IEOF area, precipitation volume, infiltration, and overland flow volume. Alternation of basin shape, size, and distance due to different map projections would greatly affect the physical shape, distance, area, and direction of soft geo-objects. It would also affect calculations of total infiltrated precipitation onto ground surface, change of physical soil parameters (soil porosity, conductivity, path of subsurface flow, return flow) with different soil types, and amount of direct runoff generated in the study area. The modelling performed, however, does not consider factors in the water balance equation such as evapotranspiration losses, percolation, return flow, groundwater flow, and shallow and subsurface flow. Different physical characteristics of the MRSO and

Cassini-Soldner projections cause different values of infiltrated storm water, overland flow volume, and flow direction by volumetric soft geo-objects.

26.6 Concluding Remarks

This paper discusses the definition, mathematical expression and representation of 3D dynamic simulation of GIS based IEOF modelling using the volumetric soft geo-objects approach driven by the Green-Ampt method. We use this approach to estimate the potential locations of IEOF area, infiltration, and overland flow volume. The influence of the MRSO and Cassini-Soldner projections results in deformation of volumetric soft geo-objects, sub-basin area, slope angle, flow direction, and stream network paths. The spatial layers of soils, land use, precipitation, and runoff coefficient are all important sub-basin parameters that result in significant changes of infiltration and overland flow volume at various locations under different map projections.

3D dynamic simulation of IEOF within GIS is important for better understanding of observed phenomena as well as the representation and management of all steps of the process. Performing volumetric soft geo-objects simulation offers the possibility of visualizing affected areas, as well as of reducing economic and social losses. A thorough understanding of physical geography in hydrological process and determination of GIS properties such as map projections, scale, and coordinate systems are required before runoff modelling and data processing can be performed. A map can be drawn at any scale, but it is unclear to what extent existing hydrologic models can be applied at different map projections and scales.

To obtain higher-accuracy computation of IEOF areas, infiltration and overland flow volumes, further investigation is needed for possible new criteria. This investigation may take the form of examining volumetric soft geo-objects data structure for other locations and conducting validation using combinations of GIS and hydrologic algorithms. Although the method involves some identifiable sources of uncertainty, the results nevertheless provide an initial indication of the importance of considering map projections, scales, and grid resolutions for an actual GIS application in a region. The results obtained would benefit the relevant agencies, such as the Department of Irrigation and Drainage (DID), Department of Environment (DOE), Department of Town Planning (DOTP), and Department of Minerals and Geosciences (DMG) in determining flood risk zones, areas of prompt to produce large direct runoff volumes, careful monitoring of NPS runoff pollutant loading, proper development plan and constructions, and monitoring of water quantity and quality of river networks.

References

1. Allan, C., Roulet, N.: Runoff Generation in zero order Precambrian Shield Catchment: The Stormflow response of a Heterogeneous Landscape. Hydrological Process 8, pp. 369—388 (1994)
2. Bonell, M.: Applications of Hillslope Process Hydrology in Forest Land Management Issues: The Tropical North-East Australian Experience. http://www.unesco.org.uy/phi/libros/manaos/5.html (1986)

3. Dingman, S, L.: Physical Hydrology. Macmillan, New York (1994)
4. Dingman, S, L.: Physical Hydrology. Prentice Hall, New Jersey. pp. 607 (2002)
5. Drummond, J., Billen, R., Joao, E., Forrest, D.: Dynamic and Mobile GIS. CRC Press, Taylor and Francis Group. pp. 1--299 (2007)
6. Dunne, T., Leopold, L.B.: Water in Environmental Planning. W H Freeman and Co, San Francisco (1978)
7. Forrest, D.: Cartographic Education and Research in the U.K. The Cartographic Journal, 40 (2), pp. 141--146 (2003)
8. Freeze, R.A., Cherry, J. A.: Groundwater. Prentice-Hall, Englewood Cliffs, New Jersey (1979)
9. Galati, S.R.: Geographic Information Systems Demystified. Artech House, London. pp. 61--151 (2006)
10. Garbrecht, J., Ogden, F.L., DeBarry, P.A., Maidment, D.R.: GIS and Distributed Watershed Models I: Data Coverages and Sources. Journal of Hydrologic Engineering. pp. 506--514 (2001)
11. Gong, J.Y., Cheng, P.G., Wang, Y.D.: Three-Dimensional Modelling and Application In Geological Exploration Engineering. Computers & Geosciences (30). pp. 391--404 (2004)
12. Goodchild, M, F.: Geographic Information Science and Systems for Environmental Management. Annual Revision Environment Resources, 28. pp. 493--519 (2003)
13. Green, W.H., Ampt, G.: Studies of Soil Physics Part I - The Flow of Air and Water through Soils. Journal of Agricultural Science (4) pp. 1--24 (1911)
14. Isenbies, M.H., Aust, W.M., Burger, J.A., Adams, M.B.: Forest Operations, Extreme Flooding Events and Considerations for Hydrologic Modeling in the Appalachians – A Review. Journal of Forest Ecology and Management (242) pp. 77--98 (2007)
15. Kadir, M., Shahrum, S., Kamaludin, O., Ghazali, D., Abdullah, H.O.: Geocentric Datum GDM2000 For Malaysia: Implementation and Implications (2003)
16. Loxton, J.: Practical Map Projection. John Wiley & Sons Ltd. pp. 131 (1980)
17. Maidment, D.R., Robayo, O., Merwade, H.: Hydrologic Modeling, in Maguire, D, J., Batty, M., Goodchild, M, F. (eds) GIS, Spatial Analysis and Modeling, Redlands, CA : Esri Press, pp. 319--332 (2005)
18. MSMA.: Urban Stormwater management Model for Malaysia. Vol. 1 – 20. Department of Irrigation and Drainage, Malaysia (2000)
19. Pilouk, M., Tempfli, K., Molenaar, M.: A Tetrahedron-Based on 3D Vector Data Model for Geoinformation. Advanced Geographic Data Modelling, 40. pp. 129--140 (1994)
20. Rana, S.: Topological Data Structures for Surfaces - An Introduction to Geographical Information Science. John Wiley & Sons Ltd. pp. 1--183 (2004)
21. Shen, D.Y., Takara, K., Tachikawa, Y., Liu, Y.L.: 3D Simulation of Soft Geo-objects. International Journal of Geographical Information Science, (20) pp. 261--271. (2006)
22. Smemoe, C.M.: The Spatial Computation of Sub-basin Green and Ampt Parameters (1999)
23. Snyder, J.P.: Map Projection Used by the United States Geological Survey. Geological Survey Bulletin 1532. pp. 313 (1983)
24. Tarboton, G.D.: Rainfall-runoff Process. Utah State University (2003)

25. Wan, A.Z., Md, Nor.K., Mustofa, D.S.: Map Projection: Second Edition. University Technology Malaysia, pp. 103 (1998)
26. Ward, A.D. and Trimble, S.W.: Environmental Hydrology – Second Edition. Lewis Publishers, pp. 464 (2004)
27. Worboys, M.F.: Event-oriented Approaches to Geographic Phenomena. International Journal of Geographical Information Science, 19 (1) pp. 1--28 (2005)

Keyword Index

3

3D city model, 84, 111, 365, 398
3D continuous K-NN query, 272
3D Data Models, 79
3D Delaunay tetrahedralization, 48
3D geo database, 176, 385
3D geometric network, 284
3D Geometry Engine, 162
3D GeoNet, 286
3D internal structure, 288
3D Navigable Data Model (NDM), 48
3D navigation systems, 109
3D Spatial Engine in Oracle, 155
3D topological querying, 355

9

9-intersection framework, 128, 356

A

Adjacent operator, 49
Advance Simulation of Evacuation of Real Individuals (ASERI), 285
Aerial image, 215
Aerial image enhancement, 218
Affine Transformation, 223
Agent Analyst, 285
Agent-based modeling, 285
AHN-2 (Actueel Hoogtebestand Nederland), 112
Airborne Laser Scanning (ALS), 213, 320
Altimetric models, 214
Appearance, 17
Application Domain Extensions (ADE), 25
ArcGIS, 387
ArcGIS Build operator, 289
Architecture Engineering Construction (AEC), 26, 81
Augmented Quad-Edge (AQE), 48
AutoCAD, 385

B

Bezier class curves, 309
Binary B-Rep, 137
Binary topological relationships, 128
Bitmanagement Internet Explorer, 392
Boundary peeling, 291
Boundary primitives, 133
Boundary Representation (B-Rep), 84, 128, 159
Building Feature Service (BFS), 403
Building Information Modelling (BIM), 26, 79
Building surface texture, 257
BuildingSMART, 82
Built environment Data Integration System (BDIS), 406
Bundle adjustment, 227

C

CAD, 79
CAD tools, 398
CAD-GIS data transformation, 80
Canny edge detector, 220, 236
Catchment, 419
CIMSteel Integration Standards, 81
CityGML, 16, 80, 175, 373, 400
CityServer3D, 373
Climatic data, 327
COLLADA, 24
Combinatorial Data Model (CDM), 288
Complementary OpenLS, 110
Connectivity, 196
Constrained Delaunay Triangulation (CDT), 113

Constraints, 365
Coplanarity, 238
Corner detector, 220
CPA Geo-Information, 181
Cross Correlation Coefficient, 222

D

Data structure, 128
Database, 250
Delaunay tetrahedron, 51
DEM accuracy, 342
DEM surface interpolation, 345
Digital Elevation Model (DEM), 301, 342
Digital terrain model (DTM), 4, 18, 109, 323
DTM interpolation, 323
Dual graph construction, 48
Dual graph theory, 286

E

EDISON edge detector, 236
Egress movement, 296
Emergency services, 284
EParticipation, 365
Equality, 196
Equidistant-based Cassini-Soldner projection, 413
ESRI, 116, 284
Euler angles, 307
Euler Operators, 49
Eulerian model, 5
European Petroleum Standard Group's (EPSG), 160
EXPRESS-X, 82
Exterior of objects, 129, 131, 134, 135
Extruded 3D geometries, 84

F

False correspondences, 227
Finite Element Method (FEM), 113
Floating point arithmetic unit, 197
Flood risk, 424
Floor-space index, 368
Flux Studio, 387
Framework, 81

G

Gauss-Markoff model, 237
GDI-3D project, 110
Geo-entity, 358
Geographic Information Systems (GIS), 4, 127, 283, 342
Geography Markup Language (GML), 21, 155, 176
Geometric Relationship Functions, 163
Geo-registration parameters, 306
Geospatial database, 303
Geo-Virtual Environment (geoVE), 379
Gizmo, 367
Global Positioning System (GPS), 277
Google Earth, 100, 392
Google Sketchup, 235
Green-Ampt infiltration model, 419
GRIFINOR, 103
Ground sample distance, 216
Guided matching, 263

H

Half-Edge data structure, 48
Harris corner detector, 220
Harris operator, 259
Hash-keyed structure, 196
Hausdorff distance algorithm, 305
Hexahedron cell, 356
Hierarchical Network Structure (HNS), 288
Homeomorphic, 200
Homologous features, 303
Hough Transform, 236, 260
Human wayfinding, 274

I

ICP matching process, 304
IEEE 754, 197
IEKF (Iterated Extended Kalman Filtering), 214
Image smoothing, 323

ImageModele, 235
Import and export of CityGML, 183
Indoor navigation, 62
Industrial Foundation Class (IFC), 26, 80, 104, 402
Infiltration Excess Overland Flow (IEOF), 415
InfraWorld, 99
Intelligent Cities (IntelCities) project, 399
Intensity image, 216
Interactively, 365
International Standards Organization (ISO), 129
Interpolating DEM heights, 308
Interpolation algorithm, 307
Interpolation, 342
ISO 19100, 18
ISO 19109, 18

J

Java Architecture for XML Binding (JAXB), 184
Jordan curve, 356

K

KML, 27, 155, 392
k-nearest neighbor query (k-NN), 272
k-order six neighborhoods-based 9-intersection model (K6N9-I model), 358
Korean Land Spatialization Program (KLSP), 35

L

Lagrangian model, 5
Landmark, 272
Landmark-based navigation, 275
Landsat Thematic Mapper (TM), 8
Landscape models, 16
LandXplorer, 384
Laser scanner, 247, 400
Least residual RMS, 238
LEDA library, 203
Level of detail (LOD), 6, 18, 84, 250, 402
Light Detection And Ranging (LiDAR), 213, 302, 319, 387
Linear blending, 266
Linear interpolation, 323
Linear slider, 367
Local Authorities, 398
Location Based Services (LBS), 63, 272
LOD0 to LOD4, 17

M

Managed Objects (MOs), 97
Master plan, 365
Matching, 303
Matlab, 228, 328
Medial-axis transformation (MAT), 286
Mesoscale meteorological model PSU/NCAR MM5, 5
Microsoft .NET, 98
Microstation, 401
Milenkovic normalisation, 199
Minimum Bounding Volumes (*MBVs*), 164
Minkowski distance metrics, 277
Mobile clients, 273
Mobile objects, 273
Model view, 82
Modified Binary B-Rep (MBB-Rep), 139
Monte Carlo simulations, 342
Mosaicking, 258
MOST data model, 273
MULTIPLE objects, 129

N

N6 neighborhood, 357
Neighbourhood, 348, 352
Newton's acceleration equation, 294
Newton-Raphson method, 115
Node-relation structure (NRS), 288
Normalisation, 199
NURBS, 154

O

Oblique surface patches, 242
OGC OpenLS Route Service, 116
OGC Styled Layer Descriptor (SLD), 112
OGC Web Processing Service (WPS), 118
OntoNav system, 62
Open Geospatial Consortium (OGC), 175
OpenGL, 6
OpenSceneGraph (OSG), 6
OpenStreetMap (OSM), 110
Orthophoto, 214
OsgGIS, 7
Outer surface, 195

P

ParameterizedTextures, 24
Parametric LSM, 223
Parser thread pool, 186
Particle system, 6
Personal Navigation Systems (PNS), 61
Pixel-to-pixel matching, 258
Planarity, 198
Planning phases, 389
Poincaré Duality, 63
Point cloud segmentation, 239
Point clouds, 248
Pollutants, 6
Pollution dispersion models, 5
Polyhedra, 195
Polywork, 401
PostGIS, 116, 119
Product Lifecycle Management, 80

Q

QTmodeler, 228
Quad-Edge data structure, 48
Quaternion, 307
Queries on SDO_GEOMETRY, 160
Query engine, 155

R

RANSAC method, 263
Rauble's method, 280
RDBMS, 21
Reference grid, 216
Registration, 303
Registration process, 250
Relational schema, 181
REWERSE project, 63
RFID sensors, 62
RFID tags, 63
RGB color space, 265
RGB texture, 24
Rigid geo-objects, 418
Rounding, 202

S

Safe Software FME, 387
Scanning accuracy, 248
SDI-GRID project, 119
SDO_GEOMETRY data type, 157
Semantic 3D city models, 15
Shadow casting, 328
Shuttle Radar Topography Mission (SRTM), 116
SIG 3D, 18
SINGLE object, 129
Site occupancy index, 368
Skeletonization algorithms, 64
Skyline, 330
SmallWorld, 99
Smart Terrain (ST), 111
Snowflake Software, 181
Soft geo-objects, 418
Solar irradiance, 320
Solar radiation, 327
Spatial Data Infrastructures (SDIs), 111
Spatial indexes, 160
Spatial resection process (SR), 227
Spherical Linear Interpolation (SLERP), 307
Structured Query Language (SQL), 128
SUSAN operator, 259
Sweeping, 80

Swept, 84
Switch ball, 368
SYSUM, 5

T

Terrestrial images, 258
Thematic model components, 178
Thematic models, 178
Thin plate smoothing spline approximation, 347
Three-3D polyline trajectories, 277
Through and Next operators, 49
TIGER (Topologically Integrated Geographic Encoding and Referencing system), 110
Topological correctness, 22
Topology, 17, 128
Triangulated Irregular Network (TIN), 112, 157, 176, 305

U

U-Government, 33
Underground Facility Sensor Network (UFSN), 39
Union, 195
Urban planning, 379
Urban renewal, 380

V

Validity, 199
VEPs project, 365
Version and history managemen, 177
Vertical surface patches, 242
Viewer, 389
Virtual 3D city models, 15
Virtual Earth, 100
Virtual Environmental Planning Systems (VEPS) project, 399
Virtual Geographic Environment (VGE), 4
Virtual Reality (VR), 4, 399
VirtualPlanetBuilder (VPB), 7
VirtuoCity, 382
Visual materials, 380
Visual Studio 2005 .NET, 8
Volume rendering algorithm, 7
Volumetric soft geo-objects, 415
Voronoi cell, 51
Voronoi diagram-based thinning, 290
Voronoi-Delaunay duality, 51
Voronoi-Diagram based MAT, 286
VRML, 27, 296

W

Wallis filter, 258, 264
Wayfinding, 271
Web Feature Services (WFS), 111
Wi-Fi access points, 62
Winged-Edge data structure, 48

X

X3D, 24
X3D/VRML, 392
XML, 183
XML-based data interchange format (XPlanGML), 366

Z

Zoning map, 365

SPRINGER NATURE

GPSR Compliance

The European Union's (EU) General Product Safety Regulation (GPSR) is a set of rules that requires consumer products to be safe and our obligations to ensure this.

If you have any concerns about our products, you can contact us on ProductSafety@springernature.com

In case Publisher is established outside the EU, the EU authorized representative is:

Springer Nature Customer Service Center GmbH
Europaplatz 3
69115 Heidelberg, Germany

FSC
www.fsc.org
MIX
Papier | Fördert
gute Waldnutzung
FSC® C083411

Kolibri 360 GmbH
Ferdinand-Jühlke-Straße 7
99095 Erfurt, Deutschland
produktsicherheit@kolibri360.de